Designing, Developing, and Facilitating Smart Cities

Vangelis Angelakis · Elias Tragos
Henrich C. Pöhls · Adam Kapovits
Alessandro Bassi
Editors

Designing, Developing, and Facilitating Smart Cities

Urban Design to IoT Solutions

Springer

Editors
Vangelis Angelakis
Communications and Transport Systems, Department of Science and Technology
Linköping University, Campus Norrköping
Norrköping
Sweden

Elias Tragos
Institute of Computer Science
Foundation for Research and Technology—Hellas
Heraklion
Greece

Henrich C. Pöhls
Institute of IT-Security and Security Law
University of Passau
Passau
Germany

Adam Kapovits
Eurescom GmbH
Heidelberg
Germany

Alessandro Bassi
Alessandro Bassi Consulting
Juan les Pins
France

ISBN 978-3-319-44922-7 ISBN 978-3-319-44924-1 (eBook)
DOI 10.1007/978-3-319-44924-1

Library of Congress Control Number: 2016948601

Printed on acid-free paper

This Springer imprint is published by Springer Nature
The registered company is Springer International Publishing AG
The registered company address is: Gewerbestrasse 11, 6330 Cham, Switzerland

Preface

The trend of cities leveraging Information and Communication Technologies (ICT) to sustain their growth and offer additional services to citizens became mainstream in the past few years. This has been increased with the emergence of technologies, such as the Internet of Things (IoT), that allow a digital representation of physical environments. Many predictions are made regarding the way IoT will change everyone's lives. To understand the implications and instead of making more predictions, this book combines in-depth explanations of the latest results from IoT research and technology with societal, economical and legal perspectives. This interdisciplinary approach makes this book unique as it covers a broad spectrum of smart city aspects. We believe that this broad view is necessary to maintain citizens' well-being and privacy in a "Smarter" city.

We have not found many reference books covering the topic of smart cities from motivation to enabling technologies and use cases in such a broadened approach. The book provides a comprehensive guide to selected topics of research, both ongoing and emerging, in the broad area of ICT enablers for smart cities, using a pedagogical approach. Technologically, the areas of cloud computing and IoT are considered highly relevant to be understood in their limitations as well as in their offerings to successfully build an ICT enabled smart city. The contributions to this book come from worldwide well-known and high-profile researchers in their respective specialties; selected topics covered in this volume are related to assumptions (Chaps. 1–4), enabling technologies (Chaps. 5–10), and experiences of solutions designed for smart cities. (Chaps. 11–14). We have prepared this book hoping that it proves itself to be a valuable resource, dealing with a broad spectrum of smart city aspects. And we hope that it will be a helpful source and reference for instructors, researchers, students, engineers, scientists, city managers, and industry practitioners in this exciting new field.

Book Overview and Features

The book is organized in 14 chapters, each written by well-recognized experts in the area covered. Chapters have been integrated in a manner that renders the book as a supplementary reference volume or a textbook for use in both senior undergraduate and graduate courses on smart cities. Each chapter has an expository, but also a scholarly or survey style, on a particular topic within the book scope.

Chapter 1 proposes an analysis of past urban developments in an attempt to predict how future developments may look like, and how the technology could be used to overcome current spatial cities' barriers towards a sustainable vision for future generations, on both social and economical levels.

Chapter 2 investigates who the assumed user in the contemporary smart city is. Building a critical framework using examples from contemporary smart cities projects, it reflects on the characteristics of the assumed user, and which users may have been unwittingly overlooked during design and development.

Chapter 3 discusses how the onlife manifesto principles can be turned into guidelines for smart city frameworks and IoT development leveraging experience from research and innovation projects.

Chapter 4 identifies and analyzes some of the expectations and possible long-term effects of big data and the Internet of Things. Without reducing the importance of success stories achieved so far it points out areas that need further attention in the future, such as security at large and proposes to trigger a wider discussion about the impact on society and unforeseen side effects.

Chapter 5 recognizes that it is quite common in the last few years for large cities to form strategic agendas for "smartification" through IoT technologies. It provides an overview of the challenges, a methodology, and shortly summarizes the latest attempts for manifesting security and privacy protection already in the IoT architectures for smart city environments.

Chapter 6 provides an overview of the challenge of personal data protection within smart cities and shows how to formulate technological requirements that meet legal requirements. It also shows how to address the challenge to implement Privacy by Design and gives an example on how to achieve data minimization in a smart transport scenario.

Chapter 7 gives a primer on general information security, its main goals, and the basic IoT security challenges in the Smart City. Built upon the basic IT security goals for confidentiality, integrity, and availability, this chapter additionally addresses security and privacy problems. Thereby it awards recognition how the latter may be a key acceptance factor of smart city ICT solutions and introduces the reader to the technical privacy goal of unobservable communication.

Chapter 8 gives an overview of the main IoT-based communication technologies which can enable services for smart cities, further commenting on the main advantages, disadvantages and open challenges involved in applying each technology to the smart city ecosystem.

Chapter 9 presents the Cloud-IoT architectural vision, as a key service provisioning technology for the smart city. The chapter exposes open challenges and points to a set of different research initiatives that aim to address them. A promising architecture for enabling Cloud-IoT services in smart cities is presented together with a case study that reveals the cloud's high potential in the context of smart cities.

Chapter 10 introduces a framework encompassing FIWARE and the IoT-A to develop innovative IoT platforms and services, and include generic IoT devices that are independent of connectivity modes and are not coupled to specific IoT protocols.

Chapter 11 gives an overview of different mobility data sources and their characteristics and describes a framework for utilizing the various sources efficiently in the context of traffic management.

Chapter 12 focuses on the energy sector advocating how ICT and signal processing techniques can be integrated into next generation power grids towards increased effectiveness in terms of electrical stability, distribution, improved communication security, energy production, and utilization. The chapter also features big data analytics for demand response and serious games as a tool to promote energy-efficient behaviors from end users.

Chapter 13 is about each building getting "smarter" to serve the inhabitants needs better. It takes the approach to incorporate user involvement, which has been successfully applied in many areas, to the engineering process of smart buildings. Smart buildings will no longer be built just as a shelter for people and a more user driven building sector suits the smart city.

Chapter 14 is motivated by the need to restructure healthcare services through a platform to empower patients to actively engage in the management of their well-being. The chapter discusses challenges of developing smart home technologies for health and care breaking down the various facets of the home and the diversity of its residents.

Overall this book has a set of unique features:

- It is designed, in structure and content, with the intention of making it useful at a broad range of learning backgrounds and levels.
- The book's authors are prominent academics, researchers, and practitioners, with solid experience and exposure on the domain.
- The authors of this book are distributed in a large number both of disciplines and countries and most of them are affiliated with institutions of worldwide reputation. This gives this book a strong international flavor.
- The authors of each chapter have attempted to provide a comprehensive bibliography of their topic, which should provide the reader with at least a good starting point to further dig into each area covered.
- Throughout the chapters of this book, core research topics from a wide range of technologies and science and engineering domains have been covered; making the book particularly useful for city managers and practitioners working directly

with the practical aspects behind enabling the technologies for city smartification.

- We allowed the authors, who are experts in an interdisciplinary range of fields, to provide in-depth inside to their views on the topic from their different disciplines and backgrounds; our goal was to transport this expertise rather than trying to achieve perfect homogeneity, but we have attempted to make the different chapters of the book as coherent and synchronized as possible.

Intended Audience

We have attempted to design the overall structure and content of the book in such a manner that makes it useful for all learning levels. The book is written to primarily target students, of a broad spectrum of disciplines. This includes students of senior undergraduate and graduate levels, in the broad range of skills employed towards turning a city smarter. A secondary audience for this book is the researcher and practitioners' community, in academia, the ICT and IoT industry, and city planning. To this end we have taken into consideration the need for getting insights not only of the practical significance of the topics, but also need to discover how this scope of knowledge and the ideas are relevant for applications and technologies enabling for smart cities.

Acknowledgments

We are deeply thankful to the 70 authors of the 14 chapters of this book, who have worked hard to bring this unique resource forward for helping the students, researchers, and smart cities community at large. We note that as the individual chapters of this book are written by different authors, the responsibility of the contents of each of the chapters lies with the concerned authors.

We would like to thank Springer and Ms. Mary James, Senior Editor, who worked with us on the project from its inception, for her professionalism and support. We also thank Mr. Brian Halm, who tirelessly worked with us and helped us in the publication process.

Norrköping, Sweden — Vangelis Angelakis
Heraklion, Greece — Elias Tragos
Passau, Germany — Henrich C. Pöhls
Heidelberg, Germany — Adam Kapovits
Juan les Pins, France — Alessandro Bassi

Contents

Part I Motivation/Scene Setting

1 **Looking at Smart Cities with an Historical Perspective** 3
Alessandro Bassi

2 **Who Is the Assumed User in the Smart City?** 17
Katherine Harrison

3 **Making Onlife Principles into Actionable Guidelines for Smart City Frameworks and #IoT Policies** 33
Nenad Gligoric, Christine Hennebert, Srdjan Krco, Carmen Lopez, Ignacio Maestro, Colin Ő Reilly, Michele Nati, Antonio Skarmeta, Rob van Kranenburg, Nathalie Stembert and Alberto Serra

4 **Factoring Big Data into the Business Case for IoT** 49
Anastasius Gavras

Part II Technologies

5 **Designing Secure IoT Architectures for Smart City Applications** 63
Elias Tragos, Alexandros Fragkiadakis, Vangelis Angelakis and Henrich C. Pöhls

6 **Privacy and Social Values in Smart Cities** 89
Leonardo A. Martucci, Simone Fischer-Hübner, Mark Hartswood and Marina Jirotka

7 **Security and Privacy for the Internet of Things Communication in the SmartCity** 109
Ralf C. Staudemeyer, Henrich C. Pöhls and Bruce W. Watson

8 **IoT Communication Technologies for Smart Cities** 139
Matteo Cesana and Alessandro E.C. Redondi

9 **Cloud Internet of Things Framework for Enabling Services in Smart Cities** 163
Dimitrios Kelaidonis, Panagiotis Vlacheas, Vera Stavroulaki, Stylianos Georgoulas, Klaus Moessner, Yuichi Hashi, Kazuo Hashimoto, Yutaka Miyake, Keiji Yamada and Panagiotis Demestichas

10 **Future Internet Systems Design and Implementation: Cloud and IoT Services Based on IoT-A and FIWARE** 193
Stelios Sotiriadis, Kostantinos Stravoskoufos and Euripides G.M. Petrakis

Part III Use Cases

11 **Traffic Management for Smart Cities** 211
Andreas Allström, Jaume Barceló, Joakim Ekström, Ellen Grumert, David Gundlegård and Clas Rydergren

12 **Smart Grid for the Smart City** 241
Riccardo Bonetto and Michele Rossi

13 **The Significance of User Involvement in Smart Buildings Within Smart Cities** 265
Mervi Himanen

14 **SPHERE: A Sensor Platform for Healthcare in a Residential Environment** 315
Przemyslaw Woznowski, Alison Burrows, Tom Diethe, Xenofon Fafoutis, Jake Hall, Sion Hannuna, Massimo Camplani, Niall Twomey, Michal Kozlowski, Bo Tan, Ni Zhu, Atis Elsts, Antonis Vafeas, Adeline Paiement, Lili Tao, Majid Mirmehdi, Tilo Burghardt, Dima Damen, Peter Flach, Robert Piechocki, Ian Craddock and George Oikonomou

Erratum to: Making Onlife Principles into Actionable Guidelines for Smart City Frameworks and #IoT Policies E1
Nenad Gligoric, Christine Hennebert, Srdjan Krco, Carmen Lopez, Ignacio Maestro, Colin Ő Reilly, Michele Nati, Antonio Skarmeta, Rob van Kranenburg, Nathalie Stembert and Alberto Serra

Index 335

Contributors

Andreas Allström Communications and Transport Systems, Department of Science and Technology, Linköping University, Campus Norrköping, Norrköping, Sweden

Vangelis Angelakis Communications and Transport Systems, Department of Science and Technology, Linköping University, Campus Norrköping, Norrköping, Sweden

Jaume Barceló Department of Statistics and Operations Research, Universitat Politècnica de Catalunya, Barcelona, Spain

Alessandro Bassi Alessandro Bassi Consulting, Juan les Pins, France

Riccardo Bonetto University of Padova, Padua, Italy

Tilo Burghardt Faculty of Engineering, University of Bristol, Bristol, UK

Alison Burrows Faculty of Engineering, University of Bristol, Bristol, UK

Massimo Camplani Faculty of Engineering, University of Bristol, Bristol, UK

Matteo Cesana Dipartimento di Elettronica, Informazione e Bioingegneria, Politecnico di Milano, Milan, Italy

Ian Craddock Faculty of Engineering, University of Bristol, Bristol, UK

Dima Damen Faculty of Engineering, University of Bristol, Bristol, UK

Panagiotis Demestichas Department of Digital Systems, University of Piraeus, Piraeus, Greece

Tom Diethe Faculty of Engineering, University of Bristol, Bristol, UK

Joakim Ekström Communications and Transport Systems, Department of Science and Technology, Linköping University, Campus Norrköping, Norrköping, Sweden

Atis Elsts Faculty of Engineering, University of Bristol, Bristol, UK

Xenofon Fafoutis Faculty of Engineering, University of Bristol, Bristol, UK

Simone Fischer-Hübner Karlstad University, Karlstad, Sweden

Peter Flach Faculty of Engineering, University of Bristol, Bristol, UK

Alexandros Fragkiadakis Institute of Computer Science, Foundation for Research and Technology—Hellas, (FORTH-ICS), Heraklion, Greece

Anastasius Gavras Eurescom GmbH, Heidelberg, Germany

Stylianos Georgoulas University of Surrey, Guildford, UK

Nenad Gligoric DunavNet, Novi Sad, Serbia

Ellen Grumert Communications and Transport Systems, Department of Science and Technology, Linköping University, Campus Norrköping, Norrköping, Sweden

David Gundlegård Communications and Transport Systems, Department of Science and Technology, Linköping University, Campus Norrköping, Norrköping, Sweden

Jake Hall Faculty of Engineering, University of Bristol, Bristol, UK

Sion Hannuna Faculty of Engineering, University of Bristol, Bristol, UK

Katherine Harrison University of Copenhagen, Copenhagen, Denmark; Lund University, Lund, Sweden

Mark Hartswood University of Oxford, Oxford, UK

Yuichi Hashi Hitachi Solutions East Japan, Ltd., Sendai, Japan

Kazuo Hashimoto Waseda University Tokyo, Tokyo, Japan; Kokusai Kogyo Co., Ltd., Tokyo, Japan

Christine Hennebert CEA, Grenaoble, France

Mervi Himanen Relate Partnership, Digital Living International Ltd, Espoo, Finland

Marina Jirotka University of Oxford, Oxford, UK

Dimitrios Kelaidonis Department of Digital Systems, University of Piraeus, Piraeus, Greece

Michal Kozlowski Faculty of Engineering, University of Bristol, Bristol, UK

Rob van Kranenburg Resonance Design/Council, Tilburg, Netherlands

Srdjan Krco DunavNet, Novi Sad, Serbia

Carmen Lopez University of Cantabria, Santander, Spain

Ignacio Maestro University of Cantabria, Santander, Spain

Leonardo A. Martucci Karlstad University, Karlstad, Sweden

Majid Mirmehdi Faculty of Engineering, University of Bristol, Bristol, UK

Yutaka Miyake KDDI R&D Laboratories Inc., Fujimino, Japan

Klaus Moessner University of Surrey, Guildford, UK

Michele Nati Digital Catapult, London, UK

George Oikonomou Faculty of Engineering, University of Bristol, Bristol, UK

Colin Ő Reilly University of Surrey, Guildford, UK

Adeline Paiement Faculty of Engineering, University of Bristol, Bristol, UK

Euripides G.M. Petrakis Department of Electronic and Computer Engineering, Technical University of Crete, Chania, Greece

Robert Piechocki Faculty of Engineering, University of Bristol, Bristol, UK

Henrich C. Pöhls Institute of IT-Security and Security Law, University of Passau, Passau, Germany; Department of Informatics and Mathematics, University of Passau, Passau, Germany

Alessandro E.C. Redondi Dipartimento di Elettronica, Informazione e Bioingegneria, Politecnico di Milano, Milan, Italy

Michele Rossi University of Padova, Padua, Italy

Clas Rydergren Communications and Transport Systems, Department of Science and Technology, Linköping University, Campus Norrköping, Norrköping, Sweden

Alberto Serra sociotal.eu, Guildford, UK

Antonio Skarmeta University of Murcia, Murcia, Spain

Stelios Sotiriadis Department of Electronic and Computer Engineering, Technical University of Crete, Chania, Greece

Ralf C. Staudemeyer Institute of IT-Security and Security Law, University of Passau, Passau, Germany; Information Science, Stellenbosch University, Stellenbosch, South Africa

Vera Stavroulaki Department of Digital Systems, University of Piraeus, Piraeus, Greece

Nathalie Stembert sociotal.eu, Guildford, UK

Kostantinos Stravoskoufos Department of Electronic and Computer Engineering, Technical University of Crete, Chania, Greece

Bo Tan Faculty of Engineering, University of Bristol, Bristol, UK

Lili Tao Faculty of Engineering, University of Bristol, Bristol, UK

Elias Tragos Institute of Computer Science, Foundation for Research and Technology—Hellas, (FORTH-ICS), Heraklion, Greece

Niall Twomey Faculty of Engineering, University of Bristol, Bristol, UK

Antonis Vafeas Faculty of Engineering, University of Bristol, Bristol, UK

Panagiotis Vlacheas WINGS ICT Solutions, Athens, Greece

Bruce W. Watson Information Science, Stellenbosch University, Stellenbosch, South Africa

Przemyslaw Woznowski Faculty of Engineering, University of Bristol, Bristol, UK

Keiji Yamada Kokusai Kogyo Co., Ltd., Tokyo, Japan

Ni Zhu Faculty of Engineering, University of Bristol, Bristol, UK

Part I
Motivation/Scene Setting

Chapter 1
Looking at Smart Cities with an Historical Perspective

Alessandro Bassi

1.1 Smart or Dumb, What Is a City?

The concept of "smart city" is often used implying that the reader has a clear and common notion of what it means. However, in the current literature it is very hard to find a precise definition. What is even more interesting, it is not so easy to find a precise definition of what a city is.

In France, the French National Institute for Statistics and Economic Research (INSEE) uses as a criterion the number of inhabitants: a City is an agglomeration of 2000 or more people. As all arbitrary numbers, this one is not exempt by critics, as just a growing number of inhabitants allow to pass from rural to urban zone, while they are clearly two different realities. Furthermore, every country has different limits: while in Denmark, for instance, an agglomeration of 250 people is enough, in Egypt we need 11,000 people, while in Japan 30,000. In the United States the number set is 2500. The United Kingdom has a different way to define a settlement in a city: only the King (or Queen) has this power, and the appellation is given without a specific criterion, although usually it matches the diocesan cathedrals.

In some cases a different concept is used. The definition of urban unity is based on the habitat continuum, not more than 200 m between 2 constructions and at least 2000 inhabitants.

The definition of Urban or Industrial Zones respond to a deeper concept that takes into consideration the level of daily migration between workplace and home, the percentage of the population not involved in agricultural work and the number and size of industrial, commercial, and administrative buildings. The dynamics of these spaces is linked to the proximity of one or more urban areas.

It is possible to notice, therefore, that the notion of city itself does not have a unique administrative definition. Furthermore, the concept of city goes beyond its

A. Bassi (✉)
Alessandro Bassi Consulting, 3 Avenue de Cannes, 06160 Juan les Pins, France
e-mail: alessandro@bassiconsulting.eu

V. Angelakis et al. (eds.), *Designing, Developing, and Facilitating Smart Cities*,
DOI 10.1007/978-3-319-44924-1_1

mere administrative domains, and evokes a way of living which is typical of a certain amount of population living in close proximity. We would then use the adjective *urban* rather than the word *city* in order to avoid the administrative definitions and rather concentrate on the lifestyle and way of living typical of a large number of people living in proximity. In this context, urban refers not only to the mere administrative facts but, more important, to a specific culture and mentality.

1.1.1 City Growth

The urban population is growing constantly. Looking at simple numbers [1], it's easy to get convinced and to understand the amplitude of the phenomenon. In 1960, the population living in cities was around 1 billion; in 1986, it doubled, and in 2005 was 3.2 billions. The increment of urban population is growing constantly, and constantly quicker. Following this pattern, our planet will see in 2030 5 billion urban people, leaving rural areas more and more inhabited, as the world population growth is estimated around 1 % per year, while the urban population is growing at a rate of 1.8 %.

According to UN data, the percentage of population living in urban environments cross the 50 % in 2007. It is interesting to notice that the strongest part in this evolution will happen in the regions of the world which are currently underdeveloped. Within these zones, in 1975 there were 815 urban areas, while in 2005 the number grew to 2252. In the same time period, in the more developed areas, the growth has been limited (from 701 to 898). In percentage, we have 2.2 yearly for the first ones and 0.5 for the second one.

1.1.2 European Trends

In Europe, the urbanization trend is very important: more than 75 % of the global population lives in towns, using 80 % of resources and contributing 85 % to the European GDP. Demographically speaking, when we refer to Europe as the old continent, we can see that the expression is rather correct: in 2009, the median age of the population was 40.6 years, and forecasts show that it should reach 47.9 in 2060.

From a macro-economic point of view, the EU 28 region is the world first economic power. The GDP of EU countries generated more than 16 Trillions USD in 2008, with an average GDP per person above 30,000 USD. However, these indicators hide huge disparities. The GDP can vary by one order of magnitude, between Estonia and Germany for instance. Differences between EU countries can be found at any level of economic dimensions, from unemployment rate to public deficit or inflation rate.

Growth of the European population (in the EU 27 countries) grew by 100 million, from 400 to 500 million, between 1960 until 2009. Until the eighties, the demographic growth was mainly due to the rate of natural increase. However, the ratio is constantly decreasing since the 1960s, and the lowering birth rate and increasing life expectancy will directly translate into a sensible increase of the average age of the population. It is possible to observe, though, that starting from the nineties, international migrations became the main cause of population growth. This factor can be a solution against the average aging, with particular regards to issues related to the workforce, as most migrants are young adults. In 2010, 9.4 % of the population living in the EU was born outside Europe, and by 2060 it is expected that at least one third of the EU population will have an ancestor born outside Europe, while an even bigger percentage will constitute the active part of the society. While the 2015 migrant crisis showed that the EU countries are still struggling to integrate important migratory trends, demographic shows that the smooth integration into the workforce is likely to be the only viable solution to keep a certain level of prosperity and social benefits.

1.2 Different Theories on Smart Cities

> The European cities are all different; but they are facing similar challenges and are looking for shared solutions [2].

A domain where the research is particularly active during this past few years is at the crossing between technology and society. The current world situation calls for a progressive but radical change. This evolution has been smoothen by the policies of the European Union, but today we see a quick acceleration of tis trend because of economical and environmental concerns. Therefore, the future of our towns is dependent to the way we will manage to work out the economical, social, and environmental developments in synergy.

Within this context, it seems interesting to state the ambitions of the EU for the coming decade. The strategy called "Europe 2020 [3]" aims to revive the economy and is the development of a smart, sustainable and inclusive growth. These priorities, which are mutually reinforcing, must allow to the Union and its Member States to ensure high levels of employment, productivity and social cohesion. This should happen by relying on greater coordination between national and european policies. In other words, each Member State will be required to follow the European directives and support the common objectives through an harmonization of local legislation. Given the growing euro-skepticism, the adoption of common policies by member states might not be obvious.

The main actions are the following:

- Smart growth, developing an economy based on knowledge and innovation. Between now and 2020, an estimated 16 million more jobs will need a high level

of qualification, while the low-skilled asset demand is expected to fall by 12 million. The improvement of the initial training is paramount—as well as the means to acquire and develop new skills during a career.
- Sustainable growth, which promotes a better efficiency energetics as well as a greener and more competitive economy.
- Inclusive growth, which supports high employment rate and a strong social and territorial cohesion.

The targets for 2020 are

- Three quarters (75 %) of the population between 20 and 64 years should be employed, (the average of the EU 27 is now 69 %).
- Reduce the poverty rate of 25 %, which means 20 million people out of poverty.
- Reduce to less than 10 % the population between 18 and 24 years leaving school without a diploma, and raise to at least 40 % the percentage of the population between 30 and 34 year with an higher degree.
- 3 % of European GDP invested in Research and Development, combining private and public sectors, which is a point higher than the current rate (compared to 2.6 % of GDP invested in R&D in the USA and 3.4 % in Japan).
- objective "20/20/20" climate change, a 20 % reduction of greenhouse gas emissions, compared to 1990 levels, raising 20 % the energy efficiency and reach 20 % of energy production through renewable sources.

All these measures should allow the creation of 1 million jobs in Europe. These objectives are linked and, at least theoretically, they are reinforcing each other. Progress in education matters will improve the capabilities of the labor pool, reducing the risk of impoverishment. On top of it, the increase in the average skill level will fuel the growth of a knowledge economy based on innovation, research and development. The European economy will have the chance of improving its competitiveness, creating wealth and jobs, closing a virtuous circle—at least, on paper. Furthermore, all these improvements will bring the opportunity to develop a fully "green economy," making our societies more environmental-friendly, and therefore more profitable, as the side effects of a development not following environmentally sustainable practices are likely to result in very expensive containment measures.

One of the themes which is common to all these dimensions is technology. Much of the progress made in the recent past in the field of Information Technology and Communication (ICT) allow an holistic design for the city of the future, which is often linked to the concept of Smart City. Within the huge number of essays on this topic a few elements are recurring. They will serve as a basis for identifying key concepts of urban form of the future.

In general way, the conceptualization of Smart Cities follows from what we explained earlier on. The economic and technological changes that relate to globalization constitute the fabric of this domain. Cities find themselves facing the need to combine economic competitiveness and urban development, in a sustainable manner and style, preserving—or by creating—an outstanding quality of life. The concept of Smart City brings together all major current concerns.

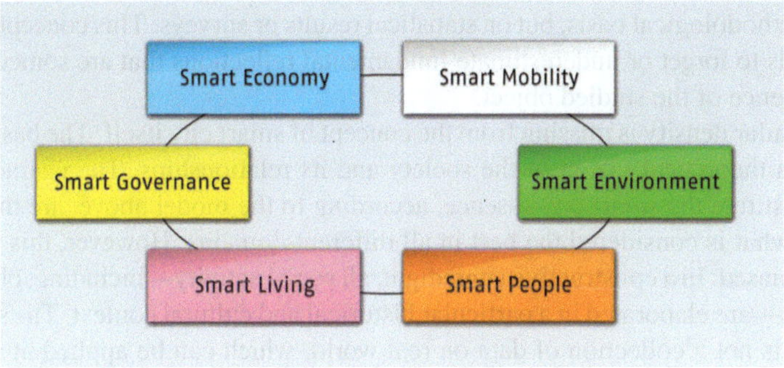

Fig. 1.1 The 6 characteristics of a smart city (*Source* european smartcities)

However, there is a specific issue while studying this theme. The literature on this new city concept comes for the most part from engineers or urbanists. In general, for humanities and social sciences this theme does not seem to resonate, and it is not yet a common object of research. As a matter of example, the diagram in Fig. 1.1, quite popular, is from the website "European Smart Cities." The research team was constituted by members of the Regional Centre of Science of the Vienna University of Technology, the Institute for Research on housing and urban mobility of the Technical University of Delft and the Department of Geography of the University of Ljubljana.

The basic model, found in many publications, promote a taxonomy with 6 domains, 31 sub-groups, and 74 indicators.

The result is a rather technocratic vision of the city, which is -at best- hard to apply. While it can represent an idealtypen, it is hard if not impossible to apply in the real world. It is a holistic approach that pretends to understand everything and explain and master everything through a mathematical formula. However, understand and manage are two very different things: although knowledge and reason are the foundation of the modern world since the eighteenth century, this taxonomy goes too far in the direction of the Reason.

Science and its applications are supposed to give The Answer to everything. This assumption might not be fundamentally wrong if science was replaced by knowledge and wisdom; however, it is impossible not to notice the return of a positivist conception with regards to analysis of the world aspects. The point is not to contest the importance that scientific observation and factual analysis can bring; but one thing is to base the reasoning on the facts, one is to develop a research as a disciple of A Comte (1798–1857). It seems like the "hard science," where everything can be quantified, and that define social laws as immutable, is taken as the cornerstone of every discussion and possible development. In this vision, the smart city concept goes between Supreme Theory and Abstracted Empiricism, two derivatives that CW Mills attributed to the sociology of the 1950s. The Supreme theory claims that purely formal studies can provide an analytical framework to the study of society. The abstract empiricism suggests that knowledge production is not based on a solid methodological basis, but on statistical results or surveys. This conceptual atrophy leads to forget or underestimate fundamental reflections that are sometimes the very essence of the studied object.

A similar density is missing from the concept of smart city itself. The basic model lies on a theoretical vision of the society and its relationships. The characteristics that constitute the smart city essence, according to the model above, are the reflection of what is considered the best in all different domains. However, this vision is clearly biased. In a constructivist paradigm, all social activity—including, obviously, sciences—are elaborated in a particular historical and cultural context. The scientific domain is not a collection of data on real world, which can be applied at any time and anywhere. Rather, it is a discussion built on a certain number of actors (the "scientific community") in a precise historical, technical, political, social, economical moment.

As a consequence, while the scientific origin of a concept is a socially valid assumption, it needs to be put into perspective. We cannot agree more that an absolute relativism, that postulates that all knowledge, scientific or not, has the same level of importance and truth, is truly negative, but we sustain an approach that can keep a different perspective on the same objective data. While the notion itself of smart city is the result of ICT research, based on certain economic assumptions, nothing is stopping from considering human and social sciences to bring a strong contribution to this subject and feed it with a different set of considerations. Understanding the cities of the future with a socio-historical approach will definitely help in fixing some epistemological shortening. This allows, from one side, to define the borders of the current urban shape, on which the future cities will be built. On the other one, to draw attention on the social and logic relationships that are characteristic of the urban way of living. Without the full understanding of these perspective, any theoretical and practical construction in this domain is bound to fail, as it would be a mere exercise of style.

1.3 A Step Back: Urban Sociology

any town is a socio-economic product

Le Corbusier, probably the most influential urbanist and architect in our era, had a very precise mantra. It stated that "the city must allow people to live, to work, to move, and to have fun." While it may sound logical, this urban utopia helped a technocratic vision of the city: very often, the urban space is conceived as a functional organization in which different sectors have specific and complementary functionalities (areas focused on residential, touristic, commerce, industrial …). However, this "clearly defined zone" politics, such as separation between living and working activities, is not always—if not, seldom—a success.

In other words, the peripheral and residential areas of towns are de facto on the outskirts of the economical and social dynamics of cities. What media usually call "urban violence" has roots in these areas. A good number of social problems are, first of all, spatial problems. Ideally, it would be better to avoid building such areas, that tend to be segregated from the rest. Social integration of a town has to go through a mix of activities and spaces, and therefore be the opposite of Corbusier's functional town. The zoning and independence of areas should be replaced by a more organic vision of quarters where different spaces would be inter-dependent and multifaceted. As Jacques Donzelot [4] says, "that social problems are concentrated in certain parts of the urban fabric prove that there is a problem in the town but not of the town."

In any case, the vision of the "urban issue" as explained above is not new, as the heart of that analysis is based on the wrongdoings of the functional urbanism. Urban issues are linked with the loss of quality life following the submission of the urban fabric to the logic of production.

1.3.1 The Heritage of the Urban Issue

In general, social issues in town are perduring even in our post-industrial economy, not based any longer on production of goods. We have therefore to upgrade the classical urban social issues, linked to the industrial town context. It is though very important to refresh this classical conception before moving beyond it.

Both process of industrialisation and urbanization are to be considered. Western towns are developed around an administrative and commercial center. Max Weber conceptualised in a very clear way the genesis of the urban shape in the western world. According to this german sociologist, the emergence of the urban phenomenon goes hand in hand with the advent of legal rational power, based on a bureaucratic apparel. His taxonomy of towns is based on five factors of urban cohesion.

The first and foremost of these factors is the economy; the heart of towns is the marketplace. A town does not exist if there is not any regular exchange of goods, and those are an essential part of the habitants existence. Weber distinguishes the towns where the production or the usage of goods is paramount; respectively, they base their income on industry on one side, and on services and commercial activities on the other. This distinction today seem anachronistic, as the European economy is mostly post-industrial, and the highest part of the value creation is on activities linked to the tertiary sector and not any longer to the production of goods.

The second factor is the security. Any town, as a marketplace, would be insignificant if it could not assure protection. The ideal kind of transformation of the western town is the "fortress town." This unified continuum of a safe, secure and commercial place guarantees both the commercial and the military peace, which are necessary conditions for the long-lasting of this organizational form.

A third factor is freedom. Weber also believes that the city air makes people free. This freedom is first of all applicable to properties.

The forth point is the brotherhood. Any city inherits from the city state ideal. This has five characteristics: fortifications, market place, a tribunal, specific laws and associations, allowing an independent administration. Therefore, any citizen is a member of a brotherhood: of place, of right, and of goods.

As a counterpart, the citizen must be able to defend himself from the city, which induced a interdependency between the social layers constituting an urban continuum. This relationship drives a fifth and last factor: the legitimacy conflicts. As the interdependency relation is a necessity for the functioning of the different parts of a city, it can also be a theater for social fights and conflicts, having as a main focus the legitimacy of the power and of the possession of the resources. The management of a city is usually the business of a very small number of elected people. Furthermore, corporations try to claim the legitimacy of their work according to some rationale, more often than not economic.

This is often linked to the fact that the industrialisation process is often the cause of urbanization. The development of large industrial areas brings as a necessity the expansion of the urban population, draining a huge number of the population from rural areas. This massive population increase brought a development of the urban shape, both regarding the density and the space.

The social consequences of this development are a piling up of heterogeneous populations.

The appearance of social issue is natural within this context. The once administrative and market centers became industrial and popular. At the same time, conflicts linked to work conditions and survival are brought into the cities. Social issues became urban issues. Working classes became dangerous classes, and the original urban population, once upon a time composed only by the noble and middle-class, is afraid of this heterogeneity and all the problems they bring. The undesirable effects of this concentration of people started to be perceived; hygiene worries became closely linked to cities. Therefore, the trend at that time was to isolate the working sections of town from the administrative ones and put the workers under the direct control of the patronage.

The transformation of Paris made by Haussmann is a clear example of the above-described trends. The new laying out of the center of Paris bring a specific social, political, hygienic and aesthetic perspective. The objective is to unlock the town and to make it a safer place, through the creation of large boulevards and pushing certain classes outside the center; this is also functional of making the city more monumental, showing a clear representation of the running power.

This logic of moving the working class away from the center of cities had been followed for a long time. After World War 2, the necessity to re-construct cities in Europe fast, in large quantities and in a modern way, in order to have an economy of scale, was paramount. The developments of that era were not only the result of a necessity—the need of housing—with a strong constraint—a fragile economy: they were presented as modern jewels, and meant to unify different social classes. Besides, these new housing were extremely comfortable (in terms of bathrooms, central heating, lifts, …) in comparison to old buildings.

Yet, very soon it became evident that these kind of habitat had several problems. In 1973, for instance, France stopped all construction of large housing sets, as they were considered far from the expectations of the inhabitants and promoting an active segregation, as this housing complex were build in the outskirts of large towns.

Several researches showed that a relatively heterogeneous population changed within a few years, reducing to a mainly low-income one. This is mainly due to the distance from the urban center and the difficulty of getting there: whoever could, left the suburbia for relocating in a more convenient place, and the low cost of housing drove the ones that could not afford to be in a more central location. When the first issues with unemployment rose, these areas were the worst hit, as the population was composed almost exclusively by low-qualified workers.

Therefore, the majority of urban issues in the western world are linked to this historical trend to separate the different classes, and a fortiori of the segregation that follows.

The concept of "civil society," bearing its own power separated from the state, rises from social movements which are the effect of the tensions due to the separation. The process of industrialization and urbanization led to the creation of "urban society" which, according to H. Lefebvre, draws its essence from key elements of the historic city: the centrality, the public space, the street. Praxis—or practice of the city—that is an effect of it, cannot be assimilated to other perspectives. On the contrary, it takes all these different viewpoints and transforms them all. Through this and the "right to the city," Lefebvre argues that urban society can survive and reverse the industrial era that created it. According to his work, urban planning hides the capitalist strategy in which the user of a city disappears in favor of its market value; the user is therefore marginalized vis-a-vis of a consumer. The mercantile vision, though, would necessarily lead to the extinction of sociability in favor of market exchanges. This eminently political perspective is clear from this text: "[...] We had to denounce urban planning both as a mask and as instrument: mask for the State and political action, instrument of interests hidden in a precise strategy. Urbanism does not seek to shape the space as a work of art, neither for technical reasons. Urbanism shapes a political space" [5].

From a larger perspective, the urban problems covers not only what happens in a city but "a set of actions and situations typical from everyday's life strictly the progress and characteristics of which are depending by the general social organization" [6].

In other words, certain life or social conditions are intrinsically urban. As well, social practices such as the culture or the consumer habits are at least in part explainable by the social position of a person.

Now, how these problems have an impact on the development of new technologies, or, conversely, how the technological development can help in mitigating the outstanding issues that the urbanization is developing since centuries?

1.3.2 Spatial Distances Reflecting Social Differences

The phenomenon described beforehand is not new and even less unheard of. If we go back to feudal times, social and spatial separations are flagrant. The political, economic, administrative, and military kernel is concentrated in the fortress. Around it, there are the market towns which, in exchange for a relative military and merchant peace, can grow and prosper. While this example may seem too distant to be useful, things did not really change in our era. During the industrial revolution, housing estates were carefully built on the margins of bourgeois cities, as mentioned earlier. Today, popular residential areas follow the same spatial logic. Whether we look at ghettos in the US, banlieues in France, favelas in Brazil or periferie in Italy, all countries are facing these social and spatial relegation.

But if the phenomenon is historically and geographically recurrent since centuries, why there is a problem today? The answer is not in the forms of housing in itself, but in the global society in which these forms are realized. Indeed, the polarization of urban housing highlights a first problem of poverty and social exclusion. To avoid getting lost in the maze of a comparative approach that would not bring much to the analysis, we will develop briefly the French case. As already mentioned above, in France urban problems are often related to the theme of the banlieues. These habitat areas usually cover suburban complex and multifaceted realities. We will not focus on the "typical" and recurrent trends, as the goal here is not to make a case study. But in order to draw a general picture, without distorting reality, we need to identify the elements that make these neighborhoods particular spaces. First of all, these are areas of spatial concentration of social inequalities. This goes back to what we have already mentioned on the polarization. Yet, other variables must be added. Apart of the lower income, compared to other neighborhoods, banlieues concentrate a younger than average population. In addition, there are more employees and workers than elsewhere. Finally, the unemployment rate is often higher than the national average. As an added statistical fact, there is a higher presence of immigrants, or people that are culturally and/or by birth foreigners to the perceived French "orthodox values." This series of factors combined are leading to a delicate situation that the French state is struggling to manage. Indeed, it is an aggregation of structural, social, immigration, and urban planning problems, to the point that it became hard to say if it is more a people's problem or a spatial one. Anyway, it is a fact, sadly: the situation is deteriorating. The concentration of social unrest in these places make the ones who can, to leave these neighborhoods, weakening the diversity and its inherent dynamics. According to Pierre Bourdieu, there is a close link between places and social position. "The structure of social space is shown in the most diverse contexts, in the form of spatial oppositions, the living space (or appropriate) working as a kind of spontaneous symbolization of social space" [7].

In any case, this covering by the social to the spatial domain is more or less blurred by an effect of naturalization. In other words, historical and social phenomena can be understood as implicit in the very nature of things. Yet, this physical show of social logics contributes to objectify these struggles among different social groups.

As well, conflicts are often linked to specific places, as different kind of profits are associated with them. These can be about the location (close to scarce goods such as cultural infrastructure, health, education …); position or rank (prestigious address …); occupancy (size of the owned space). The social dimension permeates the relationship with the space. Indeed, possession of capital—economic, cultural, social, or symbolic—determines the ability to dominate and own space, either physically or symbolically.

Moreover, the stakes in terms of location is part of a twofold logic of being close and moving far away. On one side, the search for proximity to rare and desirable goods and services; and secondly with the distancing—or exclusion—of people and unwanted things. Constitution of homogeneous groups based on spatial difference became the norm, helped by the state and its politics.

This segregation is strengthened by a particular phenomenon, that J. Donzelot call "affinity urbanism." First of all, it is possible to observe an increasingly widespread suburbanization. This implies that the city grows outside its historical functional limitations. Second, as individual mobility increases, the link between a territory and a population gets more loose. The weight of the neighborhood decreases and the residence becomes selective. The places where people choose to live are not functionally related and prescribed as in the old industrial city. This process is gearing towards a phenomenon, where the distance is chosen and selective, based on considerations relating to lifestyle, entertainment or security. Clear examples of this trend are the gated communities, where individuals benefiting from certain economic and social resources choose to live in a "among-pairs-group" away from the global society, often in a spatially separated area. For instance, in some countries, such as Brasil or South Africa, it is common to see estates protected by concrete walls and security at the entrance.

Without getting to this extreme point, the current problem of cities is to be split between these antagonist ghettos logic. The result is a double polarization: towards the "low-end," where new forms of marginalization and inequalities take place, and towards the "high-end," where the cultural, economical and political powers tend to unify and separate from the larger population.

1.4 Towards New Trends

The usage of digital technologies to enhance the quality and attractiveness of the city, to provide services to the inhabitants and tourists and to improve operational costs became a mainstream trend in this decade. In the recent past, there has not been a single city that did not use the word "smart" for labeling some of their initiatives. However, how can the use of digitalisation impact the life of citizens, given the sociological and historical perspective illustrated above?

In any given city there we can distinguish three distinct factors:

1. Aspects that do not change, or evolve with a speed which is by far slower than human life. It is the case of history, for instance, or geography, or climate. The Coliseum is in Rome; the Statue of Liberty in New York. The average rainfall in Tokyo in November is 100 mm; Marseille is on the sea, Stockholm is on an archipelago, and Paris is on the Seine river. Now, while in course of centuries this can change (Pisa, when founded, was on the sea, while today it is around 10 Km from the coast because of sediments brought by rivers), the pace is extremely slow and we can consider these parameters are "stable."
2. Aspects that change slowly, and require a lot of effort and commitment. Cultural aspects, for instance, or major urban modifications. Jordaan district in Amsterdam, for instance, was a few decades ago a working-class neighborhood; nowadays is arguably the most expensive area in Netherlands. Detroit population dropped by 60 % since 1950, and 25 % since 2000. Always in Amsterdam, the construction of the north-south underground line started in 2002, and it is supposed to be finished not before 2018. Therefore, while these characteristics are often slow to move, specific trends can have a strong impact on those, in particular social and economic trends tend to draw a different picture depending on the historical period.
3. Aspects that can be changed easily. These aspects, which are often "cosmetic," may nevertheless have an impact on the quality of life in a specific town. Use of NFC payments for public transport, for instance, or specific traffic restrictions, or else laws allowing (or disallowing) specific behaviors like smoking in public places. However, these hardly modify the structure of society and the spatial issues as described above.

As discussed earlier, any city is a economical and social product. Urban spaces are often conceived as a functional organization in which different areas have specific functions (residential areas, commercial ones, industries, ...). This politic, however, does not necessarily lead to good results. The city peripheries (hinterlands) are often de facto on the margins of the city social and economical dynamics.

It is important to notice is that Smart City projects as they are often advertised are addressing only the third category. A common example is a service which seems to be widely used to indicate the smartness of a city: parking sensors with a dedicated app showing the available space at real time. While the usefulness of such developments can be debated, and even positively argued, it does not tackle any of the city issues at its roots, but rather promote a further digital divide exacerbating the existing separation between different realities within the same city.

On the same line, some advertisement of these smart city projects are even clearly showing their beliefs and their intentions. When we read headlines like this

> Our Cities are rapidly becoming both more populated and more complex. Because of this, people's security needs are constantly changing. That's why Hitachi is developing solutions to keep people safe in their communities [8].

It is flagrant that—at least—some of these planned solution for smart cities do not address the trend to ghettification, trying to solve historical problems, but rather promote an even stronger segregation of inhabitants, keeping them safe *in their communities*, not beyond …

In our opinion, digital technologies are extremely powerful as they virtualise the notion of space. This can clearly overcome the spatial separation developed for centuries, and allow a "brand new start" as the digital space add another dimension, which is still mostly unchartered.

Furthermore, not all is lost. New trends in urban planning are focused on environmental-friendly cities which are eco-sustainable, and several experiments are sprouting, such as the vertical gardens or agricultural spaces within the city limits. These urban utopias, fully belonging to the Smart Cities phenomenon, conceptualize certain strong elements that could be the building blocks of the cities of tomorrow. For instance, a breakage of spatial barriers: mixing agricultural spaces within the urban territory will allow a different symbiosis between nature and urban society. Relevant work in this area has been already made (by C.J. Lim, for instance).

Therefore, with a little imagination and hope, it is possible to consider this current trends in a positive way. We briefly discussed about the fragmentation of society, that transforms a city creating "self-segregation" zones based on attraction/repulsion process. However, some current development allow people with a different logic, to coexist and to share the same space. We are talking here about the eco-neighborhoods that reconcile economy and ecology, and often also social links. This type of habitat was very marginal and rural rather than urban until a short time ago. The passage from the countryside to the city is due to an evolution of mentalities and legislations favoring a more environmentally friendly living, This trend can therefore be seen as extremely positive as it tackles not only the cosmetic aspects of cities, but leverages technological advances developing a sustainable vision for future generations, on both social and economical level.

References

1. UN: World Urbanisation Prospects (2005). http://www.un.org/esa/population/publications/WUP2005/2005wup.htm
2. EC: Towards a new culture for urban mobility (2007). http://ec.europa.eu/transport/themes/urban/urban_mobility/green_paper/doc/2007_09_25_gp_urban_mobility_memo_en.pdf
3. EC: Europe 2020 strategy (2016). http://ec.europa.eu/europe2020/index.htm
4. Donzelot J (2009) La ville à trois vitesses. 2915456488. Editions de la Villette, Paris
5. Lefebvre H (1970) La revolution urbaine. 2070352161. Gallimard, Paris
6. Castells M (1975) Luttes urbaines et pouvoir politique. B018WKFA9K. La découverte, Paris
7. Bourdieu P (2015) La misere du monde. 2757851527. Points, Paris
8. Hitachi: Hitachi social innovation business (2016). http://www.hitachi.com/businesses/innovation/solutions/index.html

Chapter 2
Who Is the Assumed User in the Smart City?

Katherine Harrison

every city wants to be a smart city nowadays.
March and Ribera-Fumaz [14], p. 1.

Much hyped and funded to the tune of billions of dollars annually, smart city projects claim to solve a range of contemporary urban problems including "air pollution, traffic congestion and assisted living for the elderly" [20], thus providing cleaner, safer, more energy-efficient cities of the future. Local and national governments, technical companies and researchers are keen to emphasise how the integration of various kinds of technologies (for example, big data analysis, Internet of Things, wireless sensors and cloud computing) are key to tackling the challenges posed by an ever-increasing urban population. The mission statement from the recently created Institute of Electrical and Electronics Engineers (IEEE) Smart Cities Initiative, for example, reads:

> we see an opportunity for IEEE to assist municipalities in managing this transition to urbanization. This would include raising awareness of the benefits and downsides of technology and help guide the appropriate uses of technology [23].

However, at the same time, smart cities have also been criticised for not taking into account the needs/customs of local inhabitants or for increasing social divides. This latter critique has been directed particularly at brand-new smart cities. Laveesh Bhandari, for example, talking about the push to create more smart cities in India such as the currently-under-construction Gujarat International Financial Tec-City (GIFT), argues that such cities will potentially create "special enclaves" where only the economically secure will be "enjoying the privileges of such great infrastruc ture" (quoted in [18]. Much of the current debate about smart cities in both popular media and the social sciences thus centres round the role of the inhabitant. To what

K. Harrison (✉)
University of Copenhagen, Copenhagen, Denmark
e-mail: tjx856@hum.ku.dk

K. Harrison
Lund University, Lund, Sweden

V. Angelakis et al. (eds.), *Designing, Developing, and Facilitating Smart Cities*,
DOI 10.1007/978-3-319-44924-1_2

extent are existing inhabitants being enrolled in the process of "smartification" or what kinds of future inhabitants are envisaged occupying future cities such as GIFT? In the hype about technological advances, is the diversity of human needs and behaviours at risk of being reduced to that which smart city designers think is efficient/sustainable/"smart"? Or, more precisely, to what extent are these cities designed with the technology (rather than the citizens) at the forefront of the designers' minds? To what extent is the diversity of potential citizens' and their needs considered? These are important questions because smart cities *are* being hyped, *are* being funded and *are* being built at an increasing rate. The WHO predicts that 70 % of the world's population will live in cities by 2050 and this means that the cities where future generations will live are likely to be "smart" to some extent, and the design of these cities has a very real impact on what kinds of lives can be lived in them.

This chapter thus contributes to the existing scholarship that takes a more critical perspective on smart cities, particularly in relation to considering current or future inhabitants. Much of this book focuses on the technological developments that have made the smart city possible. Instead, in this chapter the focus is primarily on the everyday user of these technologies, both the future user as the designers imagine her/him to be and the existing users who are already living in and interacting with smart cities. The overall aim is not to discard the idea of the smart city, but rather to encourage greater thoughtfulness as to how smart technological solutions might be developed which can take into account a range of different lived experiences by city dwellers. This approach is often called the "bottom-up" approach, and usually refers to projects that try to ground themselves in users' everyday experiences and which will be widely adopted by the urban population. The recently launched Google Sidewalk Labs initiative, for example, plans to assemble "small teams of experts to brainstorm ideas and launch experimental projects that have the potential to catch on virally with large numbers of city dwellers" [22]. Commenting on this initiative, Carlo Ratti of MIT praised this people-centric approach:

> This kind of "bottom-up" approach has the potential to bring rapid change at low cost (…) in contrast to the more centralised, "top-down" tech projects that cities have used in the past to reduce costs or improve services [22].

The "bottom-up" approach appears to offer not only a more equitable model for urban development that recognises a diverse range of users, but also contains the promise of widespread adoption by the local residents. Taking careful account of the diversity of human needs has the benefit of improving chances of new technologies being adopted by the whole community because it starts from the users' experiences rather than from the technologies. As Madeleine Akrich notes: "the success or failure of innovations frequently depends on their ability to cope with dissimilar users possessing widely differing skills and aspirations" [1]. This is not to say that existing solutions have ignored the user. Rather, that the existing technology-centric paradigm concerns itself primarily with opportunities and challenges connected to smart city technologies, instead of reflecting on how these technologies may assume certain users behaving in certain ways—at the expense of other users or behaviours. Existing solutions assume certain users, but this assumption may be somewhat limited.

Starting from the user experience prompts an important and different set of questions about smart city technologies. This chapter starts with the premise that the design and development of a smart city "artefact" (a term which is here used to refer to a system or tool or technological object) is a process that assumes certain users and usages [1]. The assumed user of an artefact shapes the affordances and limitations of the artefact; designers and developers strive to create an artefact that balances their understanding of the needs of the assumed user and the limitations/affordances of the technology itself. Starting from this point, this chapter asks: what is an assumed user? Who is the assumed user in the contemporary smart city? And, why might reframing smart city technologies as "artefacts" help us to think more critically about the balance between people and technology in urban environments? These questions will occupy the first half of the chapter.

In the second half of this chapter, we will turn our attention to some examples from contemporary smart cities projects. Here we will explore characteristics of the assumed user in particular cases and subsequently ask how limitations in the understanding of the assumed user may lead to the needs of some groups being overlooked during design and development—precisely because they do not "fit" the profile of the assumed user. This half of the chapter therefore poses the question: which citizens' needs are not adequately addressed because they do not constitute the assumed user(s)? and what does this mean for the success or failure of smart cities initiatives?

I conclude with some questions for my readers, who I assume are from more technical disciplines, about how to develop an approach that balances both technical affordances and human requirements.

2.1 Assumed User(s) and Artefacts

This chapter on is based on the premise that what counts as valuable knowledge in one scientific discipline may be different from what counts in another, and that you and I (as author and reader) may have quite different ideas about this. So, to get us on the same page.... A researcher or a student of a technical discipline might be more accustomed to thinking about smart cities from, for example, the perspective of creating "dependable, reliable, and secure networks of heterogeneous smart objects" [29]. What counts as valuable knowledge here is the affordances and limitations of the technology itself. This might include considering how efficiently or securely a technology functions, or whether it can be easily implemented in a particular environment. In contrast, social science researchers use their critical perspectives to "unpack, contextualise and make theoretical sense of smart city rhetoric and initiatives" [10] and are thus more interested in when, where, how and what kind of activities city inhabitants take part in, and whether these are accurately reflected/responded to in any smart city initiatives proposed for implementation. In particular, within the social sciences, there is a well-established tradition of taking context into account; understanding the broader socio-historical context is essential in order to understand the development, use and understanding of any technology. This promotes a lively questioning of the "common sense" or "naturalness" of an artefact.

As a researcher from the social sciences, specifically science and technology studies, I am going to use the term "artefact" a lot in the following chapter. I use "artefact" to refer to material objects, facts or ideas. It is a word which can be applied to a wide range of "things" but which fundamentally understands them all as socially constructed, not as "natural" or neutral [17]. When we look at "facts", for example, it is tempting to think that we are seeing something objective, which was discovered as part of a linear process of experimentation, and which is possible to present in a transparent way. Or, as Rob Kitchin puts it:

> Smart city advocates imagine themselves as creating technologies, techniques and visions that are scientific, objective, commonsensical and apolitical [10].

The term "artefact" can be used to bring our attention to "facts" as the results of complex negotiations between different interest groups (this could include—but is not limited to—competing scientific paradigms, political agendas or budgetary constraints). In this chapter, I am using the term "artefacts" when referring to smart city technologies in order to emphasise that the objects and ideas being created and implemented by smart city designers, developers and engineers emerge within particular contexts that affect the artefacts themselves. This emphasis on context can, for example, help to reveal the particular political decisions that shaped which technologies were (or were not) funded. Other examples of important contextual information might include organisational infrastructures that determined what information was considered "important" to capture in a database and which could be discarded [8], or the terminology used to describe the artefact that implicitly suggests it as a tool for a particular person [3].

Artefacts (be they material objects like bicycles or less tangible objects like facts) take a particular shape, thanks to the questions we pose, the limitations/affordances of the tools we use, and the ways in which results are interpreted. In other words, human beings and our personal, professional and social contexts shape the artefacts, including how we understand the "smart city" itself, the technologies that are needed and the inhabitants who populate it. The creation of any artefact thus takes place within a particular context by particular people, and it contains within it particular ideas about how it will be used and by whom, and these may be more or less clearly articulated. In a commercial setting, for example, when a new product or service is developed this process involves trying to imagine who might want to buy it. In so doing, a picture of this potential customer is built—either through explicit techniques (such as marketing surveys or consumer testing) or through implicit techniques (such as the designer's personal experience or from looking at the existing users for similar products) (see [1, pp. 169–175] for a more detailed outline of explicit and implicit techniques). Similarly, the development of any smart city artefact (whether intended for sale or not) contains a number of explicit and implicit ideas about how it will be used and by whom. This is what can be called the "assumed user". In the case of smart cities, the "assumed user" could be a list compiled by the developers of the different kinds of people who might want to use the night buses in a town, or the municipal organisations who would like to know more about air pollution at particular intersections.

However, if the idea of the assumed user does not accurately match the end user there is a risk that the project will not address a community's needs, will increase social divides by only serving some groups in the community and will not be adopted long-term. Thus, reflecting on assumptions about the users when developing a smart city technology can significantly improve the everyday lives of users and the process of implementation. In short, I identify two aspects of the current, technology-centric approach to smart cities where incorporating critical, citizen-centric perspectives can make a big difference:

1. To stop assuming that technologies are objective or apolitical, and
2. To get a better understanding of the diversity of user needs/behaviours.

The call to think more critically about smart city technologies is clear in both research and policy recommendations. For example, the recent report by innovations charity, Nesta, titled "Rethinking Smart Cities from The Ground Up" draws attention to this:

> 'Smart cities' offer sensors, 'big data' and advanced computing as answers to these challenges, but they have often faced criticism for being too concerned with hardware rather than with people.

Advocating greater use of collaborative technologies, one of the reports' policy recommendations is to "take human behaviour as seriously as technology" (Nesta policy recommendations for smart cities, [15]. Meanwhile, Rob Kitchin, in his recent research article titled "Making sense of smart cities" draws attention to how:

> Left untouched are issues such as panoptic surveillance, technocratic and corporate forms of governance, technological lock-ins, profiling and social sorting, anticipatory governance, control creep, the hollowing out of state provided services, widening inequalities and dispossession of land and livelihoods [10].

With this in mind, how can we engage with emerging smart city technologies to address the concerns being raised by researchers and policy makers? In this chapter I suggest that one entry point involves paying closer attention to who the assumed user is when designing or implementing smart city artefacts. In the following section, I suggest some ways to try and identify the "assumed users", and problems that might occur for those who do not "fit" the profile of the assumed user.

2.2 The Assumed User in the Contemporary Smart City

Early ideas about smart cities tended to stress the technological innovation that would characterise these environments [10] or the financial competitiveness that would be the result of these fast-paced, well-managed urban spaces [13]. These early imaginings of smart cities presented a visually stimulating, almost science fictional image of the buzzing, technology-dominated city where success was to be measured in terms of economic success and technological advancement. At best, this vision assumed technologically savvy citizens who were active producers and consumers, financially contributing to society. Absent from this picture are concepts such as "wellbeing" or "work–life balance" or "sustainable living". Also absent are discussions of how these technologies might serve members of the population who do not "fit" the idea of the assumed user; for example children, the elderly, those with disabilities, non-native speakers or those less comfortable using certain technologies.

Visions such as this did not die out with the 1980s, big hair and shoulderpads though. The Center for Innovation, Testing and Evaluation (CITE) is a contemporary billion-dollar experiment in building an unpopulated, test city where new technologies such as smart devices can be trialled while "sidestepping those pesky humans" [4], see also [5]. The managing director of the company leading this project went so far as to say: "It will be a true laboratory without the complication and safety issues associated with residents" (as quoted in [4]. Experiments such as CITE show that for many the thrill of developing innovative technologies remains the most interesting aspect of smart cities. Smart cities where the emphasis has been on using technologies to maximise efficiency have also been criticised for their ambience (or lack of it), as in the case of Brasilia (see Fig. 2.1).

Inevitably, when we design something, our own perspective and experience plays an important role in shaping how we imagine the technology might be used. It is much harder to imagine the different challenges or potentials experienced by another person using the same tool. And yet, focusing primarily on the technologies or not taking the diversity of human experience into account can result in low levels of adoption or even complete failure of a project. The world of Information and Communications Technologies for Development (ICT4D), for example, provides some striking examples of technologies designed and built in the Global North for the Global South that failed because the designers had not understood the local perspective. In his book *Smart Cities: Big Data, Civic Hackers and the Quest for a New Utopia,* Anthony M. Townsend recounts the example of the Lincos project (designed by MIT engineers 1999–2001) which aimed to pack into a single shipping container everything necessary to connect a rural community to the Net [24]. This container would then be airdropped into local communities in the Dominican Republic and Costa Rica. The project failed spectacularly in the Dominican Republic because "(g)overnment officials apparently found the container design, so revolutionary for the MIT engineers, a symbol of poverty. Dominicans wouldn't be caught dead walking into one" [21]. The Lincos example shows what happens

Fig. 2.1 Brasília. Completed in 1960, it was heralded by boosters as the first jet-age city. But critics later mocked its sterile architecture and desolate streets

when the assumed user does not match the real user; in the Lincos case a lack of accurate real user knowledge based on understanding of the specific use context for the technology significantly contributed to failure of the project. While the Lincos project exemplifies transnational challenges, it is worth remembering that these gaps in understanding can also occur closer to home. Differences in physical abilities, age, gender, ethnicity or educational background, for example, affect how comfortable a person is when moving through urban space or engaging with technologies—both of which may affect adoption of smart city services.

Both commentaries about smart cities and the types of smart technologies show an evolution of the idea of the "smart city" that incorporates increasing awareness of social divides and sustainability issues. However, an overwhelming focus on technology and a need for more "local" knowledge remains prevalent and much of the hype around smart cities continues to be connected to innovative technologies. The assumed user of a smart city therefore often remains less well defined than the technologies being proposed for their use.

> the smart growth agenda may have progressive potential, it is also in danger of being used as a means to discipline cities and their populations, reducing sustainability and the urban question to a technical discourse [7].

This problem is not specific to "smartification" and has been noted in other aspects of urban planning. There is a well-established body of research, for example, showing how the layout of urban spaces may affect the freedom of movement of women (poorly lit streets or limited-use thoroughfares may cause women to fear being attacked in these spaces and thus limit their movement through urban space particularly after dark, see for example [11]. The lessons learnt from urban planning are also applicable to the integration of technologies into cities with the aim of making them "smarter". The ways in which we design smart cities have an effect on how people live their lives; where and how citizens can move, and who feels welcomed or excluded in particular spaces become questions of what the technologies permit or prevent.

In order to respond to a wide range of citizens' needs it is necessary to reflect on the assumed user who lies behind design and development of smart cities, and to ask whose needs are not currently being addressed. Current wisdom advocates the "bottom-up" approach as the best way to achieve this. Usman Haque of Connected Environments gave a nice example of this in 2012 in a specially authored piece for *Wired* magazine:

> A citizen-led air quality monitoring system would see measurements taken at a much higher resolution in places (e.g. at the height of a children's stroller) that the official network just doesn't reach. Children could learn which side of the park to play on. People could decide to walk different routes to work. They could measure the specific impact of their own cars [9].

Haque's argument makes a connection between empowered citizens and technologies. It requires data being shared with citizens in a user-friendly way that allows them to make everyday decisions that affect their quality of life and reflect on how their own behaviour might be contributing to pollution levels. It also includes children and those who do not use transport (public or private) for their daily commute. These may be the less obvious behaviours, which do not contribute directly to the economic success of an urban environment, but which form part of a vibrant, diverse community. However, it also assumes users who are sufficiently tech-savvy to be comfortable interacting with an air quality monitoring interface (however rudimentary), and able-bodied citizens who are able to go to the park or walk to work. The above example of a citizen-led air quality monitoring system thus starts to model a balance between technology and citizen, and attempts to broaden the range of human experiences of the assumed user, when compared to the early imaginings of the smart city and its assumed user.

Smart city artefacts are designed within particular contexts with a set of assumptions about who is going to use them and in what ways. Some of these assumptions may be made explicit, for example the need for long-range communication between devices. Some of these assumptions may be ideas that developers have about users without really realising them. In the following sections, I illustrate this with examples of two smart cities, one new venture in which smart technologies are being embedded into the infrastructure of the city from its inception, and the other an example of the "smartification" of a well-established city. In each case, I am focusing on the assumed users of the smart city artefacts.

2.3 The Case of Palava

Palava claims to be the first smart city in India and, as such, has been dubbed "The Future City" by the company developing it. It is scheduled to be complete by 2025, but thousands of housing units have already been sold and built. Located in the Mumbai region, one of its main selling points is its proximity to established transport and business hubs. Palava is a commercial venture, developed as a partnership between Lodha (a real estate group) and a number of international companies, including IBM. It is also one of a wave of smart cities currently being built in India, supported by the government.

Palava is an interesting case study because it is a brand-new city, envisaged from the start as a smart city and owned by a commercial firm, rather than a municipality. A significant amount of the city has already been constructed, and a well-developed media and marketing strategy is in place outlining key selling points to attract businesses and residents. This means that a number of promotional and informational materials are available online about the city. From these online materials, which include films on YouTube, the Palava website and various press releases, we can learn a lot about the assumed users of this city and how developers understand the concept of a "smart city". A number of themes can be identified in these materials. In the following, I examine two of these materials: the "Palava Megacity" short promotional film from Lodha (available on YouTube) and the Palava website [26]. The promotional film is 6 min 35 s in length and dwells on different aspects of the city, illustrating these with photorealistic images, plans for the city and maps, accompanied by a soaring soundtrack and polished voiceover in English. It gives a flavour of the city, rather than providing many specific details. The website provides more detailed information, including housing prices, city transport, energy and water supplies, information about Lodha and the Palava City Management Association. Of the many aspects of Palava sold on the website and in the film, there are two themes that dominate: innovation and economic success.

The promotional film for Palava uses "vision" as an overarching theme [25]. This term is used repeatedly in the voiceover and is connected to the idea of it being a futuristic, aspirational city. The promotional film uses expressions such as "largest ever private, planned self-sustained city", "technologically advanced and futuristic", "to wash away the old" and "the city of the future". These expressions are accompanied by aspirational imagery of breaking waves and soaring skyscrapers, a sweeping soundtrack and concluding with the tagline "Palava—The Future City". As such, Palava positions itself as being a radical evolution from existing cities in India, one characterised by: (i) the latest technologies and (ii) the scale of the development.

The theme of innovation is developed on the website through a combination of both aspirational text such as "Palava isn't just a new place to live, it's a new way to live" [27] and details of the telecommunications and transport infrastructure built into the city. Smart cities in other countries are mentioned as inspiration or models for aspects of Palava. However, it is worth noting that Palava positions itself as a leader within India, comparing its offering both explicitly and implicitly with other Indian cities, and focusing particularly on technological artefacts not widely available within India such as smart cards, surveillance, solar panelling and rain-water capture. Palava is framed as offering "innovation" within a specific national and cultural context. To many in an international audience familiar with such amenities, 24/7 electricity may be less innovative and more expected.

One of the other main selling points of Palava concerns its location and resources for businesses. This takes the form of Special Economic Zones, IT Park, offices and a central business district, all of which are featured in the promotional film. The website provides more information about this in various sections, including the "City Advantages" section of the site. Here there are details of the low operating costs offered by Palava, accompanied by graphics, which claim to show annual cost reductions for business over a number of metrics such as lower utility bills, walk-to-work benefits and communications infrastructure. In a neighbouring section, the different groups of customers (businesses, citizens and tourists) envisaged as the basis for Palava as a high-growth market are detailed, and once again accompanied by graphics. Here plain text, specific figures and use of tables and graphs work to support Palava's claim to be a "City of Opportunity".

Financial success as a characteristic of Palava is conveyed not only through its framing as a location for businesses, but also as "a destination for the rich and famous". Detailing the different types of housing, the word "luxury" is repeatedly used, together with glossy images of house interiors and pictures from the sales events. In each case, the voiceover informs us how many "units" were sold within the initial sales period, framing this as a commercial real estate success. Palava is thus simultaneously framed as offering good opportunities for businesses due to its lower cost options, and as a location for luxury housing. Its ideal users are thus skilled workers/business people or those who are already wealthy.

Palava appears to be a city that draws on many aspects of European cities as an aspirational model. As such, it sets up a power dynamic in which local culture, skills and techniques are designated second best. The contact details at the top of the home page give numbers for the Indian, US, UK and UAE offices, suggesting that Lodha see many of their potential customers as an international group, possibly drawing on the economically successful Indian diaspora. The film does not use the term "smart city" although it refers to environmental and traffic management technologies commonly associated with smart cities. The website however, calls Palava "India's First Smart City" and in its consistent focus on technological innovation, sustainability and financial growth is in line with much of the prevailing rhetoric on smart cities.

By fitting itself with the dominant model of a smart city in which technology is used to facilitate economic growth and efficiency, Palava to some extent fails to take into consideration its citizens. The city is primarily defined by buildings and facilities, rather than by people; in the promotional film there are many images of buildings, roads and transport that are often taken from overhead, so the people are reduced to scuttling ants. On the few occasions when people are shown in close-up in the film, it is worth noting that they are often light-skinned, able-bodied adults expensively dressed in outfits of the Global North. As such, they epitomise the active producers/consumers imagined as the users of early smart cities.

The only evidence of citizen engagement in the city appears in the section of the website that describes the Palava City Management Association, a body comprised of "citizens of Palava, expert city administrators and urban planners" which will be responsible for the running of the city [28]. The involvement of citizens therefore takes place after the city has been constructed, rather than during its design and planning stages. Overall, Palava is promoted as a private city, only available to those who can afford to live there or who wish to subscribe to the particular luxury lifestyle on offer. Indeed, some commentators have suggested that cities such as Palava will be "more fortresses than places of heterogeneous humanity, because they are meant only for specific classes of people" (Pramod Nayar quoted in [18]. This may contribute to a growing divide between haves and have-nots as these private enclaves draw natural resources from the region with the blessing of the municipal government and under the guise of boosting the economy. Something of this can be seen in the use of surveillance and security technologies in Palava such as the "intelligent security and monitoring system that includes everything from electronic access systems and fire alarms to CCTV surveillance and street-level panic buttons" as well as smart identity cards for every citizen. Use of systems such as this to create and police the boundary between haves and have-nots was noted by academic and author Pramod Nayar who wrote that:

> Smart cities will be heavily policed spaces (…) where only eligible people—economically productive consumers (shoppers) and producers (employees)—will be allowed freedom of walking and travel, while ambient and ubiquitous surveillance will be tracked so as to anticipate the 'anti-socials' (as it appears in [18].

2.4 The Case of Barcelona

Since the early 2000s, Barcelona has been working to reinvent itself. The most recent manifestation of this was the announcement in 2012 by the newly appointed mayor of his intention to transform Barcelona into a global model for the smart city. This included a number of projects, some of which have been realised, and some of which are ongoing. One of the most discussed is the project to transform a large run-down area of the city, Sant Marti, into a new knowledge hub, now known as 22@Barcelona. Aims of the project included regeneration of the area to bring together industry, university and technology transfer companies, as well as boosting employment (a large percentage of which was envisaged to be an international workforce). The project also incorporated housing and social amenities. Located in an area near the centre of this well-established and bustling city, 22@Barcelona was thus quite a different proposition to Palava. The existing infrastructure and population meant that this was more a project of "smartification" of an existing urban space, rather than being able to build something entirely new. In common with

Palava, however, was the goal to develop simultaneously multiple aspects of the space to create a work-live environment: "the execution of a strategy that integrates, economic, physical and social regeneration with investment in economic and social programmes as much as in property development" [12].

It is widely acknowledged that the 22@Barcelona project had ambitious goals, and even the most positive commentaries on the smartification of Barcelona acknowledge (albeit briefly) that the smart city initiatives which took place in Barcelona faced a variety of challenges and that there have been problems with adoption. The reason for this is often suggested to be a lack of awareness of the broader context by designers and planners. This takes several forms, including the broader economic context, the labour market, and organisational cooperation. In their study of Barcelona, Bakici et al. suggest that "Barcelona faced certain challenges such as providing exact and appropriate infrastructure, deployment and management of wireless networks, creation of triple helix, networks, clusters and collaborations" [2] and tentatively admit that "local engagement and collaboration across departments could be challenging sometimes" [2].

Leon goes further in his analysis, identifying five major challenges that this project faced, most of which centre around the lack of suitable labour and industry in the 22@Barcelona area, namely "The human capital was originally not aligned with the needs of industry clusters" and "The incipient level of local entrepreneurship was very low", as well as a lack of early stage venture capital funding and the absence of large firms with headquarters in the region [12].

This lack of attention to the existing context could also be seen when, shortly after the 2012 announcement about the latest phase of regeneration, two of the sites identified for development were discovered to be unavailable for use due in one case to being occupied by squatters (who then needed to be relocated), and in the other case to having been turned into a food bank by the local population. Lack of awareness of local populations and their needs here highlight a gap between idealistic smart city planning and lived realities. These, it has been suggested, may have contributed to difficulties in adoption of the smart city initiatives:

> …flagships of the new urban model in Barcelona, have already encountered the opposition of neighbourhood associations, and new urban infrastructures are far from being examples of democratic participation [14, p. 10].

Both Bakici et al., and March and Ribera-Fumaz draw attention to the particular economic challenges faced by Barcelona, and emphasise that this aspect cannot be underestimated by planners of future smart city initiatives:

> As new urban smart interventions are being designed and applied, little has been explored about how they are inserted into a wider political economy and ecology of urban transformation [14, p. 8].

This can be seen in other critiques of the transformations effected in Barcelona in order to transform it into a smart city, most particularly the lack of engagement with private citizens. Like Palava, the transformation of Barcelona was driven by a clear vision. In this case, the vision of chief architect, Vicente Guallart, who re-envisaged the city on the same model as the Internet, a flat network of connected hubs. At the heart of this vision was the use of ICTs and technologies to improve environmental efficiency. Guallart did stress that such technologies were only of use if they improved people's lives. However, as the above examples suggest, this vision of smart technologies that improved people's lives failed to be realised in a way that took into account local conditions, and which ultimately was technology-centric rather than people-centric.

2.5 Conclusion

> In truth, competing visions of the smart city are proxies for competing visions of society, and in particular about who holds power in society [16].

In this chapter, I have used the concepts of the "artefact" and the "assumed user" to suggest how assumptions and biases may be unwittingly included in the design and implementation of smart city initiatives. In line with recent calls to make such initiatives more people-centric or "bottom up", I argue that a critical reflection on assumed users is a useful way to avoid thinking of these technologies as unproblematically "scientific, objective, commonsensical and apolitical" [10]. Rather, we need to recognise that smart city technologies are "artefacts", constructed in a particular socio-historical context and thus shaped by politics, budgets and local culture as much as by the material limitations or affordances of the technology itself (such as processing power, bandwidth, etc.). Understanding smart city technologies as "artefacts" highlights the human aspect of design and use, something that may be lacking in "top-down" approaches. I have illustrated this with two examples—one from a brand-new smart city, and one from an established city in the process of smartification. Both cases support March and Ribera-Fumaz's critique that: "the Smart City is a rather empty and ambiguous concept that is being deployed more on an imaginary and discursive level, rather than materially" [14].

Examining promotional materials for Palava helps us to identify the financially successful producer/consumer who is the assumed user for this particular smart city, and consequently reveals those members of society who are missing from this glossy picture of urban living—an absence that commentators suggest will lead to long-term problems within Indian society. Reading accounts of the smartification initiatives in Barcelona reveals how a mismatch between the developers' assumptions about the users and the actual users of a space contributed to poor adoption of the processes. Palava and Barcelona are examples of the different ways in which smart city technologies and visions are being currently implemented. They have important differences (for example, commercial vs. municipal leadership, and smart-from-inception vs. smartification) but both show a primary concern with technologically led change rather than citizen-led change. Furthermore, in both cases it is possible to see how these initiatives may aggravate existing social divides as segments of the population are erased from the vision of the city.

As the above discussion suggests, it is clear that it is essential to think critically about the assumed user both at the planning stage and during use for reasons of both equality of opportunity and successful adoption of smart city initiatives:

> …it is of special importance to find ways of ensuring that certain user representations—which would otherwise not be considered by the innovators and the entrepreneurs—are taken into account. (…) In addition, public authorities should identify, create and/or use a number of mediators between innovators and end-users to redefine the demand and thus allow new user representations to be considered [1].

How then to do this? How can designers, developers and engineers enhance their existing practices and pay closer attention to the kinds of assumed users implicit in their technologies?

One way forward might be to have more interdisciplinary development teams that include those whose interest is primarily with the users and citizens, who have expertise in human behaviours. For example, in a recent article in *The Guardian* newspaper about the Future Cities summit, the writer highlights the following solution put forward by one of the summit delegates:

> One sceptical observer of many presentations at the Future Cities Summit, Jonathan Rez of the University of New South Wales, suggests that "a smarter way" to build cities "might be for architects and urban planners to have psychologists and ethnographers on the team." That would certainly be one way to acquire a better understanding of what technologists call the "end user"—in this case, the citizen. After all, as one of the tribunes asks the crowd in Shakespeare's Coriolanus: "What is the city but the people?" [16].

Indeed, interdisciplinary collaboration between technical and social sciences seems key to developing smarter "smart cities". By integrating these two distinct perspectives, a more complex understanding of the citizen–technology relationships in contemporary and future smart cities might be reached, one that guarantees access and inclusion for all, and a better chance of adoption by the citizens themselves. This kind of collaboration might seek to engage with the following kinds of questions:

How can you include a wider range of user experiences, needs and behaviours into the technical design and development process? How might including more experiences into the technical development stage make the solution more robust and more likely to have a smooth adoption process by citizens? To what extent do technical developers and designers have a responsibility to consider diversity and equality in their work in smart cities?

References

1. Akrich M (1995) User representations, practices, methods and sociology. In: Rip A, Misa TJ, Schot J (eds) Managing technology in society: the approach of constructive technology assessment. Pinter, London, pp 167–185
2. Bakıcı T, Almirall E, Wareham J (2013) A smart city initiative: the case of Barcelona. J Knowl Econ 4(2):135–148

3. Cohn C (1987) Sex and death in the rational world of defense intellectuals. Signs 12 (4):687–718
4. Daly J (2015) Is this planned ghost town the city of the future? Wired Mag http://www.wired.com/brandlab/2015/05/planned-ghost-town-city-future/. Accessed 23 Oct 2015
5. Dobinson L (2011) Technology company builds desert ghost town. The Telegraph 18 September 2011. http://www.telegraph.co.uk/technology/news/8768847/Technology-company-builds-desert-ghost-town.html. Accessed 23 Oct 2015
6. Foley A, Ferri BA (2012) Technology for people, not disabilities: ensuring access and inclusion. J Res Spec Educ Needs 12(4):192–200
7. Gibbs D, Krueger R, MacLeod G (2013) Grappling with smart city politics in an era of market triumphalism. Urban Stud 50(11):2151–2157
8. Harrison K (2015) 'No thought of gender': bodily norms in Swedish rescue services incident reporting. Gender Work Organ 22(3):211–220
9. Haque U (2012) Surely there's a smarter approach to smart cities? Wired, 17 April 2012. http://www.wired.co.uk/news/archive/2012-04/17/potential-of-smarter-cities-beyond-ibm-and-cisco. Accessed 23 Oct 2015
10. Kitchin R (2015) Making sense of smart cities: addressing present shortcomings. Cambridge J Regions Econ Soc 8:131–136
11. Laws G (1994) Oppression, knowledge and the built environment. Polit Geogr 13(1):7–32
12. Leon N (2008) Attract and connect: The 22@Barcelona innovation district and the internationalisation of Barcelona business. Innov: Manag Policy Pract 10:235–246
13. Logon J, Molotch H (1987) Urban fortunes. The political economy of place. University of California Press, Berkeley
14. March H, Ribera-Fumaz R (2014) Smart contradictions: the politics of making Barcelona a self-sufficient city. Eur Urban Region Stud:1–15
15. Nesta (2015) Rethinking smart cities from the ground up. http://www.nesta.org.uk/publications/rethinking-smart-cities-ground. Accessed 23 Oct 2015
16. Poole S (2014) The truth about smart cities: in the end, they will destroy democracy. The Guardian, 17 December 2014. http://www.theguardian.com/cities/2014/dec/17/truth-smart-city-destroy-democracy-urban-thinkers-buzzphrase. Accessed 23 Oct 2015
17. Pinch TJ, Bijker WE (1984) The social construction of facts and artefacts: or how the sociology of science and the sociology of technology might benefit each other. Soc Stud Sci 14(3):399–441
18. Ravindran S (2015) Is India's 100 smart cities project a recipe for social apartheid? The Guardian, 7 May 2015. http://www.theguardian.com/cities/2015/may/07/india-100-smart-cities-project-social-apartheid. Accessed 23 Oct 2015
19. Strengers Y (2014) Smart energy in everyday life: are you designing for resource man? Interactions 21(4):24–31
20. Temperton J (2015) Bristol is making a smart city for actual humans. Wired, 17 March 2015. http://www.wired.co.uk/news/archive/2015-03/17/bristol-smart-city. Accessed 23 Oct 2015
21. Townsend AM (2013) Smart cities, big data, civic hackers, and the quest for a new utopia. W. W. Norton, New York
22. Waters R (2015) Google's smart city ambitions take shape. FT Weekend 20 June/21 June 2015, p 15

Web Resources

23. IEEE Smart Cities Initiative 'About'. http://smartcities.ieee.org/about.html. Accessed 23 Oct 2015
24. Lincos project. http://www.media.mit.edu/unwired/center.html. Accessed 23 Oct 2015
25. "Palava Megacity" promotional film. https://www.youtube.com/watch?v=qtDNRoMrQ1A. Accessed 23 Oct 2015

26. Palava 'Home'. http://www.palava.in/. Accessed 23 Oct 2015
27. Palava 'Overview'. http://www.palava.in/overview/introduction. Accessed 23 Oct 2015
28. Palava 'City Management'. http://www.palava.in/citymanagement/pcma. Accessed 23 Oct 2015
29. RERUM 'Home'. https://ict-rerum.eu/. Accessed 23 Oct 2015

Chapter 3
Making Onlife Principles into Actionable Guidelines for Smart City Frameworks and #IoT Policies

Nenad Gligoric, Christine Hennebert, Srdjan Krco, Carmen Lopez, Ignacio Maestro, Colin Ő Reilly, Michele Nati, Antonio Skarmeta, Rob van Kranenburg, Nathalie Stembert and Alberto Serra

building the raft while swimming—Rober Madelin.
Since action is the political activity par excellence, natality and not mortality, may be the central category of political, as distinguished from metaphysical, thought (Arendt 1959, p. 11).

Before the Cloud appeared around 2000, Internet of Things or #IoT was present as smart gadgets, smart offices and smart connectivity in demo form only as the data coming from the applications could be stored and analyzed only at huge costs. The EU facilitated large research programs. In this text an attempt to integrate two policy frameworks informing responsible innovation is made—Onlife Manifesto—and IoT development in research and innovation projects.

The original version of the chapter was revised: Author name has been updated. The erratum to the chapter is available at https://doi.org/10.1007/978-3-319-50094-2_15

N. Gligoric · S. Krco
DunavNet, Novi Sad, Serbia

C. Hennebert
CEA, Grenaoble, France

C. Lopez · I. Maestro
University of Cantabria, Santander, Spain

C. Ő Reilly
University of Surrey, Guildford, UK

M. Nati
Digital Catapult, London, UK

A. Skarmeta
University of Murcia, Murcia, Spain

R. van Kranenburg (✉)
Resonance Design/Council, Tilburg, Netherlands
e-mail: rvk@theinternetofthings.eu

N. Stembert · A. Serra
sociotal.eu, Guildford, UK

V. Angelakis et al. (eds.), *Designing, Developing, and Facilitating Smart Cities*,
http://doi.org/10.1007/978-3-319-44924-1_3

3.1 Early EU Research

The need for nontechnical research in the area of machine-to-machine communication and Internet of Things as the developments got closer to market and everyday lives of citizens was acknowledged in the 1996 EU Call for Proposals of the i^3: Intelligent Information Interfaces, an Esprit Long-Term Research initiative. The aim of i^3 (pronounced "eye-cubed") was to develop new human-centered interfaces for interacting with information, aimed at the future broad population.

This approach was also the starting point and rationale for the EU-funded proactive initiative "*The Disappearing Computer*" a cluster of 17 projects by interdisciplinary research groups. Its mission was "to see how information technology can be diffused into everyday objects and settings, and to see how this can lead to new ways of supporting and enhancing people's lives that go above and beyond what is possible with the computer today." [1]. The third research iteration of this approach was Convivio (2003–2005), a thematic network of researchers and practitioners developing a broad discipline of human-centered design of digital systems for everyday life. The coordinator of Convivio stated that human-centered design "still has little influence either on governmental and super-national policies or on industrial strategies. As a result, it also has little impact on the quality of ICT in public and private life." [2].

In i3magazine 2003, Jakub Wejchert of Future and Emerging Technologies Unit, European Commission, project officer of i3 and Disappearing Computer Research Initiatives, said:

> In the long term, as we move towards a 'knowledge-based society', we have to ask: "Have we been supporting only one kind of knowledge up to now"? and "How can we best support different kinds of knowledge"? To do this we need to support a diversity of things we understand by 'knowledge'—ranging from the cognitively abstract, through to knowledge that is 'at hand' and embodied in our physical everyday world, to sequences of past events, and our memories of past experiences…This will involve rethinking what we mean by 'knowledge representation', constructing new forms of 'flows' between content and context, and exploring the balance between the 'global' and the 'local'. Perhaps in the future we will look back to our preindustrial roots as inspiration—back to a reverence for 'place', 'location' and the importance of the 'being within the world' and the 'here and now'… Perhaps in the future, we will live in a world that is more 'alive' and more 'deeply interconnected' than we can currently imagine?

In his 2003 text Lessons learnt from LIVING MEMORY @ 1:3—listening to and developing technology for ordinary people, Steve Kyffin of Philips, involved in the LiMe project that build an interactive table to be used in the neighborhood and be placed in community centers, bus stops and other local meeting spaces, says:

> ¥ USEFUL … listening to and developing technology for ordinary people sums up what we might refer to as Co-Creative design. Involving the end user in a core and proactive manner at all stages in the product or system creation process.
>
> ¥ RELEVANT … listening to and developing technology for ordinary people is so relevant because the 'ordinary… ness' is the issue. Much of what we concentrated on in Living Memory was the means by which people interact with each other (the technology) and the interfaces to that technology, which we offer.

¥ IMPORTANT … listening to and developing technology for ordinary people. The world of stuff is not enough… the design discipline is changing fast. Design must now respond to an economic model which supports the provision of converged and connected solutions, such as LIME, combining products and services to suit individual needs.

¥ SUCCESSFUL … listening to and developing technology for ordinary people. The extent to which the project was successful is the subject of the complete review and validation process, which we conducted. Design is often seen as an applied discipline, where fine art may be regarded as the pure discipline. We believe that this is not the case and that it is possible to: research the 'pure' Design discipline in order to develop its future role in differing contexts; to use Design as a research tool to help us better understand and contribute to the changing nature of People, Culture and Society and in turn assist in the integration of emerging and future technologies into the lives of 'ordinary people'; to find more effective tools and research areas for Design to consult and investigate in order to provide more holistic and relevant propositions within our commercial practice.

¥ INSPIRATIONAL… listening to and developing technology for ordinary people. Certainly, in all the ways mentioned above… inspired us to find NEW KNOWLEDGE: NEW ROLES FOR DESIGN: ways to integrate SYNTHETICAL and ANALYTICAL research: NEW IP: seeds for NEW OPEN PRODUCTS-SYSTEMS-SERVICES…

¥ DIFFERENT … Research is by definition, DIFFERENT, especially when it is this trans disciplinary, collaborative, human focused, artistic and scientific driven end results in such enriched experiences for people…

The industry has been able to capitalize on the early ubicomp and ambient EU projects, enabling them to actualize the idea of the new more participatory user and user-centered design. The EU #IoT programs and projects up to Horizon 2020 however have not integrated this vision as end users—citizens—have not been involved in building use cases in FP7 IoT projects. The renewed focus on impact and business in Horizon 2020 aims to change this.

3.2 Internet of Things

Why would we want an Internet of Things? We want it because it can offer us the best possible feedback on physical and mental health, the best possible resource allocation based on real-time monitoring, best possible decision-making on mobility patterns and the best possible alignments of local providers with global potential. Operationally this means that we can define Internet of Things as the seamless flow between the BAN (body area network): wearables, LAN (local area network): smart home, WAN (wide area network): connected car, and VWAN (very wide area network): the smart city. Key to this flow is having control of the data. That is why Google is offering a Glass and a Lens so you can synchronize your health data into the NEST and the Google Car throughout the smart city applications of Google.org. The idea is that in consumer applications and services you never have to leave the Google Cloud. The products are gateways linking up the networks.

Internet of Things is a new beginning. In our current architectures we are used to dealing with three groups of actors: citizens/end users; industry/subject matter experts (sme); and those involved in governance/legal matters. These all are characterized by certain qualities. In our current models and architectures we build from and with these actors as entities in mind. The data flow of IoT will engender new entities consisting of different qualities taken from the former three groups diminishing the power of the traditional entities.

In his seminal text *The Social Order of a Frontier Community*, Don Harrison Doyle, wrote that "social conflict was normal, it was inevitable, and it was a format for community decision-making." [3]. Sociologist Lewis Coser also advised that, instead of viewing conflict as a disruptive event signifying disorganization, "we should appreciate it as a positive process by which members of a community ally with one another, identify common values and interests, and organize to contest power with competing groups." [4]. The new environment of the IoT will resemble these "frontier communities" because of their seeming disorganization where conflict will be the norm.

It needs an ethical framework to inform decision making about decision making on what kind of #IoT system architectures to support. The Onlife Manifesto might be relevant in this context.

3.3 The *Onlife Manifesto*

The deployment of information and communication technologies (ICTs) and their uptake by society "affect radically the human condition, insofar as it modifies our relationships to ourselves, to others and to the world. In order to acknowledge such inadequacy and explore alternative conceptualisations, a group of scholars in anthropology, cognitive science, computer science, engineering, law, neuroscience, philosophy, political science, psychology and sociology, instigated the Onlife Initiative, a collective thought exercise to explore the policy-relevant consequences of those changes." [5].

The *Onlife Manifesto* is a positive contribution to rethinking the philosophy on which policies are built in a hyper-connected world. It attempts to describe, explain and give a policy framework to the *Digital Transition*, which can be understood as pervasive digital presence in everyday and institutional practices, the real-world IT context enabled by the Cloud.

The Digital Transition is a change in ontological foundations, questions that run deep into what it means to be human. The Onlife Manifesto therefore states that "redesigning or reengineering our hermeneutics, to put it more dramatically, seems essential, in order to have a good chance of understanding and dealing with the transformations." It addresses the widespread impression that our current ICT conceptual toolbox is no longer fitted to address new IoT, Internet of Things-related challenges: "We grasp reality through concepts. When reality changes too quickly and dramatically, as it is happening nowadays because of ICTs, we are conceptually wrong-footed."

There are strong links between Onlife and CAPS. The initiative "Collective Awareness Platforms for Sustainability and Social Innovation" (CAPS) aims "at designing and piloting online platforms creating awareness of sustainability problems and offering collaborative solutions based on networks (of people, of ideas, of sensors), enabling new forms of social innovation. CAPS are expected to support environmentally aware, grassroots processes and practices to share knowledge, to achieve changes in lifestyle, production and consumption patterns, and to set up more participatory democratic processes" [6]. Practically this translates in the objectives of the DCENT project. By "enabling multi-agent systems and opening new possibilities for direct democracy, ICTs destabilize and call for rethinking the worldviews and metaphors underlying modern political structures" which is exactly the focus of DCENT [7].

The *Onlife Manifesto* states that the impact of the Digital Transition is due to at least four major transformations; "the blurring of the distinction between reality and virtuality, the blurring of the distinction between human, machine and nature, the reversal from information scarcity to information abundance, the shift from the primacy of stand-alone things, properties, and binary relations, to the primacy of interactions, processes and networks."

It states that ICTs "are not mere tools but rather environmental forces that are increasingly affecting: our self-conception (*who we are*); our mutual interactions (*how we socialise*); our conception of reality (*our metaphysics*); and our interactions with reality (*our agency*). In this new *conceptual* space we can co-create notions of solidarity (economics), privacy (self), security (trust), assets (potentials), risks (resilience) and threats (competition), tailored to a reality of today."

The Onlife Manifesto builds on two drivers that have been articulated by Hannah Ahrendt; *plurality and natality.* Arendt recognizes that **plurality** can best be experienced *at city level.* "*The larger the population in any given body politic, the more likely it will be the social rather than the political that constitutes the public realm*" (Arendt 1959, p. 39). With big numbers, plurality degenerates into mere and unendorsed interdependence, while natality and its inherent openness and unpredictability are perceived only under the categories of uncertainty and risk.

Arendt describes plurality as "the coexistence of equality, specificity and reflectivity. The threefold understanding of plurality (equality, specificity, and reflectivity) undermines radically an omniscience-omnipotence utopia's worldview."

In the perspective of mortality, "the future is coloured with the certainty of our eventual death, while in the perspective of **natality** it is coloured by the recurrent remembrance of the '*infinite improbability*' (Arendt 1959) of our birth and conducive to *confidence* and *wonder*. The philosophy of Arendt is anchored in the praise of beginnings."

The Onlife principles that have as a basis Ahrend's positive and life-affirming yet critical view on technology are able to address real-world issues in an actionable context.

Carl Schmitt distinguishes between der *Wirkliche Feind* and the *Absolute Feind.* The latter is '*die eigene Frage als gestalt.*' The absolute enemy is the inability to change convictions, alliances and opinions. The absolute friend is always very near to you, consisting of everyday routine skills; it is your blind spot. The real enemy can differ from time to time and period to period. Each historical situation demands the capabilities to define as those real enemies the ones that can redefine all that you hold normal, dear, and take for granted. The relation self describes a situation where there is a productive balance between the energies directed at these two types of existential questions. *Onlife* believes that everybody needs *both* shelter from the public gaze *and* exposure: "The public sphere should foster a range of interactions and engagements that incorporate an empowering opacity of the self, the need for self-expression, the performance of identity, the chance to reinvent oneself, as well as the generosity of deliberate forgetfulness."

Onlife considers the distinction between private and public to be more relevant than ever: "Today, the private is associated with intimacy, autonomy, and shelter from the public gaze, while the public is seen as the realm of exposure, transparency and accountability. This may suggest that duty and control are on the side of the public, and freedom is on the side of the private. This view blinds us to the shortcomings of the private and to the affordances of the public, where the latter are also constituents of a good life."

Ethics has become a matter of social being(s), much in the sense of enaction of Varela. He claims citizens by definition possess an ethical sense and sensibility. Responsibility for the effects brought about by technological artifacts are thus no longer only attributed to their designer, producer, retailer or user, but to an ever-increasing semi-autonomous interoperability of products and services in intranets, and an IoT ecology that globally knows no firm standardization, though there are attempts. Notions of distributed responsibility must thus inform any transaction whether between city council and vendor, B2B, B2C or C2C (the Sharing Economy).

Onlife thus states that "the development of a critical relation to technologies should not aim at finding a transcendental place outside these mediations, but rather at an immanent understanding of how technologies shape us as humans, while we humans critically shape them."

This immanent understanding can only occur in dialogue and co-creation. Co-creation workshops and Internet of Things meetups are capable of readdressing technological concepts in cultural contexts, bringing real dialogue with stakeholders in the cities in which they live. In the following chapter we describe how co-creation workshops and meetups were able to give direct feedback to smart city development questions.

3.4 Co-creation Workshops in the EU Project SocIoTal.eu: Building Use Cases with End Users

At the IoT Rotterdam meetup during IoT day (April 9) [8], Ben van Lier [9] showed how the old Shannon paradigm of communication allows engineers to port 'meaning' onto a different plane, out of their immediate consideration. This explains the huge speed and convergence of efficiency intrinsic system and applications only. It also explains that we feel somehow, 'stuck' in, 'selling' platforms to citizens who cannot articulate their need and do not see the offered services as something so amazing in the age of their own daily app agency with smartphones and companies like Google, and Facebook gradually spilling over into the real-world objects. That means that only in the recent decade system engineers realized 'meaning' had to be patched 'back on' as semantic interoperability.

Maybe we should rethink the entire structure so that it reflects the ongoing connectivity as a 'new' reference framework?

SocIoTal is an EU FP7 funded STREP project addressing the objective FP7-ICT-2013.1.4 “A reliable, smart and secure Internet of Things for Smart Cities” and more specifically sub-objective a) “A reliable and secure Internet of Things.” SocIoTal designs and will provide key enablers for a reliable, secure and trusted IoT environment that enable creation of a socially aware citizen-centric Internet of Things by encouraging people to contribute their IoT devices and information flows. SocIoTal research identified as main barriers to broad IoT adoption in ‘smart’ cities:

- lack of understanding by SME’s and city councils
- lack of third-party trust providers
- lack of involvement of end users in building use cases and developing news services.

The lack of understanding is addressed in meetups, introducing research questions and listening to the local stakeholders. The lack of involvement of citizens we addressed in co-creation workshops with researchers from University of Cantabria, Santander and Novi Sad. The co-creation workshops[1] were prepared and executed by Nathalie Stembert and Rob van Kranenburg. Following lines will present the experiences with co-creation workshops within the SocIoTal project from the different points of view of attendees and coordinator.

[1]Her master thesis with the topic “participation and co-creation in the public domain,” represents the final project of the master Design for Interaction at the faculty Industrial Design Engineering, University of Technology Delft (TUDelft) and was executed for the STT Netherlands Study Centre for Technology Trends in The Hague (STT). http://cocreatetheiot.com.

3.4.1 *Brief Description of the Co-creation Workshop from the University of Cantabria Researchers*

Six people within and out of the SocIoTal project were involved in the session in Santander. After the welcome and a brief introduction by Nathalie, the first of the two main parts of the activity started, which was to create a use case (UsCa) from scratch. The creation process started from analyzing some cards in which different situations in a city were represented (e.g., a woman buying a gift or kids planting flowers). From those cards participants were asked to select those who could present situations where IoT could offer some benefits. After a bit of discussion, three cards related to the transport, one related with IoT education and other about Smart Shopping were selected. After an initial description of the use cases extracted from these cards, and after establishing the pros and cons of each of them, the Smart Shopping was elected to be analyzed deeply. The UsCa was studied trying to answer the questions *who* (who are the actors?), *where* (what are the places involved in the UsCa?), *what* (what is the evolution of the UsCa) and *why* (what necessities does the UsCa fulfil?). After that, all the descriptions were translated and represented on a Santander map, establishing the objects, the intelligence and the connections between them using different touchable pieces and totems. Also, the devices needed during the UsCa were defined and analyzed in terms of how the user would observe the use case through them (e.g., what information do I need to see in the application?). To finalize this first part of the workshop, participants recapitulated the experience highlighting the benefits that can be added to the UsCa, and the barriers that can be found.

During the second part of the session the carpooling UsCa, described by the University of Cantabria for the SocIoTal project, was explored. The same previous process was followed, observing different points of view of the UsCa, possible extensions and new ideas to enrich it.

At the end of the session, some conclusions were highlighted by the participants in terms of the usefulness of this kind of workshops:

- It is a more visual, enjoyable, and collaborative way to introduce people within the IoT and to take advantage of all their ideas to elaborate or re-elaborate the UsCas
- The structure of the session allows to guide users easily to create a complete description of a new UsCa
- In the case that the UsCa is already described, it allows to discover the point of view of the final user who could offer new requirements, and descriptions about what would be really and new valuable functionalities for them
- The touchable materials used during the process allows the users to become abstract ideas for them into reality
- Allows developers to discover users' reaction to the UsCa, acceptance and barriers. Also, it allows to explore the availability of devices which at the end

could be translated into the acceptance of a new service or the necessity of change technological aspects of the UsCa
- It is an interesting way to capture potential users in pilots and trials

As general feedback, the University of Cantabria team agreed that this workshop process could help rank the necessities—the must and should haves—with end users in the current use cases, as it will draw good and structured feedback from the real users. As a result, a second co-creation workshop was performed with final users to gather valuable requirements to be added during the development of one of the pilots of the SocIoTal project: an application to create routes for disabled citizens avoiding blocking elements along the city. All the participants contributed to the co-creation workshop in a very valuable way, due in part to the collaboration between multiple stakeholders (disabled users, developers, city council representatives) which led to more empathy between them. The city council and developers could obtain very detailed information from the target group, since the people with mobility problems provided first-hand experience and could therefore convey the experience and problems that they face in their daily lives. Consequently their ideas and suggestions were very rich and contained detailed information that a person without a disability could never imagine. During the session, the use case *accessible routes in the city* was explored in depth and led to valuable information concerning the design and the development of the application.

3.4.2 *Natalie Stembert: Notes—Workshop Santander. From a Facilitator Point of View*

From skepticism to enthusiasm! The SocIoTal researchers started a bit hesitant with the co-creation workshop, giving it the benefit of the doubt. The first step, deriving three use cases from the interaction cards, needed explanation. The reason behind the steps was therefore explained frequently during the process, moreover there was emphasized that the workshop is designed for collaboration with the target group and that several aspects are therefore simplified. Soon the SocIoTal researchers, assembled three use cases, describing and evaluating them by means of positive (incentives)/negative (barriers) aspects. After which they decided to work out the use case about smart shopping. At this point they already became more convinced about the purpose of the workshop and discussions were fruitful. When the co-creation artifacts appeared on the table to whole process and its steps became clear to the SocIoTal researchers. They mapped the activity chain of the use case, by means of the metaphorical objects and sensors, while discussing the interaction in the mean time. The data flows were visualized, the scenario was played out with the actor artifacts and the interfaces were visualized on the device templates. Conclusions about the use case were written down, to subsequently end in an enthusiastic discussion about the benefits of the co-creation workshop.

As accomplishments of the session, the following can be highlighted:

- An entirely new and feasible use case was developed.
- The use case was visualized in terms of location, objects, intelligence (sensors and network), data flow, interface input/output and actors involved.
- The use case was evaluated in terms of incentives, barriers, devices and willingness to use these devices for the purpose of the use case.
- By playing out the use case insights were gained in the interaction of the target users with the Internet of Things network.
- The workshop let to a number of requirements for the old and new use case.
- It moreover turned out to enable SocIoTal researchers to prioritize requirements above others (based on the needs of the target group).

From this experience, some recommendations for future workshops with final users were provided:

- Discussions are most important to elicit.
- Time has to be managed carefully; one use case is the maximum to elaborate in per workshop
- Participants have to be selected according to their level of technical expertise.
- Collaboration between the SocIoTal researchers and the target group is important to guide the participants (the process can be still too technical), yet also to let the SocIoTal researchers engage with the target group to uncover their needs.

In the Novi Sad workshop the core of stakeholders in any local situation was present. The Belgrade Chamber of Commerce main task is to integrate local initiatives and look out for co-financing models. The local development offices are considering to open up data from local databanks in a dual way; a particular level for free and an optimum service for more granular data. For end users it would be interesting to add user generated content, personal stories that can become part of feedback and validated information, like rating someone routes with the bonus that certain users become trusted sources of information. It becomes clear that is impossible for the local service #IoT and M2M companies to go beyond the efficiency and cost-cutting paradigm and propose innovative and meaningful applications on their own. We are facing a clear situation of a no win for the first investor; result a current deadlock for a citizen-centric IoT.

As citizens' active involvement is thus the necessary precondition of possible success, according to Ezio Manzini, we need "to take in account why and how people collaborate is a fundamental component: collaborative economies, collaborative services, collaborative consumption, collaborative innovation spaces, collaborative events are very diverse initiatives, with a common denominator: they all ask for collaboration. Following typologies can be recognized:

- Vertical collaboration: individual citizens collaborating with solution promoters. Example: Fix my street.
- Vertical and horizontal collaboration: individual citizens collaborating with solution promoters and then, collaborating among them in a p2p way. Example: Carpooling.
- Horizontal collaboration: p2p collaboration among citizens. Example: Circle of care and Collaborative housing."

Rich scenarios embed all three types of collaboration. It can be therefore investigated, as Louisa Heinrich [10] suggests,

- "Help identify patterns within communities, indicators that might help newcomers to a city or area decide where to visit or where they might want to live
- Give people modular tools that they can use to 'mark up' and monitor what's important to them—whether that's embedding history into the physical environment (personal or official), keeping track of noise and pollution levels, or planting and maintaining communal gardens

The idea of the project would be to design and prototype a small set of hardware and software tools that could be given to a community and then used by them in whatever way made the most sense."

The key elements as they were voiced during the discussions in the co-creation workshops were:

- **a mentality change**: "How can we all (ourselves included) make the switch from 'This is their building', to 'This is our building, our street, our park'?. This is a mindset change and extremely complex. Pretty much a lot of citizens are depressed. Youth unemployment is very high, much to high. There is a sense of togetherness that is missing."

- **mixing public and private responsibilities**: The funding should come partly from the government and partly from crowd funding and private donors as ownership must be taken by citizens and it should not feel as if everything is already decided. A business model could be on some basis of vouchers: I can donate time, money or can I buy a plant or tree? I have certain skills: "Can you use them? In exchange of what?"
- **not inventing the wheel**: use for example taskrabbit.com in the idea for the portal where citizens can log in and subscribe to donate a gift—time, money, a tool to a problem or cause in the street or neighborhood.

3.5 Meetups

According to Marshall Van Alstyne, a business professor at Boston University, companies have a "really difficult time with the mental models. It's fascinating. Most companies compete by adding new features to products. They haven't been in the business of thinking of how to add new communities or network effects. In many cases, the governance models have not been established. For instance, population density can be determined by mobile phone distribution. A telecom company owns that data. How do you motivate them to share it? All these sensors are capturing data, but how do you divide the value? Those are the rules that need to be worked out, and that's the missing piece of most of these discussions about the Internet of things. You have to build economic incentives around it, not simply connectivity." [11].

Key findings from the 2014 report, The Hansard Society's Audit of Political Engagement, the only annual health check on British democracy: "Levels of knowledge and interest in politics have improved this year, but the public continue to feel powerless." They feel that they have very little influence on decision-making and that their own involvement in politics will have little effect on the way the country is run…the desire to actually be involved in decision-making, both locally (43 %) and nationally (38 %) continues to outpace their personal sense of efficacy and influence…67 % of the public say 'politicians don't understand the daily lives of people like me.'

Actionable reasoning in this new 'middle' is not formed by academic conferences or informed structured debate but in informal and open meetups where practitioners listen to each others experiences over a beer. The numbers of people engaging in IoT meetups globally is rising fast. At the end of 2014 the total number of members in SocIoTal Meetups was *431*. On August 9 2015 it is *1063* in five cities: Santander, Novi Sad, Guildford, Ghent and Grenoble. During these meetups, different speakers have talked about the Internet of Things, new projects being developed in these regions related with IoT, and possibilities it can offer to multiple stakeholders in the city. Along the journey of the meetups, a lot of interesting interactions have appeared that have helped us to learn more about the relation between society and IoT, and that have given rise to various situations to take advantage within the project, like the one described above. Engaging with local developers, businessmen, researchers, geeks, hackers, artists and citizens devoted to unique and idiosyncratic technical solutions is messy and very hard to formalize as it is in the praxis of doing that real interaction. We can discern certain types of interactions: regional coordination, informal learning and facilitating and building local connections.

regional coordination

Srdjan Krco: "One of the developers from a local SW company who attended Meetup in Novi Sad (one during which I spent time behind the screen:) came to us afterwards looking for collaboration. We helped him with some of the hardware kits and at the next Meetup (next week) he will be talking about his IoT experience—how he learned to program Galileo and use related tools."

The coordination linking up IoT incubators, conferences, startups and regional policy makers, based on the work of Srdjan Krco and Community Manager RD who was invited to Brno, Cluj, and Bled "to map countries in CEE, all stakeholders who want to work together, plan events, put in entrepreneurial orb all of the companies to synergize them and start delivering solutions on a bigger scale, beyond some fancy gadgets and apps." Damir Čaušević[2] **is** "already in Brno Smart City Committee, which has a direct influence on the whole region of South Moravia. We're now building the Smart City concept, putting in place processes and procedures (not overdoing it) and setting up the solid ground for the infrastructural projects." RD posted the view of a modular and domain specific IoT approach on Council ("If you look once more into my article Rob posted less than a month ago (http://www.theinternetofthings.eu/damir-%C4%8Dau%C5%A1evi%C4%87-entrepreneurial-orb-modular-service-approach) you'll notice how it can be devel oped."); Telco's between Damir Čaušević (Brno), Tomaz Vidonja (Bled), Srdjan Krco (Novi Sad), Radu Ticiu (Timisoara), Andree Balaci (Cluj) is set for March 11 led to further coordination during the 5th Living Bits and Things 2015 June 8th–9th, 2015, Bled, Slovenia, the event that gives attendees insights and understanding of the central and east Europe (CEE) region and opportunity to share and discuss the challenges with professionals and experts. Among the key results from the 5th IoT event »Living bits and things 2015« is that the IoT CEE Community is rising as well as the awareness in the region (especially well connected with JIC, Czech Republic), according to organizer Tomaz Vidonia.

informal learning

Srdjan Tadic wrote that the Meetup "helped me to get more structural view on things we see on every company-brainstorming." (IoT meetup #6 (a prvi u Beogradu), February 10 2015)

---

[2]Manažer startup akcelerátoru StarCube | Startup Accelerator StarCube ManagerJIC, zájmové sdružení právnických osob INMEC, Purkyňova 649/127, Brno–Medlánky 612 00.

A 2015 Vodafone study claims that local government "still remains unaware, by and large, of the opportunity presented by smart city technology, and how it can be used to deliver better and more cost-efficient public services." The survey questioned 1,624 UK adults and 629 councillors, revealed that "smart in-building energy management systems and street lighting alone could save local councils across the country £402.3 m." [12]. The study revealed that "67 % of urban councillors were not aware of machine-to-machine (M2M) technology and how best to take advantage of it, even though 77 % of people living in urban areas said they would support their council's decision to invest in the internet of things (IoT) to improve public services. *The gap in understanding*, said Vodafone, would explain why smart city technology has not yet been widely deployed—beyond a few test beds in tech-heavy locales such as Bristol and Milton Keynes—to improve lighting, rubbish collection, traffic, public transport management and so on."

The interview with Ana Milicevi Senior Advisor, Association of IT activities, Belgrade Chamber of Commerce (www.kombeg.org.rs) shows the local awareness of this gap and the attempts to bridge it, validating the meetup as an important positive factor. She points two important factors. First, the importance of the host. Ana M recalled how she felt quite lost personally in the 6th SocIoTal Serbian Meetup (Belgrade, February 10) as she is used to more formal arrangements where someone introduces the participants and is used to meeting CEO's and CTO's of companies, not the coders. Not knowing what to expect she felt confused and surprised to see the makers, the coders, not the business men. At that particular Meetup WP6 leader RD was present, saw that she was struggling and introduced her to Srdjan and the others. For the project this means that social skills in the meetup are very important even if—as it is hosted mainly by engineers and computer scientists—these skills are not the primary skills of coders. The main outcome of recognizing this diversity is that Ana M realized that she had to find a way to include the coders themselves in the activities of the Chamber of Commerce, that there is a special value in these informal meetings, of new way of creating knowledge together, that she was surprised by the feedback in email that she received after the meetup, "actually it was the first time I had such feedback." Ana Milicevic, subsequently took part in the co-creation workshop on Open Data in Novi Sad. She is very positive about the process and format of the co-creation workshop. She realizes in her practice that #IoT is a horizontal operation that is connecting different startups scenes that she sees happening in Belgrade at the moment: what does it mean for startups in technology, advertising, design, gaming, retail, etc., to collectively think about business models from the start, not on their own? And what kind of third trusted party would be able to convince them of doing this? What kind of process is necessary to build this kind of trust?

In Santander, SocIoTal partners arranged a set of meetups with the purpose of introducing practical IoT knowledge to the attendees. During the previous sessions, speakers provided theoretical talks about IoT and how it can improve their lives through different projects which are being developed in the region. From this viewpoint, the natural next step was to teach them how to develop their own IoT devices and explain different options (platforms) to create their own services.

Positive feedback was provided by the attendees, claiming the importance of teaching nontechnical people this new technologies as well as providing them (and also technical people) with updated information about the available resources to create new and valuable services for them and society. Also, it is relevant to highlight the importance of this kind of events have to improve the IoT local ecosystems that will be described in the following lines.

> facilitating and building local ecosystems
>
> Meetups are a valuable way to produce quality relations between different stakeholders, what result in an enhancement of the local ecosystem of the city. As an example, the experience of Nigel Stirzaker from Guildford is reported: "Thought I'd find out about the kind of things your likely to look at going forward and let you know where I'm coming from. I was really nice to see your group. As I work in London (Dev for Disney in Hammersmith) I don't often think to look for groups closer to home (Woking). One of my main interest/passions is getting kids involved in tech and I'm looking to move what I "offer" (all free via schools and hackathons) to include more hardware. This is on top of my general passion for tech. I'm looking forward to the BBC MicroBit coming out with the hopes that it will give another push to IoT."

A similar account can be made for all partner cities; Ghent, Grenoble, Novi Sad, and Santander. It testifies to a strong effort of all partners involved to invest a significant amount of (extra) hours in activities that do not directly lead to concrete results but that are helping to build a local atmosphere of communication and trust. An example of this can be extracted from the reported experience of Michele Nati, organizer of Guildford Meetups:

> Everybody found it very interesting and people usually ask to present in some of the upcoming Meetups. Recently we had in April a very low attendance one (6 people). Didn't have time to properly prepare the program due to proposal submission. We ended up to sit in few persons and informally chat in the pub. Mark Hill, entrepreneur in IoT (former bank employee in London) really wanted to help this local Meetup to fly and compete against the London one. He gave some suggestion and we came up together with a strong plan. We put this in practice for the May Meetup. The result was a fully booked Meetup and a room with 70 people, sponsors, and 4 great presentations. We will definitely keep doing this. Participants always try to connect together and few relationships started among entrepreneurs, companies and borough councils. Zebra technology who gave a speech last time wants to partners as mentor for 5G activities at University of Surrey.

References

1. The Disappearing Computer II (DC) Proactive Initiative. http://cordis.europa.eu/ist/fet/dc2-in.htm
2. Letter to the Convivio community, Giorgio De Michelis, Convivio network coordinator. http://daisy.cti.gr/webzine/Issues/Issue%201/Letters/index.html
3. Harrison Doyle Don T (1983) Social order of a frontier community: Jacksonville, Illinois, 1825–70. The University of Illinois Press, Illinois
4. Coser LA (1957) Social conflict and the theory of social change. Br J Sociol 8(3):197–207, http://links.jstor.org/sici?sici=0007-1315%28195709%298%3A3%3C197%3ASCATTO%3E2.0.CO%3B2-H

5. The Onlife Manifesto. https://ec.europa.eu/digital-single-market/en/onlife-manifesto
6. Collective Awareness Platforms for Sustainability and Social Innovation. https://ec.europa.eu/digital-single-market/en/collective-awareness
7. Vercellone C, Bria F, Gentilucci E, Griziotti G (2015) Managing the commons in the knowledge economy. Decentralised Citizens ENgagement Technologies. https://www.nesta.org.uk/sites/default/files/d-cent_managing_the_commons_in_the_knowledge_economy.pdf
8. http://www.theinternetofthings.eu/iot-rotterdam-kudos-martin-pot-and-martin-spindler-short-report
9. http://www.theinternetofthings.eu/ben-van-lier-information-crucial-when-considering-future-mankind
10. http://www.louisaheinrich.com/2015/03/12/internet-of-neighbourhoods-part-1-post-19100/
11. Regalado A (2015) The economics of the internet of things. As everyday objects get connected, brace yourself for network effects, says one economist. http://www.technologyreview.com/news/527361/the-economics-of-the-internet-of-things/
12. Scroxton A (2015) Local government blind to internet of things savings Networking Editor 09 Jun 2015 16:15: a report from Vodafone says local councils could save billions by implementing smart city systems, but are ignorant to the possibilities. http://www.computerweekly.com/news/4500247803/Local-government-blind-to-internet-of-things-savings

Chapter 4
Factoring Big Data into the Business Case for IoT

Anastasius Gavras

4.1 Big Data

There is a widespread expectation that big data offers tremendous opportunities. The potential economic value to be generated from the vast amount of data is given in tens or even hundreds of billions of euros per year. However, the cost of big data is rarely mentioned. We can hardly find figures explaining the long-term cost of big data. Yet, there must be a price tag attached to the investments needed for handling big data, including operational expenses for maintaining the data and making them available in the long term.

Citing ex-Commissioner Neelie Kroes' speech at the ICT 2013 event in Vilnius [1]; *every two days we create as much information as was created from the dawn of civilisation to 2003*. In addition she claimed that big data is growing by 40 % per year, a figure that is hard to correlate to the previous statement and in fact even harder to believe if we accept that a similar law like Moore's law applies on big data and that Moore's law will likely continue to be valid until at least 2030 [2], implying that most likely the growth of big data is higher than claimed by Neelie Kroes.

4.1.1 Expectations

The traditional big stakeholders in very large database management systems see a great opportunity and a new market potential. The current hype speaks indeed about disruptiveness in how business is conducted. A large share of the value to be

A. Gavras (✉)
Eurescom GmbH, 69123 Heidelberg, Germany
e-mail: gavras@eurescom.eu

V. Angelakis et al. (eds.), *Designing, Developing, and Facilitating Smart Cities*,
DOI 10.1007/978-3-319-44924-1_4

created will come from new types of data use which are unprecedented. There are tremendous investments to be made in storage, computation and transmission capacity; and there are undoubtedly costs for keeping the systems running.

High expectations are put in non-ICT sectors that so far, although users of ICT had little exposure to big data because it was not easy and it was costly due to highly specialised solutions. Sectors often mentioned include energy, environment, agriculture, health, government and many others. The same applies to purely scientific data repositories. The sectors today are either increasingly engaged in collecting new data, or engaged in opening their archived data under the open access legislative initiatives. Current economic sectors are able to provide a cost-benefit analysis, although the benefits are rather expectations today and depend a lot on yet unknown applications that may emerge, while there is a lot of uncertainty.

It is often neglected that any data that is collected has only meaning in its context. All big data must be contextualised in its scientific or socioeconomic context. In a big enough data set one can discover unrelated correlations and the temptation is high to accept such a correlation as *truth* just because it is discovered in big data. This problem has been impressively demonstrated by the website 'Spurious Correlations' [3] and analytically discussed by Calude and Longo [4], who prove that very large databases have to contain arbitrary correlations. These correlations appear only due to the size, not the nature, of data. This phenomenon plays an important role later in the section about the necessary skills to master data sets.

4.1.2 Big Data Players

Open platforms that support knowledge exchange and sharing of largely unstructured data are a strong trend. Several initiatives and companies have picked up this basic idea and try to support organisations in managing their unstructured data and extract knowledge out of the data. However, the landscape is currently very unclear and full of buzzwords.

The fact that new data management companies emerge means that there is a certain complexity which has a cost and through which a profit can be realised. Those who find ways to efficiently manage the data complexity at low cost will be able to sustain a viable business.

Large established IT companies like IBM are offering solutions for managing big data and are advertising these by publishing related use cases [5]. The topics advertised are: Big Data Exploration, Enhanced 360° View of the Customer, Security Intelligence Extension, Operations Analysis and Data Warehouse Modernization. Taking a closer look into the use cases, I can hardly discover disruptive potential. In most cases it is about doing things better and faster which is good enough for a start.

4.1.3 Cost Versus Value

Admitting that there is a data management cost due to the sheer volume and complexity of big data, even considering the dropping cost of enabling technology, inevitably raises the question about the break-even point. The following general considerations are meant to highlight the challenge.

Figure 4.1 illustrates the increasing management cost, which roughly translates to Operational Expenditure (OPEX), versus the dropping technology cost, which roughly translates to Capital Expenditure (CAPEX) over time.

From the dawn of IT, experts are trying to calculate the optimal equilibrium state of processing, transmission, and storage costs for data. Simply put the question is when it is more economical to transmit large raw data volumes multiple times, versus replicating these data volumes at several locations, versus re-calculating a *result* and transmitting or storing only this result. Of course we know today that this is a moving target and depends on the costs trends of the different related technologies. However virtualization of technology and its use as a service is blurring its classification into the established boxes (OPEX/CAPEX). Using cloud storage induces OPEX, although in the past the purchase of storage arrays was inducing CAPEX. Clever buzzword finders have started to coin the term COPEX (standing for Capital and Operational Expenditure). As of May 2016 this definition for COPEX does not appear yet in the web's largest acronym repository [6].

As a concrete example of OPEX for big data, Google posted in 2009 [7], figures about the actual energy cost for a query. According to Google's blog article a single query accounts for 0.0003 kWh or energy. In terms of greenhouse gases, one Google search is equivalent to about 0.2 g of CO_2. Today Google is striving to reduce its CO_2 footprint to zero [8]. Of course this is achieved by large investments in renewable energy, which means a shift of OPEX to CAPEX. Newer studies are published almost daily that try to pinpoint accurate cost figures for the resource consumption of ICT.

Figure 4.2 illustrates the aggregate cost (CAPEX + OPEX) as compared to the value for society. The dotted line indicates an unanticipated scenario in which the

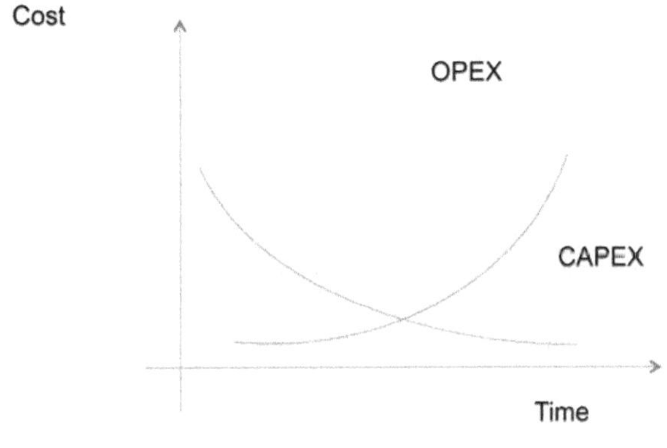

Fig. 4.1 Qualitative projection of CAPEX versus OPEX for big data

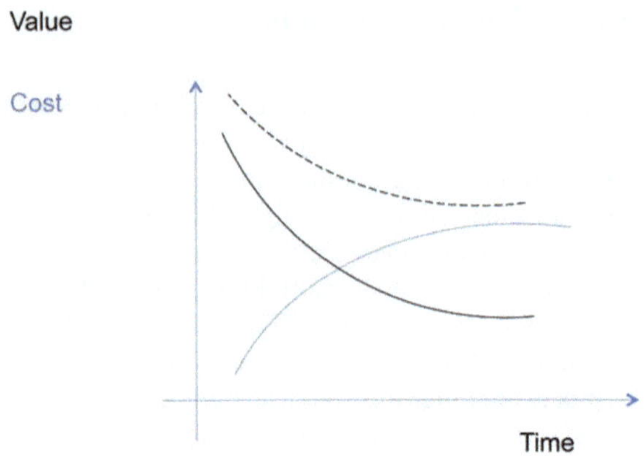

Fig. 4.2 Qualitative projection of aggregate cost versus value for big data

total cost of ownership exceeds in the long term the value provided to society or businesses. We still do not have a complete picture of the overall cost of Big Data and the Internet of Things (IoT).

4.1.4 Societal Cost

Not all costs have a direct price tag in terms of money or CO^2 footprint. Not focused on Big Data and the Internet of Things alone, the current calculation of the total environmental impact of ICT may have some defects [9], for example by not considering the rebound effect of stimulation of increased demand due to time-saving optimisation, or the software-induced hardware obsolescence and the miniaturisation paradox, which indicate that hardware is getting cheaper faster than it is getting smaller.

In an article, Helbing et al. [10] analyse the impact of advanced algorithms, artificial intelligence and Big Data on the future of society at large, pledging for the safeguarding of fundamental societal values developed over centuries such as freedom and democracy.

4.1.5 Skills and Enabling Technology

Miniaturisation and cost reduction of ICT has enabled almost anyone to collect and make accessible digital data, mainly because the technology exists and is affordable. Terabyte hard disks are nowadays in the 50 Euro range, very small scale PCs running free and open source software, such as Linux, are available for less than 30 Euros, and broadband subscriptions, even mobile broadband, are in the 20–40 Euros per month range. With these components at hand each digital native is capable of building a system that collects, temporarily stores, eventually processes and maybe makes available arbitrary and unstructured data virtually without limits. But is everyone able to assess, organise, preserve and maintain the data in the long term?

As noted earlier the knowledge in the data requires often the correct interpretation by domain experts applying fundamental scientific practices in addition to mining the big data sets, in order to avoid fallacies as discussed in [4]. Relying only on popular wisdom that 'numbers speak for themselves' is a dangerous assumption. It is a scientific skill to give numbers their meaning.

Once we have the skills to combine scientific rigour with big data analytics the relevant question about the value of the knowledge that could be extracted from the data and whether this value can be monetized to cover its cost and yield an economic profit and societal benefit should be easier to answer.

4.1.6 Streams of Data

At a certain point in time the cost of processing and querying big data stores may become economically unaffordable. At this point in time we must have found ways to intelligently distil useful knowledge out of a passing stream of unstructured data and just drop the rest. The challenge lies in identifying what we can extract from this stream of raw data that is produced by the Internet of Things, and which will provide also in the future opportunities for yet unforeseen use of historical big data.

Stream data processing is also necessary in cases where even structured data would be meaningless without their temporal characteristics, implying their processing in real time or capturing the full context in space and time. Obviously stream data processing makes most sense at the edge of the network, basically at the location where data are generated. However due to the extreme decentralisation of processing this may result in increased security problems, since *traditional* security measures—as will be discussed later—are not appropriate for securing every single data source.

4.2 The Internet of Things

The Internet of Things (IoT)—used herein in a broader scope—promises many benefits in terms of new applications and in particular new opportunities for a substantial change in societal behavioural patterns. And indeed we have witnessed many exciting new technologies and applications that are enabled by the IoT. Considering IoT from a narrow point of view and looking to the currently deployed sensors in smartphones we could actually question whether there is actually an Internet of Things that is different than what we have today. This is because it looks like we have already multi-billion sensors connected to the network, yet the vendors and operators manage to keep the networks up and running and new services emerging daily. It is time for a reality check and to ask the question about the sustainability of the current approach as well as to examine future directions for new business cases enabled by IoT.

4.2.1 Cost for Connectivity

In terms of cost for connectivity, we have not progressed much in the last few years and perhaps we have even done a few steps backwards. Today virtually all IoT applications are based on some sort of client/server principle in which there is a substantial computing capacity in a virtualised computing environment to which each single sensor and actuator is connected somehow. This means that beyond the investment needs for the IoT enabled world (CAPEX) in the frontend, there is a substantial OPEX and CAPEX in the backend support for IoT-enabled applications. Assuming an average lifetime of three to five years for the hardware delivering the computing capacity we are facing a disconnect with respect to the lifetime of other supporting infrastructures, such as networking where the depreciation time of infrastructure investments is in the order of 10–20 or more years, although shrinking. This means that the cost of supporting and serving hundreds of billions of smart objects in the long term is considerable and may not be included in the current assumptions about Total Cost of Ownership (TCO).

4.2.2 Security and Trust

In terms of security and trust, there exists to date no future proof concept; even less a concept that is economically viable at the anticipated scale. The traditional model in which IT domains are organised and protected centrally by some gatekeeper will not work for the IoT world for reasons induced by the extreme decentralisation of most IoT-enabled systems. We cannot protect each smart object individually on an economically viable basis.

In the area of trust falls also the practice of many vendors delivering smart products that collect data about the behaviour of their customers in the hope that these data may be an exploitable asset. This is a serious and to date underestimated problem. In many cases individuals may not care (even if they knew) about the practice. However enterprises care a lot and sometimes have the means to discover and resist the practice. At least in Europe the trend in legislation is to strengthen the rights of citizens and businesses with respect to the control of their data.

Furthermore, a discovered vulnerability is a product defect and must be fixed. Traditional industries have been hit very hard economically in cases where recalls are necessary (e.g. the car industry). In the ICT industry the strategy is to distribute software updates, which nevertheless puts an increasing cost to the long-term maintenance of IoT-enabled applications. The fact that many new IT devices have a short expected lifetime motivates vendors to drop older devices from their maintenance roadmaps so that these devices do not receive security updates anymore. How will such devices be protected in the future? Do we need to replace our energy smart metres, digital doors locks and smart connected cars every three or five years when the vendors stop delivering security updates?

4.2.3 Future-Readiness of IoT

In *Cyber Physical Systems* (CPS) IoT devices are directly connected to real-world artefacts and have a substantial influence on their properties. Where homes, buildings and factories are smartened, it is a reasonable expectation by the customers that the devices used in this context have an expected lifetime in the same order of magnitude to be future proof. This is in the order of 30–50 years and beyond. In many cases different generations of products, that appear every 2–3 years, have different maintenance requirements and different ways of servicing. How will vendors and suppliers be able to cope with the long-term cost of maintenance? Dumping the long-term cost on the customers will not work, because when the early adopters discover that the TCO of a smart fridge is an order of magnitude higher than that of the stupid old fridge, they will start to question the added value of the smart world.

The hype about IoT and CPS has triggered many 'innovations' in the market for which the added value for the customer is questionable. Vendors may think in terms of better maintenance and service for the benefit of the customer or about warranty tracking for the benefit of their own supply chain optimisation. But a smart coffee brewer, a smart toaster, or a smart water boiler, do not provide added value to the customer unless they make better coffee, better toast or better hot water! Certainly this observation does not devalue many useful applications of the new technology, but it is always a simple and clear value proposition that decides about the market success of an IoT-enabled product or service.

All above deficiencies culminate to an uncomfortable truth that the current business models around IoT might be broken. Of course this is a pessimistic view. An optimistic view formulates the problem such as that no one has yet found a viable and sustainable business model for the large scale. In order to progress the search for viable business models, the discussion of the broader concerns above, gives us an indication that we have to face issues in four dimensions; namely (i) technology, (ii) business, (iii) policy and (iv) last but not least customer. All four dimensions are a source for requirements that have to be satisfied at the same time. Smart city experiments around the world are a good starting point to learn to deal with all these requirements, since all dimensions are prominently present in most scenarios. However all smart city pilots that exist to date are just that: pilots! None of these experiments claims a self-sustainable operation that is ready for the long term in the city scale.

In order to develop sustainable business models, they should not be designed around the traditional understanding of value chains, but should rather be designed with the flexibility to cope with value networks that emerge in digital business ecosystems.

4.2.4 Legal Frameworks

In January 2012, the European Commission proposed a comprehensive reform of data protection rules in the EU. On 4 May 2016, the official texts of the Regulation and the Directive have been published in the EU Official Journal in all the official languages. While the Regulation will enter into force on 24 May 2016, it shall apply from 25 May 2018. The Directive enters into force on 5 May 2016 and EU Member States have to transpose it into their national law by 6 May 2018 [11].

A further regulatory reform may by triggered by the question on product liability. Traditionally product liability is limited to *products* in the form of tangible personal property. In the future the correct functioning of a device (e.g. IoT, medical sensor) includes a functioning network and backend service. Smart (connected) devices will have a far reaching impact on manufacturers, service companies, insurers and consumers, since legally a product or service may become defective upon network or service failure (even temporal) or upon discovered security vulnerabilities.

The new European data protection rules and potential evolution of the legal framework on liability will have a significant impact on how businesses deal with data, products and services and it is reasonable to expect that it will induce a higher long-term cost.

4.2.5 Cannot Sell the Solution

Many companies are developing solutions that are marketed as products, e.g. for smart cities or large corporate customers. But a city is unlikely to buy a solution which is prone to all uncertainties about the economic viability and correct functioning of the solution in the long term as described above. Of course the cities are eager to provide more services to their citizens and to find ways to more efficiently manage the city as a whole. But the value is in the information and the knowledge that can be extracted from the data a solution provides, once it is deployed and operational. Buying the information or knowledge means buying sanitised data that are extracted by specialists and who can perhaps attach a *quality label* on the data.

4.2.6 *The Value Is in the Data*

This means that the business cases for IoT and Big Data is in fact a business case for Quality Data. Quality guarantees could be of substantial value to customers, whether public customers (cities, government) or private corporations. From a financial and investment point of view such a customer would have no CAPEX and would only subscribe to a service that provides Quality Data, hence paying a subscription fee. Such a fee can much easier be assessed with respect to the value it provides to the customer and consequently a decision to subscribe could be easier and faster.

Remains the question of who should build the infrastructure for IoT and who should extract the knowledge from the Big Data. The dilemma looks like the recent dilemma of the telecoms industry for which the telecom operators are building and operating the infrastructure, however other—so called over the top (OTT)—players earn the profits. However this is not entirely true since as hinted earlier many grass root initiatives are emerging and growing, which are building or retrofitting IoT infrastructures and are offering interfaces to access the information which can be utilised to extract Quality Data. Some of the observed properties of the shared economy should be examined in this context to identify how a value network could be build that allows for a reasonable compensation of all stakeholders. This compensation may not be of financial nature in all cases.

A more traditional and therefore perhaps more credible assumption is that new entrants, in fact they should be called *market creators*, will appear in the market and who will deploy IoT and Big Data solutions directly delivering *radical innovation*. They would own and operate the solutions and take the full risk (and profit) of this approach. They will offer Quality Data as a Service (QDaaS) for anyone that is willing to subscribe to a real-time feed of Quality Data in the same way as we subscribe to a newsfeed today or subscribed to a newspaper in the past. Most importantly some market creators may augment such data with vertical sector knowledge which will increase the Value of the Data and will render them more useful for sector applications and services.

These market creators would then leverage on the current communication, computing and storage infrastructures provided by the incumbent ICT stakeholders, regardless of them being network operators or cloud providers and will provide value based on Quality Data beyond networking and services.

4.2.7 *Radical Innovation*

This subsection provides an interpretation of *radical innovation* as was used in the previous subsection. Innovation per se, can be categorised according to several

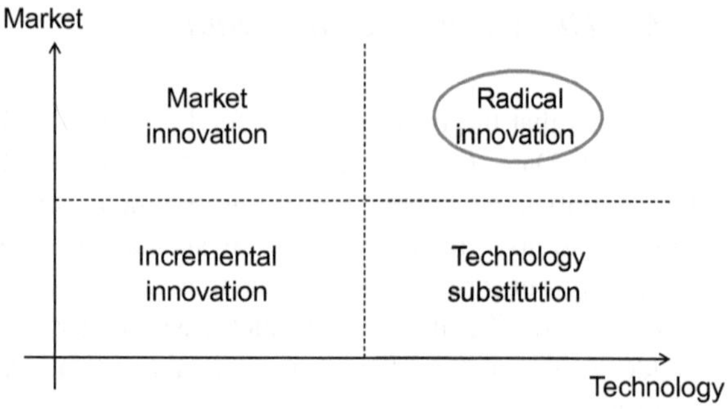

Fig. 4.3 Partitioning innovation in technology versus market dimensions

aspects. For example strategic innovation, which has a business oriented focus, and process or product innovation, which are concerned with the improvement of existing, or the introduction of new processes or products. Herein innovation is perceived along the axes *Technology* and *Market* as illustrated in Fig. 4.3. *Incremental innovation* denotes the case where existing but improved technologies are used to improve existing services and products. *Technology substitution* denotes the case where new technology is used to create new products and services in existing markets. *Market innovation* denotes the case where existing but improved technologies are introduced in new ways into the market, effectively creating a new market segment. Finally *Radical Innovation* denotes the case where new technologies are used to create new markets, effectively introducing disruptions on both axes.

Big Data, and the *Internet of Things*, taken alone, introduce relatively clear paths to either *Market Innovation* or *Technology Substitution.* However, since the first endeavours to introduce disruptive ways of creating knowledge from existing data and the latter endeavours to create new technology for the enablement of new products and services, in combination they likely lead to *Radical Innovation.*

4.3 Conclusions

This paper names and analyses some of the expectations and possible long-term effects in relation to Big Data and the Internet of Things. It spells out the hidden cost of the new technologies that are emerging very fast and fuel the imagination of entrepreneurs and investors. Certainly it does not aim to diminish the success stories achieved so far by early adopters. Its purpose is rather to point out the areas that need further attention in the future, such as security at large, or to deliver quality data, as well as to sketch a possible future in which radical innovation takes place for the advantage of the brave entrepreneurs and early adopters and the benefit of the society in the long term.

References

1. http://europa.eu/rapid/press-release_SPEECH-13-893_en.htm (Dec 2015)
2. Powell JR (2008) The quantum limit to Moore's law. Proc IEEE 96(8). http://ieeexplore.ieee.org/stamp/stamp.jsp?arnumber=4567410 (May 2016)
3. Spurious Correlations, http://www.tylervigen.com/spurious-correlations (May 2016)
4. Calude CS, Longo G (2016) The deluge of spurious correlations in Big Data. In: Foundations of science, pp 1–18. doi:10.1007/s10699-016-9489-4, http://www.di.ens.fr/users/longo/files/BigData-Calude-LongoAug21.pdf (May 2016)
5. Big Data at the speed of business; the 5 game changing big data use cases, IBM. http://www-01.ibm.com/software/data/bigdata/use-cases.html (May 2016)
6. What does COPEX stand for? http://www.acronymfinder.com/COPEX.html (May 2016)
7. Powering a Google search, Jan 2009. https://googleblog.blogspot.de/2009/01/powering-google-search.html (Dec 2015)
8. Google green, our footprint: beyond zero. http://www.google.com/green/bigpicture/ (May 2016)
9. Gavras A (2008). Why there is no such thing as "green ICT", Eurescom mess@ge magazine, Nov. 2008. http://archive.eurescom.eu/message/messageNov2008/Why-there-is-no-such-thing-as-green-ICT.asp (Dec 2015)
10. Helbing D, Frey BS, Gigerenzer G, Hafen E, Hagner M, Hofstetter Y, van den Hoven J, Zicari RV, Zwitter A (2015) Digitale Demokratie statt Datendiktatur, published Dec. 2015. http://www.spektrum.de/news/wie-algorithmen-und-big-data-unsere-zukunft-bestimmen/1375933 (May 2016)
11. European Commission, Protection of Personal Data. http://ec.europa.eu/justice/data-protection/ (May 2016)

Part II
Technologies

Chapter 5
Designing Secure IoT Architectures for Smart City Applications

Elias Tragos, Alexandros Fragkiadakis, Vangelis Angelakis and Henrich C. Pöhls

5.1 Introduction

The Internet of Things (IoT) has received significant attention lately due to the numerous potential improvements it can bring to the everyday lives of peoples, simplifying many of their daily life activities. This is mainly evident in the domains of smart buildings, smart health, and smart cities. Especially in the smart city domain, the benefits for the citizens, the society, and the economy at large scale, have transformed the IoT into a major trend, and many municipalities are trying to build their city development strategies around it. Of course, a very first point is to build an IoT infrastructure that can be quite costly. Then, the cities have to develop applications that will run on top of this infrastructure and will address the requirements and the needs of the citizens. The latter part is very important in order to motivate the citizens to "accept" and adopt the IoT technologies.

A major factor for the citizens to adopt and use the IoT technologies in this direction, is the trustworthiness and the reliability of IoT as a whole. It is reasonable to assume that citizens can be sceptical for this new technology, especially when thousands or millions of new small and "invisible" devices can be around them all the time, with the potential to monitor continuously their everyday activities. These devices are not only becoming smaller, but also more intelligent, autonomous and active and are seamlessly integrated in the smart city environments. There are

E. Tragos (✉) · A. Fragkiadakis
Institute of Computer Science, Foundation for Research and Technology Hellas, (FORTH-ICS), Heraklion, Greece
e-mail: etragos@ics.forth.gr

V. Angelakis
Communications and Transport Systems, Department of Science and Technology, Linköping University, Campus Norrköping, Norrköping, Sweden

H.C. Pöhls
Department of Informatics and Mathematics, University of Passau, Passau, Germany

V. Angelakis et al. (eds.), *Designing, Developing, and Facilitating Smart Cities*,
DOI 10.1007/978-3-319-44924-1_5

various projections that many billions of devices will be connected to the IoT in the next few years [1, 2]. This increasing number of autonomous connected devices that are not only monitoring their environment, but can also act upon it can become a worrying factor for the citizens, in terms of stealing their private information, logging their activities, monitoring their presence, or even harming their life. In order to motivate the users to adopt the IoT and use the smart city applications in their everyday activities, they have to be provided with guarantees that the IoT systems: (i) will be reliable in terms of the information they provide, (ii) will exchange the information in a secure way and (iii) will safeguard their private information.

To address the above requirements, the whole IoT system has to be secure in a cross-layer manner, starting from the physical layer and covering all layers up to the applications. Since IoT systems are based on Wireless Sensor Networks, various sensor security mechanisms can be adopted also in the IoT [46]. Security and privacy-preserving functionalities have to be embedded in the system's operation from the design phase, because post-mortem corrections and improvements can only cover some holes, but will not provide full-scale security. Thus, there is lately a move toward adopting the concepts of "security and privacy by design" when someone wants to build a new IoT system. As described in [3], the concept of "security by design" describes the need for considering all possible security issues at the conception phase, and designing respective mechanisms to prevent and react to these issues.

Although for adopting the concept of "security by design" a security toolbox might be enough, for adopting the concept of "privacy by design" a more holistic approach is required, since a set of privacy enhancing technologies cannot fully protect the users' privacy [4]. This requires embedding privacy concepts in the core of many system functionalities, e.g., onboard the sensors in order to be able to not collect identifiable information, at the gateways to be able to hide identifiable information before forwarding encrypted data, at the middleware to disallow the linking of information between different services, etc.

Apart from security and privacy, the issue of Trust in IoT is also attracting a lot of research interest lately. Trustworthy internet is defined in [5] as the Internet system that is secure, reliable, and resilient to attacks and failures. This means that the notion of Trust in the IoT has to be evaluated using metrics of: (i) reliability of both systems and data, (ii) security of the infrastructure and the provided services and (iii) ability of the system to prevent and respond to attacks and failures. An overall trust model for the IoT, since it incorporates security, and privacy, can be potentially very useful for promoting the IoT systems to end users or citizens. It is reasonable to assume that a system that is certified as "trustworthy" will be much more attractive for a user than any other system. However, in order to make a system trustworthy, its functional architecture has to incorporate all security, privacy, and trust functionalities.

In this chapter, we provide an overview of the requirements and challenges for designing secure IoT architectures. We also present previous attempts of many IoT-related projects to design and develop mechanisms and concepts for improving the security, the privacy, and the trustworthiness of IoT. The focus is on the latest IoT projects and we try to provide a brief description of the main security and privacy features that they have embedded on their architectures. The chapter will conclude with a brief discussion on the common functionalities of the projects and the open research items providing some suggestions toward the future research in the area.

5.2 Securing IoT Architectures: Challenges and Methodology

There are many challenges when designing IoT architectures, and the most important of them stem from the fact that IoT systems are assumed to consist of very large heterogeneous networks of both constrained and unconstrained devices, continuously operating mostly without any power source. From this, one can assume that the main factors that prohibit the inclusion of strong security and privacy mechanisms in an IoT system are [6, 7]:

- the heterogeneity of the involved systems and devices in terms of communication technologies, software, hardware, and capabilities,
- the constrained nature of many IoT devices, which can have very limited resources,
- the need for scalable solutions to function properly even at large-scale deployments and
- the need for energy efficient optimization both in terms of hardware and software so that the devices can operate without the need for battery replacement for long periods.

These challenges can pose severe difficulties when designing the architecture to be secure and privacy preserving, and especially when trying to embed security and privacy functionalities on resource-constrained devices. A recent report by Hewlett Packard Enterprise [8] analysed the vulnerabilities of existing IoT hardware devices and the findings were very concerning. The highlights of this report was that more than 90 % of the devices were collecting personal information, 70 % of them used unencrypted traffic, while 60 % of them used weak credentials. It is obvious that no matter how secure the backbone IoT system is, the lack of security on the devices can really compromise the whole system, and this has to be seriously taken into account in the design of secure IoT architectures.

The IoT system architectures are designed to be secure to minimize the risks of attacks or failures. Actually, there can be either human or non-human risk sources in an IoT system. The human risk sources stem from the fact that malicious users may be hacking the devices and steal information or when users' faults or accidents affect the system performance. The non-human risk occurs when there are security issues due to natural phenomena, i.e., a flood, a fire, heat or device hardware failure.

Usually, system designers consider only human risk sources and mainly intentional malicious attacks, considering all other risk sources as out of the scope. However, using IT technologies, the impact of many other risks can be mitigated too. For example, when the reliability of the data is calculated in order to discard erroneous measurements, the data gathered by a device that is affected by a failure or a fire can be identified as erroneous and not considered in the system decisions. Furthermore, when devices are affected by flood and are not sending measurements, proper monitoring mechanisms can create alerts so that the system operator can resolve the issue and revive the devices.

As described in [9, 10] toward designing a secure IoT architecture, as a first step, one has to identify the elements (or assets) he wants to protect in an IoT system. In general, the main elements to be protected can be split into IT-based and non-IT:

- IT-based elements: this category includes assets like the user/device credentials, the various types of data that are exchanged within the system (user data, sensed or actuation data, application data, control data) and the software [9].
- Non-IT elements: this category is more generic and includes elements, such as: human users, devices (leaf or intermediary), users' privacy, services, communication channel, and in general the infrastructure [10].

For the identification of risks in IoT systems and the assessment of their impact, standard methodologies such as Microsoft's STRIDE/DREAD [11] or analysis of the Confidentiality, Integrity Availability on the assets and the threats against Authentication, Authorization and Accounting, as well as the Privacy threats. For a detailed analysis of the risks and their assessment in IoT systems, the reader can refer to [9, 10].

After the identification of the risks, the next step is to identify what are the system requirements in order to be able to mitigate these risks, and especially which are the design choices that the operator or system designer has to make in order to secure and protect his system and his data. In some cases, the system designer has to take tough but important decisions, since security/privacy and functionality/flexibility/interoperability/openness can be conflicting requirements. For example, one of the key enemies of privacy is linkability of the data, which is basically required for improved interoperability of an IoT system. Moreover, strong security comes at the expense of system performance due to the high complexity for running, i.e., powerful encryption mechanisms on the devices. However, the level of security, privacy and performance can be dynamic, selected at the operational phase so that it can correspond to the device/system capabilities and the requirements of the applications that are provided by the system. The latter is a very important requirement that has to be taken into account, since the applications can have very different security and performance requirements, and a proper IoT system has to be able to adapt to these diverse requirements. For example, an environmental monitoring application may have very low security and performance requirements, while a smart health application will have strong security, privacy, and performance requirements. Thus, the IoT systems due to the requirement for interoperability and for breaking the silos must be smart enough to be able to support both these applications without any issues.

5.3 Overview of Secure IoT Architectures

This section provides a brief analysis of the most important EU projects in the area of IoT that have embedded in their architectures security, privacy, and trust functionalities. The literature review starts with the IoT-A project, which is the lighthouse project of the EU with regards to setting the foundations for the design of IoT architectures and continues with other key projects.

5.3.1 IoT-A

The goal of the Internet of Things—Architecture (IoT-A)[1] project was to set the foundations for the design of an Architectural Reference Model (ARM) for the IoT. The project aimed to create a coherent architecture that allows the integration of heterogeneous technologies and supports the interoperability of IoT systems. IoT-A outlined principles and guidelines for the technical design of IoT communication and service provisioning protocols, interfaces, and algorithms. Furthermore, IoT-A proposed an innovative service resolution infrastructure for scalable discovery of resources and entities of the real world.

IoT-A's activities were based on the concept that the key aspect of any IoT system are the "things" and the "communication" among them. On the contrary to many past approaches that used the term "things" to denote the sensor devices, IoT-A defined the "things" to be the Physical Entities (PEs) that are all around us and are being accessed through the devices. The things that are connected to the internet are, thus, i.e., the room, a pen, a fridge, a car, a city, a laptop, anything that is a physical object and is of interest for the user. These PEs are transformed into Virtual Entities (VEs) so that they can be searchable, locatable, and controllable via software. The concept of virtualization helps to conceal the heterogeneity of the devices and to hide the devices from the applications' point of view, so that the users only request data for a PE and should not have to know which devices are monitoring or controlling this PE. This service-oriented-architecture approach utilized Resources on the IoT devices that are exposed via Services that can be reusable by many applications or can be composed to provide more complex services [10].

With regards to security and privacy, IoT-A has defined many system requirements both functional and nonfunctional that assisted in the design of the ARM. The requirements were split into three main categories, as described in [12], and summarized below:

[1]www.iot-a.eu.

- System Dependability: this category includes requirements for improving the overall security of the IoT system and its infrastructure. It includes requirements for
  - Service and infrastructure availability, so that the user can invoke a service under all conditions and the infrastructure should be able to provide this service at all times.
  - Accountability, so that any failures or misbehaviours can be traced back to the responsible person/system.
  - Infrastructure Integrity and Trust, so that the provided infrastructure services are trustworthy and can operate according to their design requirements.
  - Non repudiation so that all services can be accessed by their rightful owners.
- Communication Stack: this category includes requirements for (i) the network layer and (ii) the service layer:
  - Network layer: here the requirements are related with (i) anonymization at the network level, so that users can protect their privacy through anonymity and (ii) confidentiality, so that the messages are encrypted to be protected against eavesdroppers.
  - Service layer: here the requirements are related with (i) service authentication and access control, so that only authenticated users can have access to the system services and even more to only specific services that they are allowed to, and (ii) service trust and reputation, so that only authenticated users access the system and only trusted nodes can provide measurements.
- User and service privacy: this category includes requirements either for protecting users when they are using the Services and the Infrastructure or the users cannot extract information about the data subject that is providing a specific service.

To address the previous requirements, IoT-A identified a number of security components that are required to be included in the architecture. The architecture of IoT-A, called ARM and the respective Functionality Groups (FGs) can be seen in Fig. 8.2 in [10]. Since the focus of this chapter is only for the security and privacy components of the IoT architectures, here we will be limited to the description of the respective Security and Management FGs, while the description of the rest of the FGs is omitted.

The Management FG includes among others functionalities that can improve the resiliency and the availability of the overall system. For example, the component for "Fault" can identify and correct system faults, detecting them by generating alarms and applying corrective mechanisms. The alarms can also be sent to other components that have to act in order to correct a specific fault. Another component called "Member" handles the membership and the associated information of all relevant entities of the system. It includes a database that stores information regarding ownership, rules, rights, etc., and is important for the Security FG in order to define the security and privacy policies [10, 13].

The Security FG is the main group of functionalities for ensuring the security and privacy of the system, including the following components [10, 13]:

- **Authorisation**: which handles the security policies and takes decisions for access control based on these policies. It determines if a user action is allowed or not and controls the policies for the services by adding, updating, or deleting policies according to Service Level Agreements or user preferences.
- **Authentication**: which involves authentication both for users and services. It checks the credentials and if they are valid then it allows the access to the IoT services.
- **Identity management**: which tackles privacy issues, by distributing and managing pseudonyms and accessory information to trusted subjects to operate anonymously.
- **Key exchange and management**: which provides mechanisms for secure communications between two or more system nodes. It is also responsible for distributing keys for the secure communications in a secure way, as well as for registering security capabilities.
- **Trust and Reputation architecture**: this component collects and evaluates user reputation scores to use them for calculating service trust levels. This is done by requesting reputation information from users with respect to a service and providing the reputation of the service to interested applications or users.

Overall, IoT-A set the foundations for the architecture of IoT systems and in the proposed ARM, some basic functionalities for security and privacy where included. These functionalities can be characterized as the bare minimum that an IoT system can include so that it can be acknowledged as secure.

5.3.2 iCore

The project "Empowering IoT through Cognitive Technologies" (iCore)[2] has as a vision to provide the IoT world with the necessary technological foundations to empower the IoT with cognitive technologies, so that IoT services and applications can be easily and simply widespread. iCore structured this vision toward tackling two main issues: (i) how to abstract technological heterogeneity that is inherent to the real world objects, considering the large numbers of diverse objects that impose challenges for reliability, energy efficiency and context-awareness and (ii) how to consider the requirements of different users and stakeholders aiming to support business integrity and application provision according to service level agreements [14].

Toward these objectives, iCore structured a cognitive architectural framework that includes three levels of functionality, which can be reusable to diverse

[2]www.iot-icore.eu.

applications: (i) the Virtual Objects (VOs), (ii) the Composite Virtual Objects (CVOs), and (iii) Service Level. The VOs are used to tackle the technological heterogeneity, by enabling a cognitive virtual representation of both real world (e.g., a chair, a table, a house) and digital objects (e.g., a sensor, an actuator, a device). CVOs are cognitive mashups of semantically interoperable VOs and use the services of the latter in accordance with the user or stakeholder requirements [14].

With regards to security and privacy, iCore defined the respective requirements considering existing regulatory frameworks (i.e., EU directives) and the three main use cases of the project. As a result of the analysis of these sources, five main security requirements were defined as described in [15] and summarized below:

- **Availability**, ensuring the reliability of the data sent to authorized users.
- **Confidentiality**, ensuring the privacy of the users and the protection of their data, disallowing the disclosure of information to unauthorized individuals, devices, or services. This requirement included also the requirements for anonymization and pseudonymisation.
- **Integrity**, guaranteeing the correctness of the operation of the system according to some predefined rules and the consistency of the data.
- **Authentication or authorization**, ensuring the validity of the provision of data/services verifying the authorization of an individual to receive this information.
- **Nonrepudiation**, reassuring both the sender and the receiver of information that it was sent by the correct sender and that it was received by the proper receiver.

Considering these requirements, iCore defined also some higher level requirements for the design of the architecture [15]. These requirements were related with security functions for (i) security management, (ii) scalability and (iii) multi-level security. The goal was to ensure that security mechanisms and policies are not static and can be updated on a regular basis, to avoid using obsolete credentials or hacked software. Furthermore, the system security mechanisms have to be scalable to ensure that they can perform well in heterogeneous large-scale environments. Moreover, iCore users are considered to have multiple different levels of access, credentials, and data protection, and the framework should be able to properly handle these diverse functionalities.

The iCore Architecture presented in Fig. 2 in [14] aims to differentiate the project by allowing the system to derive knowledge from the usage of context, so that applications can reach their goals and intelligently behave and organize the system's own resources. A main concept of iCore is cognition through "virtualization," which allows operations such as (i) **functional enrichment**, targeting a virtualization with highest fidelity of managed real-world objects that coexist with virtual functional spaces, (ii) **abstraction**, allowing the generalization of the functionalities do that they are applied to a larger set of situations and requirements, and (iii) **aggregation**, allowing the combination of Virtual Objects with dynamic context-awareness capabilities so that they provide higher level functionalities, close to real world demands.

To provide enhanced system security, in the proposed iCore architecture, there is a functionality group for Security Management, which includes a policy repository, an identity provider and a Policy Decision Point (PDP) [16]. At all layers (VO, CV, service) there are Policy Enforcement Point (PEP) components to intercept any request for new service, or for executing VOs or CVOs. These requests are then sent to the PDP that evaluates the requests against predefined security policies, returning the results to the PEPs. That way proper authentication and access control is enforced in an iCore system. These functionalities are provided in iCore through a Model-based Security Toolkit (SecKit) that provides the basis for security engineering, data protection and privacy. It has meta-models that support the modeling of data, identities, context, trust, risks, and policy rules.

The iCore security architecture can provide functionalities for [16]:

- Trust management, using security policies that consider trust relationships established by users with devices and operators.
- Dynamic context adaptation, so that any context changes will be considered in the security actions.
- Data privacy, through
  - data anonymization, to allow the privacy rules to ensure proper data anonymization when they are collected by the devices,
  - control of the data flow from devices, to allow users to define which and what type of data will be gathered by the devices.
- Control of the actions of the actuators, to improve the system safety.

5.3.3 BUTLER

The BUTLER project[3] aims to support the development of pervasive applications using heterogeneous devices, protocols, and standards. The applications are built in order to improve the daily activities of users in various domains, considering among others contextual information, that may be the user needs and preferences, their location or the status of the physical entities with which the users interact. The architecture of BUTLER was built on existing standards and industry initiatives such as IoT-A and FI-WARE [17]. However, the BUTLER architecture includes additional functionalities that relate with the association of Context to virtual entities.

With regards to security and privacy, BUTLER, similarly like iCore, focused on context-aware security, adapting its mechanisms using input from its environment with regard to any changes that might occur, with the target to use this information to prevent inappropriate behaviors [18]. Access control mechanisms based on

[3]www.iot-butler.eu.

context were developed, for deciding if a request for access to a resource or to some data will be granted of refused. BUTLER considered security policies based on context, so that the security responses of the IoT system are adapted to their context of use. These policies have to describe the authorized practices for a user/entity of the system at each moment in time and in each situation. Furthermore, these policies are not used only for access control, but also for communication encryption and other security functionalities. Apart from context-based security, BUTLER focused also on enhancing the users' privacy [19]. BUTLER considers that the applications may impose privacy issues in term of using the data gathered by the devices.

The BUTLER architecture is depicted in Fig. 4 in [17] where the four main layers can be seen:

- The communications layer, including all functionalities for enabling the communication between heterogeneous devices and between the various entities of the system and the users. In addition, it includes various Security Services for device and user authentication and authorization to ensure that only identified and authorized devices can join the BUTLER network.
- The data/context management layer, which handles all data and context related functionalities, (e.g., capturing and collecting data, persistent storage, data processing, context extraction, etc.) but does not include any security services.
- The System/Device management layer deals with the management and the maintenance of large numbers of heterogeneous networked devices. This layer has direct interfaces with the security services, i.e., for ensuring the secure configuration of the devices, their authorized management, etc.
- The Services Layer is responsible for describing, discovering, binding, and providing context-aware BUTLER services. It is also responsible for making the services available to the applications, providing enablers for supporting discovery and purchasing of data functionalities.

As it can be seen, the main security and privacy functionalities of BUTLER reside at the communication layer, which provides access to the users to all the other functionalities. The BUTLER security services are built around a strong authorization server. The main security services are [17]:

- Secure transport of messages between devices and the authorization server, based on Transport Layer Security (TLS).
- User and application authentication, using credentials.
- Resource registration and authentication, including information on which actions are allowed to be performed on this Resource by its consumers.
- Key management, both for encryption and authorization keys.
- End to end security between application user and Resource provider.
- Device authentication via a bootstrapping mechanism between the devices and the gateways at the sensor network level, using asymmetric cryptography and elliptic curves.

As it is obvious, the main focus of BUTLER is ensuring security and privacy at the communication layer, with many functionalities built around a powerful authentication and authorization framework. This is very important in order to ensure that access to services and private information can be allowed only to authorized persons, avoiding the disclosure of information to third parties.

5.3.4 OPENIoT

OpenIoT[4] focused to develop an open source IoT middleware for getting information from heterogeneous sensor networks, concealing the types of devices that provide this information. OpenIoT adopts the IoT-A concepts of "entities" and resources, such as sensors, actuators, and smart devices, leveraging them to build the concept of "Sensing-as-a-Service", via an adaptive middleware framework for deploying and providing services in cloud environments [20].

The OpenIoT architecture is built based on various nonfunctional requirements for improved performance, scalability, reliability, privacy, and security. The project acknowledges that the provided platform has to be secure and privacy preserving. In this respect, mechanisms for role-based authentication and authorization are developed, upon which utility-driven privacy mechanisms are provided [20].

The OpenIoT architectural components mapped to IoT-A ARM are depicted in Fig. 10 in [20]. This architecture includes, among others, a group of security components, providing mainly authorization, identity management, and authentication functionalities. As described in [21], the OpenIoT platform consists of several standalone applications, such as X-GSN, LSM, etc., which require their own security mechanisms. Due to the distributed nature of the platform, OpenIoT acknowledged the importance of designing and developing a centralized authorization server that would provide authentication and authorization services to all components of the system. OpenIoT acknowledged that a main feature of the platform is that the user credentials are only checked and maintained by this central server and then, this server performs the authorization of the applications that are running on behalf of the user. In this respect, user credentials are not being circulated among the various components, which improves the system privacy.

With respect to communication security, mechanisms like TSL or HTTPS are used for the backbone communications (between the gateways and the cloud servers). At the sensor layer, standard security mechanisms of IEEE 802.15.4 are used. For the authorization framework, OAuth is used, because it is an open standard that can be easily integrated in the open platform. Apart from these, OpenIoT developed also a framework to evaluate the trustworthiness of the sensor readings via spatial correlation of the sensed measurements. In this respect, readings from neighbor sensors are correlated to assess their trustworthiness and discarding untrusted readings [21].

---

[4] www.openiot.eu.

5.3.5 SMARTIE

SMARTIE[5] (Secure and Smarter Cities Data Management) aims to create a framework for IoT applications that are sharing large volumes of heterogeneous data. The focus is on providing end-to-end security and information trust, considering also requirements for maintaining user's privacy. Mechanisms and technologies for security and trust at the perception, service, and network layers, as well as for secure storage and access control are central to the system.

The design of the SMARTIE system architecture was based on a long list of requirements as presented in [22]. With regards to security and privacy, SMARTIE supports the processing information on trusted nodes and user anonymity for privacy protection, data confidentiality, strong access control and authentication, secure storage, user's consent, context-aware policies, location privacy, communication integrity, no tampering of devices, and system resilience.

The SMARTIE architecture is based on the IoT-A ARM, following the IoT-A methodology and the ARM functional groups to organize all the functional components of the SMARTE architecture. The architecture is depicted in Fig. 4 in [23]. As it can be seen in this figure, there are many security related functionalities, not only in the security FG, but also embedded in other FGs. For example, the "IDS Data Distribution" component scans network traffic for intruders and reports unwanted or unknown traffic to the network operator, distributing detected events to other devices. Furthermore, the "Resource Directory with Secure Storage" component provides secure access to available resources in the system, allowing also their secure storage. The "Privacy-preserving Geofencing" component offers secure location-based services using secure geofencing, aiming to avoid the disclosure of location information of the users to unauthorized persons.

In the Security FG, SMARTIE provides the following components and functionalities [23]:

- Authorization, based on DCapBAC, for ensuring that access control decisions are taken before a service is accessed. Attribute-based access control[6] using the XACML/JSON[7] framework is used, utilizing policies to express rich and fine-grained access control decisions.
- System integrity based on nodes attestation (IMASC). IMASC provides a trusted environment for most embedded devices using SmartCards, allowing the detection of malicious code or wrong measurements.
- Authentication using distributed Kerberos and symmetric cryptography.
- Encryption, based on the CP-ABE [24] schema that ciphers information based on policies.

[5]http://www.smartie-project.eu.

[6]http://nvlpubs.nist.gov/nistpubs/specialpublications/NIST.sp.800-162.pdf.

[7]http://docs.oasis-open.org/xacml/xacml-json-http/v1.0/xacml-json-http-v1.0.html

- Libraries for shortECC and LmRNG, in order to provide elliptic curves with key lengths between 32 and 64 bits and to generate cryptographic secure pseudo-random numbers respectively.

SMARTIE follows also the concept of IoT-A, splitting the IoT devices in constrained and unconstrained ones. In terms of functionalities, SMARTIE proposes in [23] a recommended set of functionalities and technologies that can be used for either constrained or unconstrained devices, justifying the proposals to make sure that only lightweight implementations of technologies are embedded on constrained devices.

5.3.6 COSMOS

The COSMOS[8] project aims to enhance the sustainability of IoT-based smart city applications, enabling things to evolve and become more autonomous, reliable, and smart. The basic concept is that things will have the opportunity to learn based on the experiences of other things via a situational knowledge acquisition and analysis. Furthermore, COSMOS aims to provide end-to-end security and privacy, with hardware-coded security and privacy on storage, introducing the concept of Privetlets for IoT services.

The COSMOS architectural framework is built to address a large number of security, privacy, and trust requirements [25]. COSMOS assumes that data must be secured against eavesdroppers, data modification attacks, identity thefts or replay attacks. It also assumes that there is secure storage on the devices to protect secret information, that the devices are booted or updated in a secure way. COSMOS splits the execution environment of the services in two parts: (i) the secure part that executes the critical services and (ii) the unsecure part for the noncritical services. COSMOS uses a secure backbone server for key management, storage, and distribution, as well as for device enrollment. Authentication and access control are also embedded in the system.

The notion of Trust is also very important for COSMOS. A Trust Model for providing data integrity and confidentiality is defined, allowing also endpoint authentication and nonrepudiation between any system entities. Furthermore, COSMOS considers the notion of Privacy as very important for protecting the private information of users. Privacy is assumed to be supported by enabling entities to maintain control over their private information and decide whether they will be gathered, used, or disclosed to other entities [25].

COSMOS defined security mechanisms across different layers. The project aimed to ensure end-to-end security and privacy with security embedded at the device level, access control, encryption and cross-application mechanisms on the data level, injecting privacy-preserving mechanisms whenever appropriate. For the

[8] www.iot-cosmos.eu.

definition of the COSMOS architecture, the IoT-A ARM was used as a basis. The functional view of the architecture is depicted in Fig. 7 in [26].

As depicted in this figure, the main security components of the architecture reside in the Security FG

- **Authentication and authorization** is used for authenticating entities and managing the allocation of their access rights.
- **Key exchange and Management** covers the generation, storage, and distribution of cryptographic keys, based on the Diffie and Hellman [27] key exchange protocol.
- **Hardware Board Communication Accountability** handles the issue of accountability, tracking access to crypto primitives for nonrepudiation purpose, computing the reputation index of the entity.
- **Cryptographic nonrepudiation** is enforced based on the hardware board communication accountability component.
- **Checksum** ensures the integrity of the data packets, detecting, and correcting bit errors.

In the rest of the FGs there are also other components that are related to security, privacy, and trust

- **Privetlets** are acting as filters to ensure that every virtual entity and every user shares only the minimum intended information, omitting all other unnecessary information in order to maintain the privacy.
- **Communication channel** between VEs or between VEs and the COSMOS system has to be secured, thus encryption of the data transferred within these channels is applied.

In general, COSMOS gives much importance on ensuring privacy via data minimization at the middleware layer, as well as on strong authentication and authorization. Encryption at the communication channel is also important for ensuring confidentiality of the data.

5.3.7 COMPOSE

The COMPOSE[9] project developed an IoT platform that eases the task of developers writing applications which are based on the Internet of Things (IoT). Following section is based on the information from the project's deliverables [28, 47] and [29]. COMPOSE abstracts from the IoT devices by assigning to them a virtual identity. Devices which are not directly connected to the Internet (e.g., a bottle of wine with a RFID or NFC tag) will need a proxy to represent them in the COMPOSE platform. IoT objects with network capabilities, but without support of the

[9]http://www.compose-project.eu.

network protocols needed for COMPOSE, such as simple sensors, will also use proxies to be able to communicate with the COMPOSE platform. Finally, there is a group of advanced devices (so-called Smart Objects, such as a Smart Phone, tablet, or an Arduino device) that are capable to communicate with the COMPOSE platform directly. All the above-mentioned physical objects are called Web Objects in COMPOSE and are represented as Service Objects. COMPOSE specifies an API by which it expects to communicate with the Web Objects, in order to obtain data from them, or set data within them (for more detailed information see [45]). A Service Object exhibits a standard API also internally towards the rest of the components within the COMPOSE platform. That API is needed in order to streamline and standardize internal access to Service Objects, which can in turn represent a variety of very different Web Objects providing very different capabilities.

The COMPOSE platform, in an effort to embrace as many IoT transports as possible, allows Web Objects to interact with their representatives in the Platform (the Service Objects) using a set of well-known protocols: HTTP, STOMP over TCP, STOMP over WebSockets, and MQTT over TCP. Out of Service Objects, COMPOSE offers to design applications. Developers locate interesting existing Service Objects or applications, and tailor-specific logic around them. In addition, Service recommendation will be made available to choose the best suitable entity based on developer needs as well as proposed composition, and recommendation based on platform knowledge, such as security related aspects.

The security architecture of COMPOSE is based on the approach of data-centric security requirements [29]. It differs from classical device-centric approaches in that the COMPOSE security framework shrinks the security perimeter to the granularity of data. It is fine granular and can be combined with static and dynamic enforcement to regain governance on devices and data without sacrificing the intrinsic openness of IoT platforms.

The framework is depicted in Fig. 5.1 and it encompasses the following concepts:

- **Security metadata** is stored together with the entities they refer to. Metadata captures security policies of users specifying the privacy level the system must maintain for them. Service-centric metadata also allows developers and providers to specify the use of the services or service objects. Finally, the metadata will be consulted to efficiently build secure applications and workflows, and to store provenance information for data generated and processed in COMPOSE, and to store reputation information about users, service objects, and COMPOSE applications.
- **Provenance information** is solely generated by the COMPOSE system. It archives the information about when, who, and how an entity has been used. It accumulates information about the applications generating specific data, the applications consuming it, and possible operations performed on this data, e.g., its combination with other data or its broadcasting to remote locations outside of the COMPOSE system. To gather precise provenance information runtime monitoring is performed during sensor update dispatching and execution of

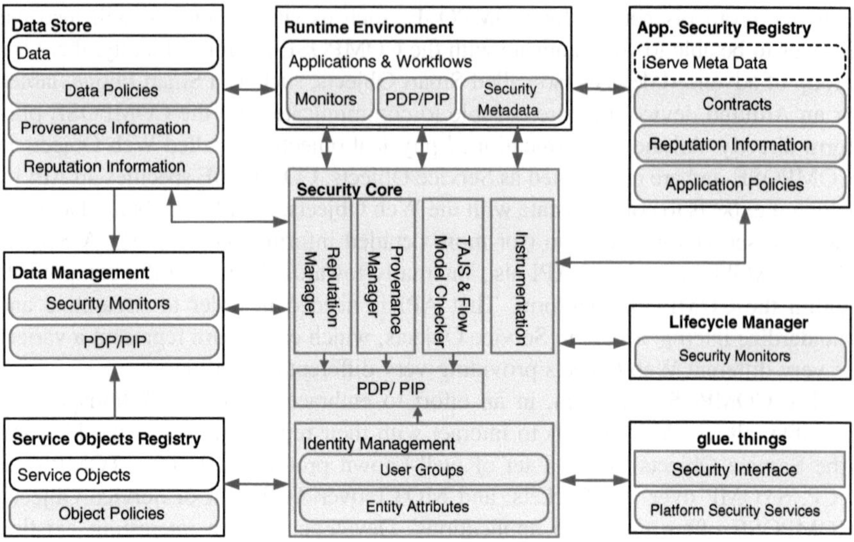

Fig. 5.1 COMPOSE security architecture overview [47]

applications. The data provenance manager tracks origins of data, the operations performed on it, and the time when operations took place. This empowers users to define policies based on data provenance, e.g., allow a Service Object to receive data only if it has been processed by a particular application. Further, visualizing the provenance of data can help users to detect when certain errors occur; for example, if several data sources are combined, but one of them is malfunctioning, the developer could examine the sources of correct values, and compare them with wrong values to isolate the malfunctioning device. Also, provenance information has an interesting potential to help to protect the user's privacy. For instance, it could eventually help to detect when particular applications harvest and correlate information from specific entities, hinting to the possibility of user profiling.

- **Reputation information** stores in the corresponding service object and service registries the collected feedback about service objects, and COMPOSE applications. Reputation information can be collected regarding service objects but also about the COMPOSE application popularity. The component called PopularIoTy[10] reflects how often a certain Service Object or application is used, i.e., invoked by other entities. Whenever data is generated, notifications are stored in the data storage. Moreover, monitors are placed in the runtime to store

[10]Parra, J.D.: PopularIoTy: http://github.com/nopbyte/popularioty-api/ (2014) and PopularIoTy Analytics: http://github.com/nopbyte/popularioty-analytics/ (2014).

notifications when an application is called. All this information is processed to calculate a popularity score. It also interacts with the contracts for applications, such that a contract compliant application will get rewarded with a positive score.

- **Contracts for applications** define conditions for applications stating when they can be executed securely, as well as flow specifications which is information about their internal, abstract data flows. While conditions specify system states which must hold before execution of an applications or which hold after its execution respectively, flow specifications provide more insights about where, e.g., to which resource, input data is flowing, and how and from where output data is generated. This information can be generated by hand describing critical security services provided by COMPOSE, e.g., the encryption or authentication of a data stream. Once an API will be published, a semiautomated process will generate contracts for these APIs. This information is used in a deployed COMPOSE environment to enable and simplify the analysis process. In general, i.e., for user-defined applications, contracts are generated by the static analysis component of the security core to save computational resources during the analysis of COMPOSE workflows.
- **Data flow enforcement** is enforced dynamically and statically. A data-centric flow policy is attached to input and output data for all components inside COMPOSE. Apart from specifying the entities allowed to access, execute, or alter a component, flow policies also describe the security requirements of data entering a component and the security properties of data leaving it. A unified policy framework can be used to avoid additional evaluation overhead, i.e., in COMPOSE the language is inspired by ParaLocks [30].
- **Identity management** is provided through a generic attribute-based IDM framework. It associates COMPOSE entities with identifiers and stores and distributes appropriate security information to also provide an authentication service. In COMPOSE an attribute-based approach allows every user to tag himself or his entities with attribute values. Once entities are tagged with attributes, e.g., the brand of the device, the tag can be used to specify security policies, e.g., accept data only from devices from that brand. COMPOSE allows users to approve attribute information depending on the group where they belong [31].
- Finally, **enforcement points** at the runtime level enforce access to data, services, or other resources based on decisions of the policy decision point (PDP). COMPOSE at its core embraces information flow security, runtime monitors (part of the PDP) detect and prevent illegal flows—as specified by the user or service object provider—during the execution of user-provided services. Finally, the PDP guides the security analysis and instrumentation components in COMPOSE whose task is the identification of potentially malicious flows in applications or workflows and their prevention or mitigation by software reconfiguration or instrumentation.

5.3.8 PRISMACLOUD

The PRISMACLOUD[11] project is a EU-funded research project developing the cryptographic tools to build more secure and privacy-preserving cloud services. To **enable end-to-end security** for cloud users and provide tools to **protect their privacy** the project brings novel cryptographic concepts and methods to practical application. The following information is based on the already available project deliverables [32, 33]. PRISMACLOUD wants to increase the pervasion of already existing and maybe nearly usable strong cryptographic primitives to the practice. It was perceived as being low at the beginning of the project and thus identified as an obstacle that withholds the use of less-trusted intermediaries (like cloud provided services, IoT gateways, or IoT middleware) in many security and privacy conscious usage scenarios, e.g., Smart Cities. PRISMACLOUD aims at bridging the divide between the needed deep cryptographic knowledge and the application requirements of cloud users in order to bring the cryptographic primitives to good practical use.

PRISMACLOUD's architecture encapsulates the cryptographic knowledge of the primitives layer inside the tools and their usage inside services. As depicted in Fig. 5.2, it is organized into four tiers. Each layer helps abstracting from the needed core cryptographic primitives and protocols, which are located at the Primitives layer, which is the lowest layer of the PRISMACLOUD architecture. Building the tools, the layer above, from the primitives requires in depth cryptographic and software development knowledge. However, once built they can be used by cloud service designers to build cryptographically secure and privacy-preserving cloud services. Thus, PRISMACLOUD's architecture levels also define connection points between the different disciplines involved: cryptographers, software engineers/developers, and cloud service architects. On the uppermost (i) Application layer are the end-user applications. Applications use the cloud services of the (ii) Services layer to achieve the desired security functionalities. The cloud services specified there are a representative selection of possible services, which can be built from the tools organized in the (iii) Tools layer. In particular, they represent a way to deliver the tools to service developers and cloud architects in an accessible and scalable way. Together the tools constitute the PRISMACLOUD toolbox.

The PRISMACLOUD architecture can be seen as one recipe to bring cryptographic primitives and protocols into cloud services such that they empower cloud users to build more secure and more privacy-preserving cloud services. In particular, it considers providing tools for secure object storage, flexible authentication with selective disclosure, verifiable data processing, topology certification, and data privacy. The services designed in PRISMACLOUD are data sharing; secure archiving; selective authentic exchange; privacy enhancing ID management; verifiable statistics; infrastructure auditing; encryption proxy; anonymization service. The tools are using the following cryptographic primitives: RDC: Remote Data

[11] http://www.prismacloud.eu.

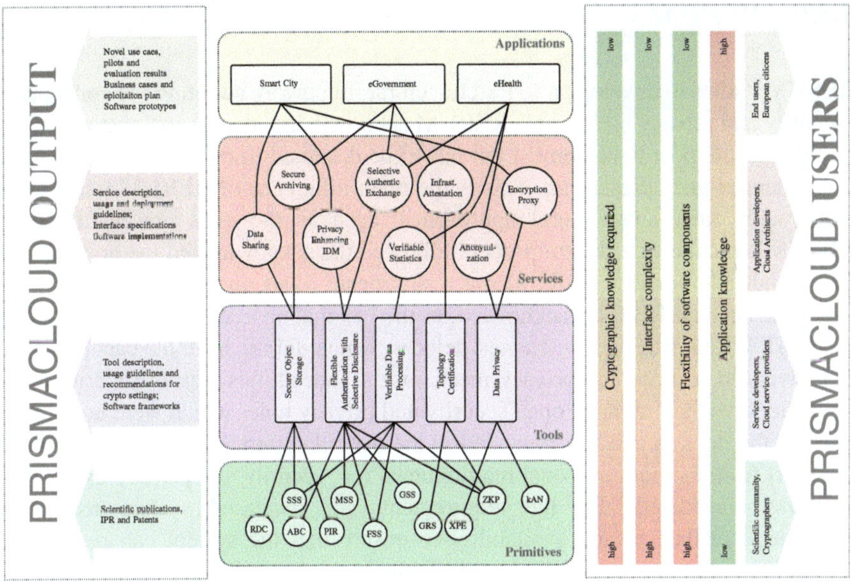

Fig. 5.2 Overview of PRISMACLOUD's 4-tier architecture [33]

Checking; SSS: Secret Sharing Schemes; ABC: Attribute-Based Credentials; PIR: Private Information Retrieval; MSS: Malleable Signature Schemes; FSS: Functional Signature Schemes; GSS: Group Signature Schemes; GRS: Graph Signature Schemes; SPE: Format- and Order-Preserving Encryption; ZKP: Zero-Knowledge Proofs; kAN: k-Anonymity (abbreviations from Fig. 5.2).

To take it to a Smart Cities example, imagine a number of IoT devices that constantly record data, but have no storage and computing capabilities to do higher level aggregation. Assuming further that this aggregation will be done on a cloud-based infrastructure then this infrastructure must be trusted to perform the computation correctly. Here the trust in this infrastructure can be removed using a cryptographic primitive called functional signatures. They allow the delegation of signature generation to other parties for a class of messages meeting certain conditions. Such schemes can be used to certify the computation done by third parties, such as untrusted intermediaries like the gateway or the middleware of an IoT stack. From this class of signature schemes PRISMACLOUD builds the Verifiable Data Processing Tool. The tool allows performing verifiable computations on data such as computing statistics on IoT data. See [32] for more information on cryptographic signature primitives that PRISMACLOUD uses.

5.3.9 RERUM

RERUM[12] aims to develop an IoT architectural framework adopting the concepts of "security and privacy by design". RERUM acknowledges that an IoT system cannot be fully secure post-development but has to be designed from its foundations to be secure and privacy preserving. This is the main difference of RERUM when compared with other architectural approaches, such as the ones presented above. RERUM's architecture design process followed the acknowledged methodology of IoT-A, but additionally, RERUM put a significant focus on the IoT devices. This was done because RERUM acknowledges the fact that up to now, the weakest point in an IoT system was the constrained devices, which did not have the capabilities to run advanced security and privacy mechanisms. Indeed, this lack of security focus on devices resulted in many open security and privacy holes which were not limited only on the devices and but expanded to the overall system [34, 35].

Apart from the nonfunctional requirements for "security and privacy by design" RERUM considered also the requirements for increased system reliability, robustness, resilience, and availability to ensure that the system can respond to attacks and that the data will be available to be provided to the applications whenever they are requested. Since RERUM's key focus is on the devices, the key differentiating factor of its functional requirements is that the developed security and privacy mechanisms were required to be lightweight and energy efficient so that they can be easily implemented and embedded even on constrained IoT devices. RERUM's requirements include among others, lightweight encryption, confidentiality, and integrity protection, authorized modification of integrity protected data, simple and strong authentication both for users and devices, attribute-based access control, user consent, data collection limitation, data minimization, accountability, secure device bootstrapping, and secure configuration of devices [36].

The RERUM architecture went through a detailed process of requirements definition and analysis in order to extract the required functional components to support the long list of requirements. RERUM defined its own functionality groups called "managers". Among others, RERUM defined the "Security, Privacy and Trust Manager" (SPT) that included a long list of functional components. These components are depicted in Fig. 5.3 mapped to the IoT-A ARM's FGs [37].

What is interesting from this mapping is that contrary to previous IoT security architectures, RERUM developed a large number of components that are assumed to be embedded on the IoT devices. For example, components for secure credential bootstrapping, secure storage, geolocation privacy enhancing techniques, integrity generator/verifier, data encryption/decryption, trusted routing, cognitive radio security, and device to device authentication are assumed to be important to be running on the devices to ensure the highest level of security of an IoT system, by strongly securing the leaf nodes.

[12] www.ict-rerum.eu.

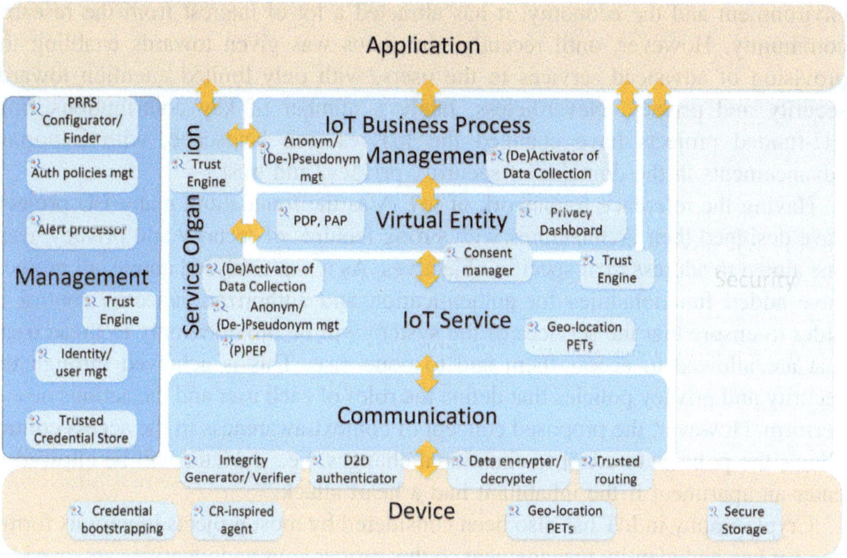

Fig. 5.3 RERUM architectural security components mapped on the IoT-A ARM [37]

Apart from that, security components for authentication and access controlled are also important in RERUM, together with secure reconfiguration of devices, creating and processing security alerts, handling user and device identities and storing securely credentials and certificates. A number of components are also defined for ensuring user privacy via anonymization and pseudonymisation, activation and deactivation of data collection based on user preferences, a consent manager to request the consent of the user for gathering or disclosing private information, a privacy dashboard to allow users themselves to handle their privacy policies, as well as Policy Decision and Enforcement Points to manage and execute the policies.

Trust is also very important in RERUM. The project has defined a trust engine to calculate and evaluate the trust ratings for devices, services and users, based on observers scattered around different system entities and which are monitoring the user actions, the data reliability as these are gathered by the devices, and the behavior of the devices in order to identify malicious or misbehaving devices so that these will not affect system's decisions. That way, the reliability of the overall system is improved.

5.4 Conclusions

The Internet of Things is becoming an important element of the everyday lives of the people, providing opportunities for simplified activities, and improved quality of life. Acknowledging the numerous benefits of the IoT for the people, the environment and the economy, it has attracted a lot of interest from the research community. However, until recently, the focus was given towards enabling the provision of advanced services to the users, with only limited attention towards security and privacy. Nevertheless, lately a number of key contributions from EU-funded projects have changed the IoT research landscape, with important advancements in the domains of security, privacy, and trust.

Having the reference framework of IoT-A as the foundation, many EU projects have designed their architectures with strong features of security and privacy, each one aimed to address their specific objectives. As it was analysed above, all projects have added functionalities for authentication and authorization/access control in order to ensure that the services of the system will be provided only to those users that are allowed to access them and to none else. This is achieved through the security and privacy policies that define the roles of each user and the actions he can perform. However, the proposed concept of context-awareness in the access control allows the policies to adapt to situational changes, i.e., a doctor will be allowed to enter an apartment if the inhabitant had a heart attack.

Cryptography in IoT has also been considered by most projects in various forms, i.e., for key and identity management so that proper keys and identities are provided to the various entities of the system. Encryption has also been supported by many projects, mainly though on the backbone communications, since the IoT devices can be very constrained to support encrypted communications. In this respect, the DTLS technology using ECC can play a significant role [38] or the novel idea of Compressive Sensing-based encryption, which performs simultaneous compression and encryption in the measurements, contributing to the minimization of the energy consumption of the devices [39–41].

Data integrity is also a very important issue in IoT, since malicious or tampered data can severely affect the decisions of the system and could potentially even harm people when actuators are involved. To ensuring data integrity, techniques such as digital signatures creation and verification must run on the devices [42]. This way the receiver can validate that no unauthorized intermediate node has tampered with the data it received. If subsequent modifications are needed, advanced techniques like malleable signatures can be applied to allow changes to parts of the data limited to authorized nodes for concealing identifiable information without breaking the validity of the signature [32, 43].

Data integrity can also be a metric of the reliability of the data, which is used for measuring the trustworthiness of the IoT system. Not many projects have worked towards trust and reputation schemes for the devices, but the focus was mainly to assess the reputation of users in order to change their access policies.

With regards to privacy, the main functionalities developed by most projects were related with the inclusion of privacy policies on the access control mechanisms. Only a few projects lately have worked towards cross-layer privacy enhancing techniques for data minimization, for ensuring unlinkability of the data, for allowing users to take control of their data and for anonymization and pseudonymisation of the data so that no identifiable information is disclosed to unauthorized third parties. Location privacy is also a key research item in some projects and this has mainly been tackled via geofencing/data minimization and anonymisation/pseudonymisation.

Overall, there have been significant advances in the areas of security and privacy in IoT in the last few years. However, as also analysed in [44], there are many solutions that are not addressed adequately so far in the IoT community. For example, embedding autonomic computing functionalities in the security mechanisms for implementing self-management based security is not addressed so far, although it can contribute to a more resilient IoT system. Research towards unobservable communications for improved privacy is also a very interesting topic, with the goal to try to avoid extracting the measurements from performing traffic analysis [4]. Traffic anonymization in IoT networks, using, i.e., Tor or I2P is yet an open research area. Anonymous credentials and identity mixers can highly contribute to improved privacy protection. Finally, trust building between devices using trust negotiation protocols requires the iterative exchange of credentials and strong cryptographic calculations, which is not applicable to constrained IoT devices. These are only some examples of open research items that can improve even more the security and privacy of IoT systems. However, the main challenge is the design of the respective functionalities in a lightweight and energy efficient way, so that they can be embedded on resource constrained devices. Only then, the IoT systems can be fully secure and the citizens will have the necessary incentives to adopt this exciting new set of technologies.

Acknowledgements This work has received funding from the European Union's Seventh Framework Programme (FP7/2007-2013) under grant agreements no 609094, 612361 and 644962.

References

1. Evans D (2011) The internet of things. How the Next Evolution of the Internet is Changing Everything, Whitepaper, Cisco Internet Business Solutions Group (IBSG)
2. Ranken, Margaret, M2M Global Forecast & Analysis 2014–24, Machina Research Strategy Report, 24, June 2015
3. Ruiz D et al (eds) (2015) Enhancing the autonomous smart objects and the overall system security of IoT based Smart Cities, RERUM Project Deliverable D3.1, 28 February 2015
4. Pöhls HC et al (eds) (2015) Privacy enhancing techniques in the Smart City applications, RERUM Project Deliverable D3.2, 2 Sept 2015
5. Blefari-Melazzi N, Bianchi G, Salgarelli L (eds) (2011) Trustworthy internet. Springer Science & Business Media
6. Vermesan O, Friess P (2014) Internet of things—from research and innovation to market deployment. River Publishers
7. Vermesan O, Friess P (eds) (2015) Building the hyperconnected society: IoT research and innovation value chains, ecosystems and markets, vol. 43. River Publishers
8. Internet of Things research study. Hewlett Packard Enterprise 2015 report. www8.hp.com/h20195/V2/GetPDF.aspx/4AA5–4759ENW.pdf
9. Mouroutis T et al. (eds) (2014) Use-cases definition and threat analysis, RERUM Project Deliverable D2.1, 31 May 2014

10. Bassi A et al (2013) Enabling things to talk. Designing IoT solutions with the IoT architectural reference model, pp 163–211
11. Howard M, Lipner S (2006) The security development lifecycle: SDL: a process for developing demonstrably more secure software. Microsoft Press (2006)
12. Gruschka N et al (eds) (2012) Concepts and Solutions for Privacy and Security in the Resolution Infrastructure, IoT-A Project Deliverable D4.2, 16 February 2012
13. Carrez F et al (eds) (2013) Final architectural reference model for the IoT v3.0, IoT-A Project Deliverable D1.5, 15 July 2013
14. Menoret S et al (eds) (2014) Final architectural reference model, iCore Project Deliverable D2.5, 2 Nov 2014
15. Baldini G et al (eds) Security requirements for the iCore cognitive management and control framework, iCore Project Deliverable D2.2, 31 May 2012
16. Neisse R, Steri G, Fovino IN, Baldini G (2015) SecKit: a model-based security toolkit for the internet of things. Comput Secur 54:60–76. ISSN 0167-4048
17. Integrated System Architecture and Initial Pervasive BUTLER proof of concept, BUTLER Project Deliverable D3.2, October 2013
18. Requirements, Specifications and Security Technologies for IoT Context-Aware Networks, BUTLER Project Deliverable D2.1, October 2012
19. Ethics, Privacy and Data Protection in BUTLER, BUTLER Project Deliverable D1.4, July 2013
20. Dimitropoulos P, Soldatos J, Kefalakis N, Bengtsson JE, Giuliano A et al (eds) OpenIoT detailed architecture and proof-of-concept specifications, OpenIoT Project Deliverable D2.3, 28 March 2013
21. Gwadera R et al (eds) (2013) Privacy and Security Framework, OpenIoT Project Deliverable D5.2.1, 27 Sept 2013
22. Azevedo R et al (eds) (2014) Requirements, SMARTIE Project Deliverable D2.2
23. Skarmeta A et al (eds) (2015) Initial Architecture Specification, SMARTIE Project Deliverable D2.3
24. Bethencourt J, Sahai A, Waters B (2007) Ciphertext-policy attribute-based encryption. In: IEEE symposium on security and privacy. SP'07, pp 321–334
25. Pitu L et al (eds) (2015) End-to-End Security and Privacy: Design and Open Specification (Updated), COSMOS Project Deliverable D3.1.2, 30 April 2015
26. Carrez F et al (eds) (2015) Conceptual Model and Reference Architecture (Updated), COSMOS Project Deliverable D2.3.2, 30 April 2015
27. Diffie W, Hellman ME (1976) New directions in cryptography. IEEE Trans Inform Theory 22 (6):644–654
28. Deliverable D1.2.2 Final COMPOSE architecture document
29. Schreckling D, Parra JD, Doukas C, Posegga J (2015) Data-Centric Security for the IoT. In: Proceedings of the 2nd EAI international conference on IoT as a Service, Rome, Italy
30. Broberg N, Sands D (2010) Paralocks: Role-based information flow control and be- yond. In: Proceedings of the 37th annual ACM SIGPLAN-SIGACT symposium on principles of programming languages. pp 431–444. POPL'10, ACM, New York, NY, USA
31. Parra J, Schreckling D, Posegga J (2014) Identity Management in Platforms Of fering IoT as a Service. In: 1st international conference on IoT as a service. Lecture Notes in Computer Science (LNCS), Springer, Rome, Italy
32. Demirel D, Derler D, Hanser C, Pöhls HC, Slamanig D, Traverso G (2015) PRISMACLOUD D4.4: Overview of Functional and Malleable Signature Schemes
33. Lorünser T, Länger T, Slamanig D, Pöhls HC (2016) PRISMACLOUD Tools: A Cryptographic Toolbox for Increasing Security in Cloud Services. In: Proceedings of a workshop on security, privacy, and identity management in the cloud collocated at the 11th international conference on availability, reliability and security. ARES'16, Salzburg, Austria, 2016

34. Pohls HC et al (2014) RERUM: Building a reliable IoT upon privacy-and security-enabled smart objects. In: Wireless communications and networking conference workshops (WCNCW), 2014 IEEE. IEEE, 2014
35. Tragos EZ et al (2014) Enabling reliable and secure IoT-based smart city applications. In: 2014 IEEE international conference on pervasive computing and communications workshops (PERCOM Workshops). IEEE, 2014
36. Cuellar J et al (eds) (2014) System Requirements and Smart Objects Model, RERUM Project Deliverable D2.2, 31 May 2014
37. Tragos E et al (eds) (2015) Final System Architecture, RERUM Project Deliverable D2.5, 4 Sept 2015
38. Capossele A et al (2015) Security as a CoAP resource: an optimized DTLS implementation for the IoT. In: IEEE international conference on communications (ICC), IEEE, 2015
39. Charalampidis P, Fragkiadakis A, Tragos E (2015) Rate-adaptive compressive sensing for IoT applications, VTC2015-Spring, Glasgow
40. Fragkiadakis A, Tragos E, Papadakis S, Charalampidis P (2014) Experiences with deploying Compressive Sensing and Matrix Completion techniques in IoT devices, IEEE CAMAD 2014, Athens, 2014
41. Fragkiadakis A, Tragos E, Traganitis A (2014) Lightweight and secure encryption using channel measurements, Wireless Vitae 2014, Aalborg
42. Pöhls HC (2015) JSON Sensor Signatures (JSS): End-to-End Integrity Protection from Constrained Device to IoT Application. In: 9th international conference on innovative mobile and internet services in ubiquitous computing (IMIS). IEEE, Santa Cantarina, Brazil, 2015
43. Pöhls HC, Samelin K (2015) Accountable redactable signatures. In: 10th international conference on availability, reliability and security (ARES). IEEE, 2015
44. Baldini G et al (2015) Internet of Things. IoT Governance, Privacy and Security Issues, European Research Cluster on The Internet of things, Activity Chain 05 Whitepaper, January 2015
45. Trifa V, Larizgoitia I (2013) Design of the object virtualization specification, Compose Deliverable D2.1.1, 30 Oct 2013
46. Fragkiadakis A, Angelakis V, Tragos EZ (2014) Securing cognitive wireless sensor networks: a survey. Int J Distrib Sens Netw (2014)
47. Schreckling D et al (2015) The Compose Security Framework, COMPOSE Deliverable D5.4.2, 15 Nov 2015

Chapter 6
Privacy and Social Values in Smart Cities

Leonardo A. Martucci, Simone Fischer-Hübner, Mark Hartswood and Marina Jirotka

6.1 Introduction

The appeal of smart cities is the exploitation of information technology to better manage and plan the utilization of resources of a urban area. The benefits of smart cities are obtained by collecting and processing data from the city public services and utility companies, such as traffic information and water consumption, and from the city dwellers and its visitors. Smart cities ideally involve real-time data collection, processing and intervention, allowing public services to adapt to new conditions and constraints as they appear, and also to be better planned. For instance, road speed limits can adapt to traffic conditions or air pollution, public transport can be better monitored, allocated and redistributed, and law enforcement officials relocated more efficiently. In this chapter, we look at the personal data collected and processed by collective adaptive systems (CAS) and the internet of things (IoT), which are key information sources for enabling smart cities. The role of the IoT is to collect data and act locally while the CAS aggregates and processes the data and allows for people and machines to complement each other and operate collectively to achieve their, possibly conflicting, goals. The goal of this chapter is to provide an overview about

L.A. Martucci (✉) · S. Fischer-Hübner
Karlstad University, Karlstad, Sweden
e-mail: Leonardo.Martucci@kau.se

S. Fischer-Hübner
e-mail: Simone.Fischer-Hubner@kau.se

M. Hartswood · M. Jirotka
University of Oxford, Oxford, UK
e-mail: Mark.Hartswood@cs.ox.ac.uk

M. Jirotka
e-mail: Marina.Jirotka@cs.ox.ac.uk

V. Angelakis et al. (eds.), *Designing, Developing, and Facilitating Smart Cities*,
DOI 10.1007/978-3-319-44924-1_6

personal data protection for smart cities by looking into the techno-legal requirements and challenges using a Privacy by Design (PbD) focused on data minimization approach.

The term "smart" in smart cities refer to the use of information technology, especially data gathering, communication and analysis, to help society by promoting efficiency in services and rational use of resources with the ultimate goal of enhancing quality of life. All the personal data processed in applications designed to support smart cities need to be handled according to (local) social and legal requirements. As computer systems, algorithms, data and devices become increasingly closely coupled to people, as individuals and collectives, significant privacy challenges arise.

In this chapter, we look at the privacy challenges in smart cities from a point of view of the data collection and processing and the CAS harmonization and coordination aspects. We list the social legal principles behind smart cities from an European-centric perspective, and list the involved challenges to privacy using a urban car pooling scenario as our case study. We illustrate the privacy challenges with a privacy impact assessment (PIA) of a car pool (ride share) application developed within the EU FP7 *SmartSociety* project (cf. [28] for general information on the project and [20] for previous *SmartSociety* work on privacy in CAS).[1]

The remainder of this chapter is organized as follows. Section 6.2 briefly introduces the background on privacy on smart cities. The application scenario on digital transport that we use throughout this chapter is introduced in Sect. 6.3. Section 6.4 outlines the legal requirements and the derived set of privacy-related technical requirements. The privacy-enhancing methods, procedures, and technologies for fulfilling the requirements are presented in Sect. 6.5. Section 6.8 discusses the limitations of the existing solutions and concludes the chapter.

6.2 Background: Privacy, Social Principles, and Smart Cities

The network-enabled sensors and actuators that constitute most of the IoT often have limited resources, such as processing power and memory. It is auto-sufficient for small-scale interventions on the local scope, such as for heating, ventilation, and air conditioning (HVAC) climate control. However, the main benefits of IoT are not on the local scope but on the global one. The IoT is a collective of interconnected data sensors an actuators which underpin larger and more ambitious projects and initiatives, including smart cities.

The increase in the number and type of devices connected to the Internet means that IoT has the potential to collect data in volumes that are many orders of magnitude greater than is possible today. This data will be increasingly intimate, as it will emerge from everyday uses of technologies leading to whole swathes of mundane

[1] http://smart-society-project.eu.

activity being newly interconnected with the digital realms. The technological foundations of IoT are blind to the nature of data that it collects and transfers, i.e., there is no distinction if the data that it handles is personal information or not. For example, if we consider two equal network-enabled sensor devices, e.g., two GPS beacons, one may process personal information (a person's location) while the other may not (a parcel's location). Hence, the context in which in the IoT devices are immersed and the purpose of the collected data are cornerstone to the question of privacy.

Threats to privacy can be very significant and aspects regarding where the data is captured, centralized, processed is of great importance for understanding the impact on privacy and the possible countermeasures, especially when the data is to be shared or forward to a government and corporate entities administering a smart city.

6.2.1 Social Principles and IoT

We already understand the potential for existing data collection and algorithmic profiling to regulate society [10]. Existing IoT applications already demonstrate how the capacity for regulation can be intensified via directly embedded norms and bridging directly between corporate interests and peoples' everyday activities. "Driving black box" technologies, like that shown in Fig. 6.1,[2] use sensors and algorithms to monitor and evaluate drivers in order to regulate individual driving patterns in exchange for preferential insurance rates. Similarly, digital medicines potentially connect peoples' use of medication directly to pharmaceutical interests [25], and smart meters connect peoples' use of domestic appliances to the interests of energy suppliers [4]. There are arguments made in each of these cases about societal benefits including safer roads, more effective medication regimes and more sustainable energy use. But by the same token there are also sinister overtones of control and important questions to answer about which norms and values are embedded in these systems, and who has a say in how these are selected.

Multiple interests may be served by IoT applications yet those who control the infrastructure and own the data have a significant advantage in embedding their interests above others. A satire and critique of the control potential of the IoT has been created by *Thing Tank*, a research project on IoT, in the form of a video of an elderly person living independently in an IoT world where his fork advises about what to eat, his bed dictates his waking and sleeping routine, and his walking stick regulates his daily exercise.[3] The video makes apparent the impersonal control non-present relatives since the IoT "assistive" technologies evidently stand proxy for the relatives' own anxieties, responsibilities and desire for control. It also shows too how these forms of non-consensual control provokes rebellion and resistance as the elderly person finds clever ways of circumventing each of these mechanisms turn. In a subtle

[2]https://www.ingenie.com/how-it-works.

[3]The "*Uninvited Guests*" is short film produced by *Superflux* and commissioned by *Thing Tank*. https://vimeo.com/128873380.

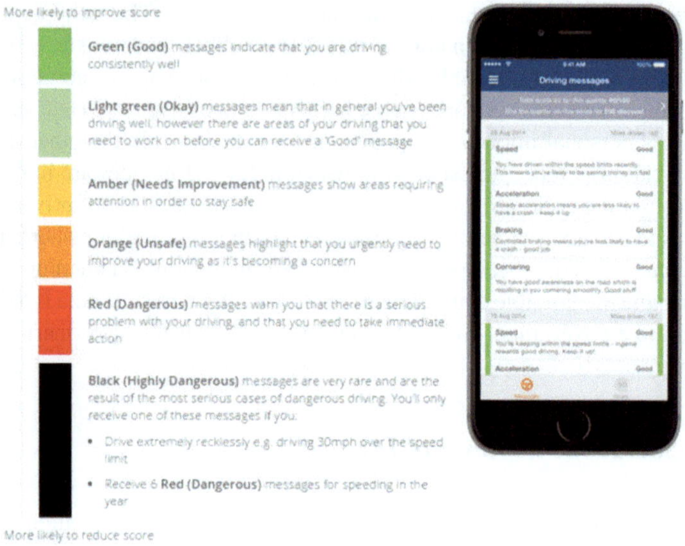

Fig. 6.1 A driving style tracker application

way, the video also expresses the value and pleasure obtained from certain freedoms that the overt regulation of the devices seems to deny. It encapsulates the conflicting values of the elderly person, his relatives and how these are entangled within wider cultural tropes about responsible lifestyle choices and personal freedoms. What is really absent in a centralized, technical and bureaucratic IoT future depicted by this video is any attention to creating space where these values can be negotiated.

6.3 Application Scenario: Urban Car Pooling

This chapter's use of digital transport as a focal example to explore privacy concerns is motivated by *SmartSociety*'s development of a car pool application, the Smart Share, to test and showcase how *SmartSociety* components can be used to build smart city applications. A privacy impact assessment (PIA) of the Smart Share application was conducted, and its architecture was conceived following a (PbD) approach focusing on data minimization. We aim, in this section, to draw lessons for Smart City IoT applications more generally. In this section, we more broadly introduce the concept of digital transport, show how real-world applications, such as Uber, can be problematic from a privacy perspective, before outlining the Smart Share, the *SmartSociety*'s own digital transport solution for smart cities.

6.3.1 Digital Transport and Privacy

Within IoT enabled smart city the vision for digitally augmented transport networks is to drive economic growth whilst mitigating problems of congestion and environmental sustainability [15]. The dynamics of real-world movements of people and vehicles are sensed, modeled and influenced via digital transport solutions comprised of sensors, algorithms and data connected by digital networks. Such solutions include, driverless cars [36], intelligent transport systems [13], personal transport advice [35], new business models, including collective utilization of spare capacity [34] and intelligent traffic management systems [19]. Alongside the undoubted social, environmental and personal benefits of digital transport there are risks too, and in this section we explore some of the risks to privacy posed by digital transport.

Travel is a vital social, economic and cultural activity so it is unsurprising that the journeys we choose or are obliged to make are also very revealing about ourselves. Our geographic location is a clue to what we are doing, and our pattern of journeys revealing of our activities and identities.

The risks of releasing our travel data is powerfully demonstrated series of media stories concerning the inappropriate use of information by the lift-sharing service Uber about journeys taken by passengers. An Uber senior executive (reportedly) threatened to reveal aspects of journalists' private lives deduced from Uber journey data as a punishment for their negative reporting of Uber. These threatened disclosures concerned journey signatures that may indicate an affair or a one-night stand. Other media stories report staff members casually accessing, circulating and commenting on journeys made by Uber passengers.[4] Subsequently Uber has stated that these practices and uses of its data are contrary to its privacy policies, and the Uber executive making the threatening remarks has apologized.[5] Although Uber could not persist with the impression of its being a playground for inquisitive employees or a vehicle for the senior executive to exact vengeance on critical journalists, these reports make clear the extent of the power attained over users from accumulating journey data. Uber's huge surveillance potential would be even further amplified should it achieve its goal of being a universally preferred option for any and every journey we might make. This spells out very clearly some of the risks to privacy of IoT in smart cities applications.

In addition to these direct implications for personal privacy, digital transport systems have a range of further risks that are linked to privacy concerns. These include the potential for new "digital divides" [37] where opting out of non-privacy-friendly systems may lead to inequality of opportunity. There are democratic risks, what were matters of public policy are transferred to private corporations. For example: supply, demand, price, quality and safety, of hire cars under Uber become (either directly

---

[4]Z. Tufekci and B. King. "*We Can't Trust Uber*". In: The Opinion Pages, NY Times, Dec. 7, 2014. www.nytimes.com/2014/12/08/opinion/we-cant-trust-uber.html.

[5]M. Isaac. "*Uber Executives Comments Leave Company Scrambling*". In: Bits, a NY Times Blog, Nov. 18, 2014. http://bits.blogs.nytimes.com/2014/11/18/emil-michael-of-uber-proposes-digging-into-journalists-private-lives/.

or diffusely) regulated within the Uber system, and no longer by local authorities. Often these types of regulation are driven by access to personal data and may work in ways to impinge on user autonomy, as with the example of "driver black boxes" given earlier. Issues around autonomy can be complicated. On the one hand, digital transport solutions promise user-centered services customized to personal need, but on the other hand they also afford delivery of finely tuned incentives to shape how transport options are chosen [12]. A simple reading is that systems with access to personal data have an equal potential to enhance or diminish user autonomy.

From a privacy perspective, exactly the same types of personal data needed for personalization may also be used to drive incentives. Thus, ommiting data to avoid incentives also restricts realizing benefit from personalization. Further types of privacy guarantee need to be built into the system, but these may lead to additional complications. For example, if individuals are (reliably) offered "opt-outs" from incentives then individual choice may conflict with the collective benefit of digital transport. Incentives need to be fairly implemented and carefully adjusted to encourage benign aims, such as carbon reduction, and avoid that those who opt out to be seen as selfish individuals undermining a common good. These types of consideration lead to important questions about those occasions where the value of privacy supports or undermines other social values expressed within the system. These are complicated questions about how privacy relates to democratic mechanism for setting system goals, how the interests of the system are made transparent and shown to be fair, and how we accept the balance between individual autonomy and collective good.

6.3.2 The Smart Share and Its Components

SmartSociety's Smart Share is a ride sharing, car pooling, application that supports drivers to fill spare capacity in their cars by enabling them to advertise the space to potential passengers. Passengers are able to search and signal their interest to participate in advertised trips. The Smart Share was designed to benefit from the following parts and components from *SmartSociety*:

- **Sensor fusion**. Use of IoT, such as sensors in mobile devices owned by the driver and passengers, to deduce information about the ride, including when it started, completed and who actually took part in it. This helps the system understand about the rides that were completed, which may feed back into reputation systems, incentives and algorithm optimization.
- **Peer profile**. It stores personal data about drivers and passengers. This includes identify information and preferences for taking rides. The release of personal data from the Peer Profile is controlled following a user defined privacy policy [22].
- **Social orchestration**. The matching algorithm that brings drivers and passengers together. "Ride Plans" are created for all the permutations of possible rides given driver and passenger constraints, with options for drivers and passengers to accept or decline rides.

- **Incentives**. They provide means of encouraging system uptake and meeting of specific goals, such as maximizing car occupancy, or use of less congested routes. It helps to meet global objectives, such as promoting sustainability, as well as improving how people meet their individual goals.
- **Reputation and provenance**. It encourages good behavior of Smart Share participants, such as timeliness and the quality of the ride. Driver and passenger reputation is visible when rides are negotiated. Provenance tracks the actions of any entity within the system, be it an algorithm or a person, to provide transparency and accountability of the actions of both algorithms and people.
- **Gamification elements**. It includes elements to make participation more enticing, such as achievement badges and the platform virtual currency.
- **Programming framework**. It provides a way of programmatically assembling the resources (including people) needed for some task to be accomplished within *SmartSociety*. In the case of Smart Share, the "task" is the ride that is collaboratively undertaken by drivers and passengers.
- ***SmartSociety* architecture**. It provides the coupling between all components in order to provide application programmers with the resources to extend the Smart Share application and other tools developed using the Smart Society platform.

6.4 Legal Aspects

In this section, we discuss the legal aspects and requirements pursuant the EU General Data Protection Regulation (GDPR) [17] concerning the processing of personal data in IoT and smart cities.[6] We first introduce the legal definition of personal data in Sect. 6.4.1, and emphasize the question around hardware identifiers, which are relevant to the discussion of personal data in IoT. Section 6.4.2 summarizes the general legal requirements, and the challenges to meetings such requirements are presented in Sect. 6.4.3.

6.4.1 Personal Data

Data Protection legislation only applies to data that classifies as personal data. The GDPR defines personal data as "any information relating to an identified or identifiable natural person (data subject)", who "can be identified directly or indirectly, in particular by reference to an identifier such as a name, an identification number,

[6]The EU GDPR was passed by the European Parliament in Dec. 2015, entered into force on 24 May 2016 and shall apply in all EU member sates from 25 May 2018. The GDPR was chosen as our reference for many reasons: (*a*) it applies to data controllers or processors located in the EU, and to any organization processing personal data of EU residents, (*b*) it reflects the basic privacy principles of the OECD privacy guidelines and (*c*) of the US Federal Trade Commission's (FTC) Fair Information Practice Principles (even going beyond them).

location data, online identifier or to one or more factors specific to the physical, physiological, genetic, mental, economic, cultural or social identity of that person."

The definition by the GDPR makes clear (as also stated previously by the Art. 29 Working Party [2]) that also unique device numbers, such as MAC addresses or RFID tag codes, can also be considered as personal data of users that can be associated with these devices (usually the device holders), with the consequence that these users could be uniquely (and secretly) profiled under these device identifiers by observers, even though the observed users may not be identifiable by name. For instance, an RFID tag in a watch that a person usually wears could be used within a supermarket to profile that user as a returning customer (cf. [2]).

The question whether MAC addresses or RFID codes constitute personal data or not can change over the lifetime of the respective devices or tags. Furthermore, in the context of IoT and smart sensing, as point out by the Art. 29 Working Party [1], individuals can often be identified with the help of data that originates from "things" and that may discern the life style of individuals and families, e.g., data generated by centralized control of lighting or heating in smart home applications.

6.4.2 Privacy Requirements

In this section, we present the legal requirements for a CAS computing platform that accommodate data protection and is designed upon a privacy preserving framework. Basic legal privacy principles, especially those by GDPR, are needed in order to identify the privacy threats as part of a Privacy Impact Assessment for a CAS computing platform and comprise the following ones listed in Table 6.1.

6.4.3 Challenges for Meeting the Legal Requirements

In the context of IoT and smart cities, fulfilling the legal requirements listed in Sect. 6.4.2 requires several challenges to be addressed, as noted by the Art. 29 Data Protection Working Party for IoT [1].

First, the current (in)security of IoT devices and platforms, which often have constrains concerning their battery and computational resources, is commonplace. IoT devices, from light bulbs and kettles to Barbie dolls, have flawed to none security mechanisms implemented.[7, 8, 9] Security is a fundamental legal requirement

[7] B. Ray. "Securing the Internet of Things–or how light bulbs can spy on you". Apr. 22, 2013. The Register. http://www.theregister.co.uk/2013/04/22/iot_security/.

[8] D. Pauli. "Connected kettles boil over, spill Wi-Fi passwords over London". Oct. 19, 2015. The Register. http://www.theregister.co.uk/2015/10/19/bods_brew_ikettle_20_hack_plot_vulnerable_london_pots/.

[9] I. Thomson. "Goodbye, Hello Barbie: Wireless toy dogged by POODLE SSL hole". Dec. 4, 2015. The Register. http://www.theregister.co.uk/2015/12/04/wireless_barbie_slipshod_security/.

Table 6.1 Privacy requirements derived from the EU GDPR

Privacy requirements	Description
Compliance with General Data Processing Principles, data protection by default (Art. 5, 25)	Key privacy principles are to be ensured, and must be enforced by the controller by appropriate technical and organisational measures by default, particularly: *Purpose specification & binding:* Personal data must be collected for specified and legitimate purposes and may later only be used for those purposes *Data minimization:* The amount of personal data and the extent to which they are collected and processed should be minimized, i.e., in particular if possible data should be anonymised or pseudonymised
Lawfulness of personal data processing (Art. 6, 7) & content	Lawfulness of processing to be ensured by an unambiguous informed consent, contract or legal obligation. The data subject shall have the right to withdraw his or her consent at any time
Lawfulness of processing special categories of data (Art. 9)	Lawfulness of the processing of "sensitive" personal data (such as data related to health, ethnicity, political opinions) must be ensured by explicit consent or special legal basis
Compliance with the right to be informed (Art. 14)	A data subject is to be provided with required privacy policy information including the identity of the data controller[a] and data processing purposes as well as the period for that the data will be stored at the time when the data is collected from the data subject
Compliance with transparency rights (Art. 15)	The data subject has the right to access their data (unless this adversely affects the privacy rights of others) and receive information about data processing purposes, data recipients or categories of recipients, the data retention period, the right to lodge a complaint with a supervisory authority as well as meaningful information about the logic involved on any automated processing including profiling, and the significance/envisaged consequences of such processing
Compliance with rights to rectification, erasure and restricting data processing (Art. 16, 17, 18)	The data subject can exercise the right to correct or delete their data, the right to restrict its processing, and the right to be forgotten in a timely manner
Compliance with the right to object (Art. 21, 22)	It must be ensured that the data subject has the right to object to the processing of their data, especially in the case of automated individual decision making, including profiling
Security of processing (Art. 25, 32)	It must be ensured that suitable security measures, including data minimization and pseudonymization, are implemented

[a] A data controller is a "person, public authority, agency or any other body which … determines the purposes and means of the processing of personal data". A data processor "processes personal data on behalf of the controller" [16].

for protecting, as presented in Sect. 6.4.2, and addressing the security problems IoT caused by flawed design is evidence to the lack of proper security testing.

Second, sensors are designed to collect data implicitly (without an explicit, case by case, consent) in an unobtrusive manner, which poses challenges to transparency. Moreover, fusion of sensor data allows further (sensitive) personal details may be derived, such as personal habits (as illustrated in the satire from *Thing Tank*, in Sect. 6.2.1), the driving style (as in the example shown in Fig. 6.1), and physical condition. The implicitly collected data and derived data can be quite diverse and with different purposes attached to each piece of information, which makes the process of obtaining informed consent difficult and cumbersome, and the end-user task of correctly setting their individual fine-grained privacy preferences for different types of data and purposes grueling.

6.5 Privacy by Design and Privacy Impact Assessment

Privacy by Design (PbD) is a framework for embedding privacy into the design and architecture of IT systems [6]. Its objective is for privacy to become an essential property of all of the components of an IT system. PbD claims a full life-cycle protection of personal information, as its guidelines advocate for all personal data to be securely collected, stored, used and destroyed. In theory, it is applicable even for evolving systems, as privacy properties are constantly analyzed and addressed following the evolutionary steps of the development of a the IT system.

PbD is based on a collection of loosely defined guiding principles, which include a proactive approach to privacy, and the promotion of user-centric systems, visibility and transparency. The absence of proper formalization allows for confusion and even intentional abuse [21]. A strategy to avoid such pitfalls is to link PbD to general privacy principles, such as data minimization.

Embedding privacy in the design of IT systems and applications requires a comprehensive evaluation of the collected personal data and its use. This evaluation is provided by the Privacy Impact Assessment (PIA), a systematic process for evaluating the effects of data processing on privacy [11]. It consists of multiple procedural and sequential steps that are related to: the characterization and use of information, retention of data, internal and external sharing and disclosure, notice, access, redress, correction, technical access to information, security aspects, and technologies involved. A PIA provides means of understanding privacy-related concerns regarding the adoption and deployment of new technologies and services, and also helps to mitigate risks to business [11].

A PIA can be summarized as five procedural step: (*a*) a check for the need of a PIA, (*b*) the identification of personal data in the application, (*c*) the identification of existing countermeasures, (*d*) the listing of the existing privacy threats, and (*e*) a recommendation for additional countermeasures.

A PIA can be tailored to specific technologies and applications. In the context of IoT, the PIA framework for RFID applications [18] is a relevant example. The framework specifies the need of the PIA, its scale, criteria and elements for assessment, including privacy goals derived from the EU Directive 95/46/EC [16] and a list of privacy risks related to RFID, e.g., collection of personal data exceeds purpose and secret data collection by the RFID operator.

The scope of smart cities and IoT applications is much broader than the scope of RFID applications, which parts are also a subset of IoT applications. Nevertheless, the RFID PIA framework offers a set of processes and guidelines that could be adapted to IoT. An IoT PIA framework targeting the privacy requirements listed in Sect. 6.4.2 identifies threats to personal data and lists the appropriate controls and mitigation measures to avoid or minimize them, such as privacy-enhancing technologies (PETs). The drawback is such an IoT PIA framework would be too general, i.e., it would include a too large spectrum of threats and countermeasures, which would make it not useful in practice as PIAs are application specific.[10]

6.6 Countermeasures: Privacy-Enhancing Technologies

Legislation offers a list of legal definitions and privacy requirements (see Sect. 6.4.2), and the fines, reparations and penalties related to the legal infringements. Legislation, however, does not offers the technological means to enforce the data protection requirements. Hence, privacy in smart cities should not rely on legal measures only, but also with the support of computer and network security tools and mechanisms to enforce legal privacy principles. These privacy tools and mechanisms are generally referred to as Privacy-Enhancing Technologies (PETs). PETs can be divided into three categories, according to their specific goals.

The *first category* comprises PETs for enforcing the legal privacy principle of data minimization by minimizing or avoiding the collection and use of personal data of users or data subjects. PETs in this class provide a subset of the following:

- *Anonymity*, which is defined as an individual not being identifiable within a set of individuals, such as a collective.[11]
- *Unlinkability*, which means that two items of interest, such as individuals, objects and actions, cannot be sufficiently distinguished if they are related or not by a third party, e.g., an attacker.[11]
- *Pseudonymity*, which refers to the use of pseudonyms as identifiers. Pseudonyms are identifiers other than an individual's real names. Pseudonyms can be classified according to their degree of linkability to the individuals holding them, from a simple substitute to an individual's name, i.e., a nickname or a mobile phone number, to short-lived pseudonyms that are used for a single transaction or operation only.[11]

---

[10]The RFID capabilities and the scope of its applications are narrow enough to produce a PIA with a (non-exhaustive) set of 15 potential threats and five groups of countermeasures.

[11]The definition of the terms *anonymity, unlinkability, pseudonymity, unobservability* in this chapter follows the Pfitzmann and Hansen terminology [30].

- *Unobservability*, which means that an item of interest is undetectable, i.e., an attacker is not to sufficiently distinguish if an item of interest exist or not, and individuals involved in the item of interest are anonymous.[11]

A PbD targeting data minimization would plan and enforce this first category of PETs by default (as postulated by the GDPR, Art. 23). PETs in this category can be further classified depending whether data minimization is achieved on the network (data communication) level or the application level. Examples for PETs for achieving data minimization at the network level are anonymous communication protocols, which are based on either specialist nodes for forwarding network traffic, e.g., Mix Nets, DC Nets and Tor [7, 8, 14], or distributed solutions, where all devices in the network forward data on the behalf of others [26, 32]. On the application level, PETs include anonymous payment schemes [9], privacy-preserving digital identifiers [5, 27], oblivious data transfer (OT) [31], and obfuscation schemes [29].

Data minimization is the best strategy for protecting privacy because it decreases or avoids personal data from being processed. Nevertheless, there are many occasions in daily life when individuals have to, need to, or want to reveal personal data. For instance, when online shopping, an individual would reveal an address for billing and delivering goods, or when people willingly disseminate personal information because they wants to introduce themselves and interact with an online audience, such as on social network. In these cases, the privacy of the individuals concerned still needs to be protected by adhering to legal privacy requirements, which are covered by the second category of PETs.

The *second category* of PETs comprises technologies that enforce legal privacy requirements, such as informed consent, transparency, right to data subject access, purpose specification and purpose binding and security, in order to safeguard the lawful processing of personal data. Electronic privacy policies are PETs in this second category. They are statements that allow to describe how personal data is processed, by whom, for what purposes, and can be mathematically and logically formalized into machine readable, purpose-specific privacy policy languages. The PrimeLife Policy Language (PPL) [33] is a privacy policy language that falls into this second category of PETs. It can be used to enhance (ex-ante) transparency for users and to derive so-called sticky policies, which "stick" to the user's personal data to define allowed usage and obligations to be enforced a the service provider requesting the data and any third party to whom the data is forwarded.

The *third category* of PETs comprises technologies that combine PETs of the first and second categories, such as identity management systems.

6.7 A PbD Case Study: The Smart Share PIA

To illustrate how PbD can be included in the design of IoT applications for smart cities, we summarize the PIA for the *SmartSociety*'s car pool application, the Ride Share, introduced in Sect. 6.3.2. To conduct the Smart Share PIA we followed the general framework for RFID applications [18] and the guidelines of the British Information Commissioner's Office (ICO) PIA code of practice [23]. In Sect. 6.5, we learned that the RFID PIA framework can be adapted to IoT and applications for smart cities. To broader the scope of the RFID PIA framework, we used elements from the general structure of the ICO PIA code of practice.

As described in Sect. 6.5, a PIA has five procedural steps. The first three PIA steps are presented in Sect. 6.7.1, the fourth step, concerning the encountered privacy threats, are summarized in Sect. 6.7.2, and last step, with the recommended additional countermeasures, is outlined in Sect. 6.7.3.

6.7.1 Personal Data and Existing Countermeasures in Ride Share

The first step in a PIA process is an initial assessment to identify the need of a PIA. For this evaluation, it is necessary to verify the objectives of the Ride Share application, and which of its component parts process personal data. The Ride Share is a car pooling application, and its explicit objective is to provide a ride sharing service. Other objectives that are less explicit are related to the use of the Ride Share as a testing platform for the *SmartSociety* components. The Ride Share processes personal data to test and evaluate its algorithms and protocols and keep records for data provenance and for its reputation system. Furthermore, Smart Share aims to release (anonymized) data sets to the general public. The collection and processing of personal data in Smart Share justify the need of a PIA.

The second step involves the identification of personal data in Ride Share, and personal data information flows in the application. This step requires a deep understanding of the system, its interfaces and its implementation details. In Smart Share, we first identified the personal data inputs, which happen either during registration phase or during operation phase. In the user registration phase, Ride Share has three mandatory fields (user name, email address, and phone number) and optional fields, such as a photo. Additional personal data collected / processed during operation phase include geographical location (departure and arrival addresses), date and time, smoking habits, tolerance for domestic animals, history of rides taken and shared, user feedback, and reputation.

We classified the personal data types collected according to their processing purpose and the source of the personal data. The personal data was organized according to the following data processing categories: (*a*) functional purpose, which means that the data is required to providing the explicit objective of the application, i.e., offer a platform for car pooling, (*b*) accountability, which includes provenance and reputation services, (*c*) statistical analysis, and (*d*)assessment and testing of the *SmartSociety* components. The possible sources of personal data were: user input, sensor data, or output from a third party application.

The third step identifies the countermeasures in place, such as PETs embedded in the *SmartSociety* components and tools and procedures that are independent of the platform, which are application specific, such as mechanisms and interfaces for deleting and exporting personal data, and for redressing incorrect or inaccurate data. This step also includes the list of the security mechanisms that guarantee that personal data is stored, processed and communicated securely. It also includes the procedures of obtaining user consent, the transparency mechanisms that allows users to check accuracy of their personal data, and contact information.

6.7.2 Privacy Threats

In this section, we present the fourth step of the PIA process, which lists the privacy threats in *SmartSociety*'s Smart Share application in Table 6.2. The privacy threats relate to the list of privacy requirements presented in Table 6.1.

6.7.3 Additional Countermeasures

When executing the PIA for Smart Share, we identified a series of common pitfalls that may occur when IoT-based applications are designed and implemented in a distributed and collaborative way.

The first three steps of the PIA (Sect. 6.7.1) demonstrated the importance of the privacy awareness concerning the definitions around personal information and the general need for application designers and software engineers to know the basics around data protection legislation, or to be supported by someone that is equipped with this knowledge. In the case of the developing team of Smart Share, it had, in general, no proper formal privacy-awareness, which led to delays in the implementation of the Smart Share, which is evident by the few identified privacy and security controls in place. In the fourth step of the PIA (Sect. 6.7.2), the list of privacy threats in relation to the privacy requirements identified in Table 6.1 were presented.

In the final step of the PIA, we present a recommendation for additional countermeasures. The non-exhaustive list of the suggested additional countermeasures included: (*a*) clear purposes for processing personal data, (*b*) well-defined consent forms, (*c*) means to withdraw consent, (*d*) a specified, limited duration for storing personal data, (*e*) data encryption (*f*) the use of pseudonyms for location, reputation, incentives and in the case of entangled data, which would allow individuals to access their personal data even if it is related to personal data of other individuals.

This list of recommendations resulted in a series of security and privacy enhancements in later versions of the Smart Share and of the *SmartSociety* components.

A variant of PPL, called A-PPL [3], was integrated to the Peer Profile component of *SmartSociety*. The Peer Profile allows users to define privacy policies to their personal data items. It also allows for semantic data obfuscation, i.e., personal data is

Table 6.2 Privacy threats in *SmartSociety*'s smart share

Privacy threats	Description
Imprecise terms in the informed consent and the right to withdraw (lawfulness of personal data processing threat)	The terms used in the Smart Share's consent form are not precisely outlined and what personal data is released to each of the different organizations involved in the Smart Share is not clearly defined. Abstract and imprecise purposes lead to the collection of much more data than strictly needed. For example, refining a purpose of "accountability" to "accountability of user profiles" would reduce the amount of personal data to be recorded There are no procedures to withdraw consent in Smart Share
Collection and storing of personal data beyond what is strictly needed (data minimization threat)	There is no limited duration defined for storing personal data collected by the Smart Share application
User profiling (data minimization threat)	Provenance makes it possible to link personal data to activities. These relationships can be used to create user profiles that include personal data beyond the original purpose/need of the application
Vague purposes and function creep (purpose binding threat)	The personal data processed is potentially excessive or irrelevant. It concerns especially data items processed for the purpose of "accountability" and "statistical analysis," which are broad and ill-defined concepts. Vaguely defined purposes allow personal data to be processed for purposes unintended at design time. For example, the purpose of "accountability" could be misinterpreted or extended to allow otherwise unauthorized parties to have access to the data
Unauthorized access to information (security of processing threat)	The *SmartSociety* platform has no access control mechanisms in place. None of the platform components encrypt data, which may allow for personal data to be read by unauthorized users
Processing of inferred personal data without consent (threat to processing of special categories of data and to lawful processing)	Information collected from sensors, including geographical location, may lead to conclusions regarding an individuals habits, life style and social connections. The personal data collected in Smart Share allows for revealing who is traveling with whom, their whereabouts, and other semantic information, such as visits to shrines, political demonstrations, etc
Obstacles to deletion of data (threat to rights to rectification, erasure and restricting data processing)	There are no automatic means to delete personal data entries. The deletion of personal data from the social orchestrator and peer manager can be performed manually upon request to the Smart Share data support team. There are no available means to delete data from the provenance and reputation servers

(continued)

Table 6.1 (continued)

Privacy threats	Description
Multiple data subjects: Obstacles to access and delete data (transparency rights threat and threat to the rights of rectification and to erasure)	Sharing rides are collective actions that involve two or more individuals. Hence, personal data from multiple individuals is entangled in Smart Share. Access requests for personal data might be denied because it may disclose data about other individuals
Limited transparency to data subjects (transparency rights threat)	There are no means to guarantee the right of data subjects to access and / or amend, to the full extent, all the personal data items that are processed by the Smart Share. The data subjects have access to personal data that is available on their user profile, and may amend their email address, but have no means to access or correct information related to past rides, or any provenance and reputation data. Furthermore, the consequences of processing personal data are often difficult to foresee. It is thus difficult for data controllers to inform individuals about all possible consequences of personal data disclosure
Obstacles to the implementation of technical measures, such as pseudonymization (security of processing threat)	The use of transaction (one time) pseudonyms hamper the *SmartSociety* incentives, provenance and reputation components, which rely on individuals' history of past interactions with the system and user profiles

semantically obfuscated after an ontology-based obfuscation mechanism [24]. Personal data in the peer manager is now stored encrypted, and the communication between *SmartSociety* components is secured using TLS 1.2. In Smart Share, it led to planning properly designed consent forms, and the inclusion of contact points and procedures for redressing incorrect data. Smart Share now provides a check-box that is mandatory for users to select to point out their informed consent. The registration page includes a link that points to the privacy policy.

6.8 Conclusions

In this chapter, we discussed the social impact and privacy aspects of smart cities that are based on IoT and CAS for collecting and processing data about individuals, taking the Smart Share application of the *SmartSociety* project as an example. A wide scope of privacy issues are raised, in particular by the nontransparent manner of data collection via IoT and challenges to secure IoT technology due to performance constraints, as well as challenges enforcing data minimization via pseudonymization due to the need to link data for the purpose of accountability, such as provenance data or reputation data. Also, in smart cities, data, such as data about reputation scores of a rater and ratee, or data about shared car rides may refer to more than one data subject, who may have conflicting privacy preferences in regard to the handling of those data.

Furthermore, we outlined the process of a PIA for Smart Share that we conducted within the scope of the *SmartSociety* project and highlight the broad scope of threats that we determined, for we had to specify mitigation measures in a subsequent step. Challenges for conducting a PIA are not only posed by the complexity of smart city applications based on CAS, but also by their inherent dynamic structures: applications based on CAS can dynamically include new types of machines as peers, which may change the type of personal data collection or processing. In such situations, a new or revised PIA may be needed.

Acknowledgements This research was funded by SMARTSOCIETY, a research project of the Seventh Framework Programme for Research of the European Community under grant agreements no. 600854.

References

1. Article 29 Data Protection Working Party: Opinion 8/2014 on the on Recent Developments on the Internet of Things (2014). http://ec.europa.eu/justice/data-protection/article-29/documentation/opinion-recommendation/files/2014/wp223_en.pdf
2. Article 29 Data Protection Working Party: Working document on data protection issues related to RFID technology (2005). http://ec.europa.eu/justice/data-protection/article-29/documentation/opinion-recommendation/files/2005/wp105_en.pdf
3. Azraoui M, Elkhiyaoui K, Önen M, Bernsmed K, De Oliveira AS, Sendor J (2015) A-PPL: an accountability policy language. In: Data privacy management, autonomous spontaneous security, and security assurance, pp 319–326. Springer
4. Borges F, Martucci LA, (2014) iKUP keeps users' privacy in the smart grid. In: CNS, (2014) IEEE Computer Society. NY, USA, New York
5. Camenisch J, Lysyanskaya A, (2002) A signature scheme with efficient protocols. security in communication networks: third international conference (SCN, (2002) Lecture Notes in Computer Science, 2576 (2003)). Springer. Amalfi, Italy, pp 268–289
6. Cavoukian A (2009) Privacy by design. White paper, Information and Privacy Commissioner of Ontario
7. Chaum DL (1981) Untraceable electronic mail, return addresses and digital pseudonyms. Commun ACM 24(2):84–88
8. Chaum DL (1988) The dining cryptographers problem: unconditional sender and recipient untraceability. J Crypt 1(1):65–75
9. Chaum DL (1992) Achieving electronic privacy. Sci Am 267(2):96–101
10. Cheney-Lippold J (2011) A new algorithmic identity soft biopolitics and the modulation of control. Theory, Culture Soc 28(6):164–181
11. Clarke R (2009) Privacy impact assessment: its origins and development. Comput Law Secur Rev 25(2):123–135
12. Deloitte: Disruptive trends for smart mobility. http://www2.deloitte.com/uk/en/pages/business-and-professional-services/articles/transport-in-the-digital-age.html (2015)
13. Dimitrakopoulos G, Demestichas P (2010) Intelligent transportation systems. Vehicular Technology Magazine 5:77–84
14. Dingledine R, Mathewson N, Syverson P (2004) Tor: the second-generation onion router. USENIX-SS 2004. USENIX Association, Berkeley, CA, USA, pp 303–320

15. Earnst & Young: Routes to prosperity: How can smart transport infrastructure can help cities to thrive. http://www.ey.com/Publication/vwLUAssets/EY-routes-to-prosperity-via-smart-transport/$FILE/EY-routes-to-prosperity-via-smart-transport.pdf (2015)
16. European Commission: Directive 95/46/EC of the European Parliament and of the Council of 24 October 1995 on the protection of individuals with regard to the processing of personal data and on the free movement of such data. Official Journal L No.281 (1995)
17. European Commission: Regulation (EU) 2016/679 of the European Council and Parliament of 27 April 2016 on the protection of natural persons with regard to the processing of personal data and on the free movement of such data, and repealing Directive 95/46/EC (General Data Protection Regulation), Official Journal of the European Union, L 119/1
18. European Union Norm: Privacy and data protection impact assessment framework for RFID applications, Appendix to the Opinion 9/2011 on the revised industry proposal for a privacy and data protection impact assessment framework for RFID applications (2011)
19. Figueiredo L, Jesus I, Machado J, Ferreira J, Carvalho J (2001) Towards the development of intelligent transportation systems. Intell Transp Syst 88:1206–1211
20. Fischer-Hübner S, Martucci LA (2014) Privacy in social collective intelligence systems. In: Social collective intelligence, pp 105–124. Springer
21. Gürses S, Troncoso C, Diaz C (2011) Engineering privacy by design. Computers, Privacy & Data Protection 14:
22. Hartswood M, Jirotka M, Chenu-Abente R, Hume A, Giunchiglia F, Martucci LA, Fischer-Hübner S (2014) Privacy for peer profiling in collective adaptive systems. In: Privacy and identity management for the future internet in the age of globalisation, pp 237–252. Springer
23. ICO UK: Conducting privacy impact assessments code of practice, v. 1.0. Technical report, Information Commissioner's Office (ICO), UK (2014)
24. Iwaya L, Giunchiglia F, Martucci LA, Hume A, Fischer-Hübner S, Chenu-Abente R (2015) Ontology-based obfuscation and anonymisation for privacy—a case study on healthcare. In: Proceedings of the 10th IFIP summer school on privacy and identity management. Springer
25. Jara A, Alcolea A, Zamora M, Skarmeta A, Alsaedy M (2010) Drugs interaction checker based on IoT. In: Internet of things (IOT), pp 1–8. IEEE
26. Martucci LA, Andersson C, Fischer-Hübner S (2006) Chameleon and the Identity-anonymity paradox: anonymity in mobile ad hoc networks. In: IWSEC 2006, pp. 123–134. IPSJ
27. Martucci LA, Kohlweiss M, Andersson C, Panchenko A (2008) Self-certified sybil-free pseudonyms. In: Proceedings of the 1st ACM conference on wireless network security (WiSec'08), pp. 154–159. ACM Press
28. Miorandi D, Maltese V, Rovatsos M, Nijholt A, Stewart J (2014) Social collective intelligence. Springer
29. Mowbray M, Pearson S (2009) A client-based privacy manager for cloud computing. In: ICST COMSWARE 2009, p 5. ACM
30. Pfitzmann A, Hansen M (2010) A terminology for talking about privacy by data minimization: Anonymity, Unlinkability, Undetectability, Unobservability, Pseudonymity, and Identity Management v.034. http://dud.inf.tu-dresden.de/literatur/
31. Rabin MO (2005) How to exchange secrets with oblivious transfer. IACR Cryptology ePrint Archive p 187
32. Reiter M, Rubin A (1997) Crowds: Anonymity for Web Transactions. In: DIMACS Technical report, pp 97–115
33. Trabelsi S, Neven G, Raggett D (eds) (2011) PrimeLife Public Deliverable D5.3.4 – Report on design and implementation
34. Trivett V, Staff S (2013) What the sharing economy means to the future of travel. Report, New York (Skift, p 7
35. Tumas G, Ricci F (2009) Personalized mobile city transport advisory system. Inform Commun Technol Tourism 2009:173–183
36. UK Department for Transport: The pathway to driverless cars. Summary Report and Action Plan (2015)

37. Velaga N, Beecroft M, Nelson J, Corsar D, Edwards P (2012) Transport poverty meets the digital divide: accessibility and connectivity in rural communities. J Transp Geogr 21:102–112

Chapter 7
Security and Privacy for the Internet of Things Communication in the SmartCity

Ralf C. Staudemeyer, Henrich C. Pöhls and Bruce W. Watson

What is a SmartCity? SmartCities face the problem of growth: with an ever growing world population living in large cities, governments and municipalities must sustain the city infrastructure and to provide public services. Therefore, the amount of information that is processed and stored to successfully, and still efficiently, manage a city's infrastructure (e.g., traffic, public transport, electricity) is growing also rapidly. To manage this SmartCities are deploying truly distributed and highly scalable information and communication (ICT) infrastructures connecting a conglomerate of smart devices or 'smart things'. These smart 'things' have self-configuring capabilities and use interoperable communication protocols to seamlessly integrate. Recently, the term Internet of Things (IoT) was coined to describe constrained systems that react via sensors to physical changes and are able to influence it via actuators. While ICT generally helps to mine that information, the IoT complements this with a direct link to sensors gathering needed data or taking immediate corrective action via actuators. Achieving some convergence of physical space and cyberspace offers manifold new possibilities for the SmartCities. Examples are traffic flow sensors measure congestion and environmental sensors measure air pollution. With this the IoT enables a fine grained monitoring and control of the city.

Why is a secure and private Internet of Things communication essential? Using the capabilities of the IoT to monitor and control the SmartCity implies numerous devices communicating data about the city and about its citizens, possibly impinging on basic privacy rights if not done correctly. The communicated data

R.C. Staudemeyer (✉) · H.C. Pöhls
Institute of IT-Security and Security Law, University of Passau, Passau, Germany
e-mail: rcs@sec.uni-passau.de

R.C. Staudemeyer · B.W. Watson
Information Science, Stellenbosch University, Stellenbosch, South Africa

V. Angelakis et al. (eds.), *Designing, Developing, and Facilitating Smart Cities*,
DOI 10.1007/978-3-319-44924-1_7

is used to make decisions that will affect many citizens and if not secured correctly, attackers (or other 'errors') could disrupt the operation of the SmartCity. This chapter provides a primer on general information security, its main goals, and the basic IoT security challenges in the SmartCity. Built upon the basic IT security goals of confidentiality, integrity, and availability, this chapter additionally addresses security and privacy problems in the communication that the SmartCity is facing. We especially focus on major issues related to private communication, as privacy is a key acceptance factor for an ICT-enabled SmartCity.

7.1 Information Security in the SmartCity

If we ask where could we apply security in the Internet of Things, and specifically in the SmartCities, the answer is: *everywhere* because a system fails first at its weakest link. Thus, we briefly introduce the overall view on security, many of these are adjacent to technical security mechanisms that we cover later in this chapter. For a deeper treatment of the fundamentals of computer security consult one of numerous excellent textbooks, e.g. [1, 2].

7.1.1 Security Management and Risk Assessment

For a start all the non-technical security functions need to be in place for the technical mechanisms to be not in vain, due to misconfiguration, misuse or easy circumvention. These organisational issues span a wide range, and can be as simple as the need to inform or train human users how to operate the system securely. Good organisational security also requires regular risk assessments to evaluate the effectiveness of implemented policies and controls and keep security updated. To effectively treat and manage risks the organisation needs a model for establishing, implementing, operating, monitoring, reviewing, maintaining and improving the protection of information assets. This is what ISO 27000 calls an Information Security Management System (ISMS) [3], and its basis is a risk assessment. In this comprehensive field, often the user becomes the weakest link [4]. Find a detailed discussion on information technology and risk management in [5].

7.1.2 Software and Operating System Security

Another viewpoint is about the design and implementation of secure software down to operating system level. It is of essential importance to understand that 'security is build in, not bolted on' [6]. This is challenging to achieve and many of the IT security problems are due to badly written software. Algorithms and especially cryptographic components must not only be well designed, they also need to be implemented to not introduce new weaknesses, like bad error handling, incorrect use of memory, broken authentication, information leakage through side-channels, prone to denial-of-service, etc. Find a detailed discussion on software security in [6–8].

7.1.3 Middleware and Trusted Systems

In the SmartCity environment, we also need to consider security of the middleware abstraction layer. In the IoT, middleware abstracts from the underlying deployment of the actual sensors and actuators (the devices) and allows applications to easily get data and communicate with the virtual representations of the physical world. So, instead of asking the temperature of a specific sensor via its previously known network address, the application can simply ask the middleware for the temperature in the area of a specified physical location (e.g. street address). Think of the middleware as an intermediary, and indeed this non-direct accessibility to the IoT devices is called 'mediated device access' [9].

From a networking security point of view, the middleware component introduces at least one additional communication partner. In the simplistic case, depicted in Fig. 7.1, the middleware intercepts all messages and transforms the messages to fulfil the abstraction. In terms of network, each communication link between two individual systems is called a hop. In Sect. 7.3.5 we will discuss in more detail what is meant by the term end-to-end security. In short, end-to-end security protects the message on its complete flow. If the data from the (sensor) device and the application is not protected end-to-end then the intermediate components and systems, like the middleware must be trusted. Note, this is even necessary when all communication links in between are hop-to-hop protected by so-called secure channels.

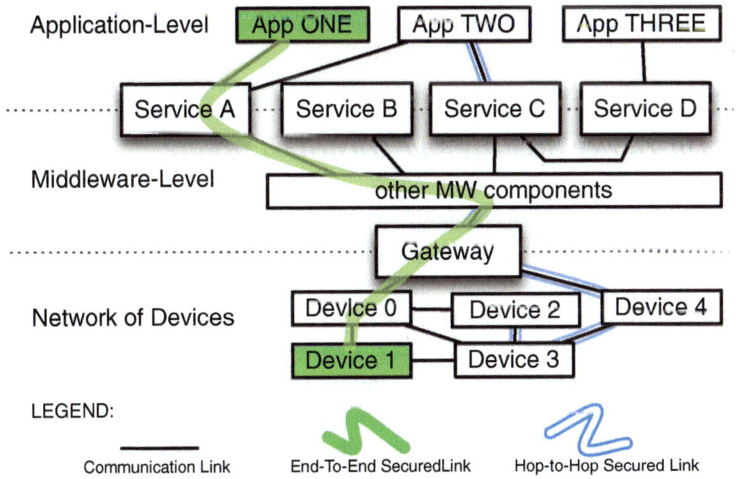

Fig. 7.1 Hop-to-hop security and end-to-end security protection in the Internet of Things levels

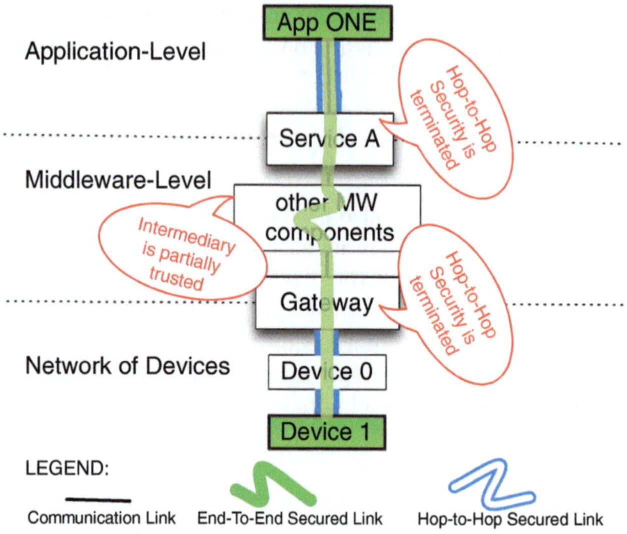

Fig. 7.2 Mixing hop-to-hop and end-to-end security for a communication between Device 1 and App ONE

In a more thought through design, the SmartCity framework shall allow important parts of a message to traverse intermediaries such as the middleware without having to trust them. In Fig. 7.2, some information (though not all) is protected end-to-end, while additional secure channels involving the middleware as one end of the communication are used to authenticate the applications on one side and the device on the other side.

Find a detailed survey on service-oriented middlewares in [10], as well as its role for end-to-end security in Sect. 7.3.5 [11].

7.1.4 On-Device Security and Trusted Hardware

We also need to discuss the general security and privacy mechanisms against a potential attacker gaining access to the device. Of course, *full protection* can only be achieved by complete physical isolation of the device, which is in a IoT context impractical. A compromise is to use *trusted hardware* which deeply embeds encryption as an intrinsic part of the processor (CPU) and memory systems—something only possible in the chip design phase, and therefore expensive. A further compromise is making the device *tamper resistant*—making it difficult or impossible to read memory contents or tap on-device data, such as the secret keys, without destroying the device. Over and above such hardware protection and tamper proofing, data on the device may be encrypted using software. In any case, the effectiveness of encryption depends on devices being able to keep keys secret, and thus on the effectiveness of measures taken to counter key extraction.

7.2 IoT Security Challenges

At first glance, IoT security challenges in the SmartCity domain (as for any IT system) could be considered already solved by applying good security hygiene, such as encrypting data end-to-end by default when communicating over the network, though they get trickier to solve in large-scale distributed systems.

We put emphasis on a selection of problems that we think are inherent to city-wide, large-scale deployments of the IoT for SmartCity applications.

Note, it is not that the underlying concept of the IoT is generally unsuitable for SmartCities, though it provides a vast number of new security challenges with indeterminate consequences. It is the large scale deployment combined with the special need for protection of the data, which is of sensitive nature, that makes this a true challenge. Those peculiarities put special focus on security problems like key distribution and confidentiality protection on numerous devices. Foremost, it creates a significant larger attack surface due to different interfaces and new possibilities to interact. We have picked on some of them and highlight their importance by giving simplified examples from the SmartCity domain.

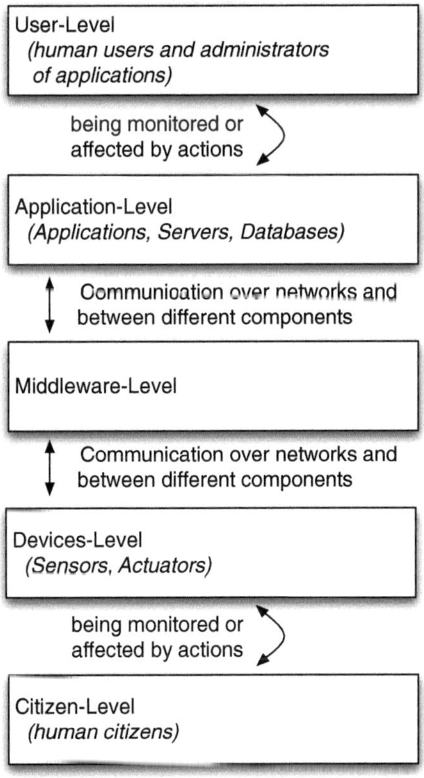

Fig. 7.3 Abstract levels of the IoT

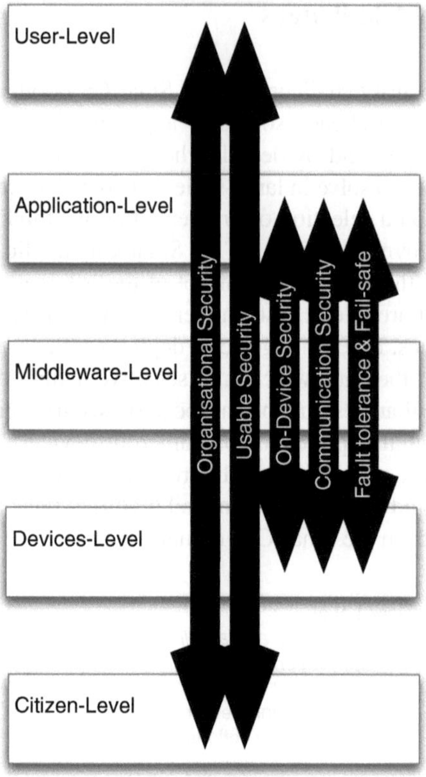

Fig. 7.4 Scope and influence of security choices

Figure 7.3 gives a very simplified layered architecture of the IoT. Via the devices, human users directly or indirectly interact in and with their physical surroundings. In this domain, users are citizens that participate in the SmartCity. At this level of abstraction the middleware is an abstraction layer. It operates, manages and unifies the plethora of IoT devices in order to allow their data and controls to be facilitated by applications. We will revisit this figure when we discuss where to place security controls, shown in Fig. 7.4.

A review of the architectural design for a secure Internet of Things communication with a focus on IP-based solutions is provided by [12]. From the legal perspective new security and privacy challenges are discussed in [13].

7.2.1 Fault-Tolerance and Fail-Safe Behaviour

Numerous SmartCity services are built assuming the ability of devices to communicate. Communication is between the devices themselves, with the middleware and finally with the application as well. Devices automatically connect to wireless networks, join the infrastructure by registering with the middleware, and become part of services. Nevertheless, IoT devices and middleware are not oriented towards human interaction during operation. Conventional security mechanisms intended to slow down automated attacks, like CAPTCHAs[1] or rate limitation, can no longer be applied. The main reason is that the devices' communication partners are not a human, but a potentially unlimited number of other devices.

Even without dedicated attacks, the connectivity to network infrastructure might not be constant and of changing quality. Transmissions can get interrupted, messages can get reordered or the data transferred can get corrupted at any time. Moreover, the context of devices might change, abruptly or over longer periods of time, whereas these changes are most probably related to device mobility. While the reason for malfunctioning might influence the design of countermeasures, the communication problems or corrupted data usually affects the application regardless of their cause.

Systems need to cope with, survive and ideally actively handle variances and errors in communication, especially given that device-to-device interactions have no human in the loop: the IoT and also the SmartCity applications built on top of it have to cope autonomously. Fault-tolerant systems might overcome problems, like using other wireless frequencies or rerouting messages via alternative networks. Regardless, they need to be able to reach a consistent state if communication fails [14].

One fall-back solution is to provide partial service until the conditions are more favourable. Under all circumstances SmartCity applications must not fail with potentially catastrophic consequences. For example, traffic lights on crossings must still enforce that once one direction has a green-light the other shows red, regardless of what an attacked or malfunctioning system for bus-preference and traffic flow tries to convince them of. If all systems fail, traffic lights shall flash yellow lights in all directions.

Note that some definitions of IT security might consider this a safety feature. We believe that it is too important to not mention it, and argue that this is linked to the problem of not being able to guarantee 100 % availability.

7.2.2 Infrastructure Integration, Monitoring and Updates

IoT is likely to be integrated into the rest of the Internet, and protocols and security mechanisms used in the Internet will be preferred and are expected to be based on IPv6 while other protocols need to bridge to them. End-to-end security needs to survive and adapt to changes in infrastructure and underlying protocols.

Malicious attacks will happen from changing attack vectors, as current trends will continue in the future. Therefore we will see the IoT being exploited by old but also by fresh, novel attack patterns. The automated collaboration between devices and services will create unforeseen threats and damages.

[1]"Completely Automated Public Turing test to tell Computers and Humans Apart".

This creates the need to make sure that the ‘health’ state of the IoT infrastructure can be monitored consistently. This cannot be a top down monitoring approach as the participating devices and services are not known a priori, but it must be an emergent property. Only if the IoT down to device-level can detect problems, it can react to changes related to malicious activity like intrusions.

A related issue is to maintain software to remove bugs and to patch vulnerabilities. This requires a specification of how patches are released to IoT infrastructure and end devices. Communication of the new software from a trusted source must reach the devices over the network—again requiring fundamental communication security.

7.2.3 Efficient Cryptography and Key Management

Many classical IT security goals are addressable with sufficiently strong and securely implemented cryptographic mechanisms. However, the SmartCities physical space is being sensed or acted upon by numerous, but potentially constrained, devices. To manage the scarce resources on constrained devices we need cryptographic algorithms optimised to save time, memory, energy and as well physical space.

Moreover, all cryptographic mechanisms depend on some sort of key material. Here, symmetric cryptography has a drawback, it requires a secret key, pre-shared securely between devices. Once a secret key falls in the attacker’s hands the added security by the cryptographic mechanisms depending on it is lost. The key will need to get revoked throughout the infrastructure; enforcing key generation and distribution afresh. Thus, capturing devices and attempting a secret key extraction are valuable attacks. To counter this, the IoT infrastructure should base its security upon public key cryptography.

Also public key cryptography has drawbacks. First, it is computationally significant more expensive than symmetric cryptography. Second, to work on a large scale we need a way to securely distribute and manage a key for each device. Distribution involves initial bootstrapping of trust and keys. Essential key management operations concern key update, revocation and recovery mechanisms.

7.2.4 Ownership and Secure Collaboration

Devices may perform operations for users, users may use devices to authenticate themselves. In this context, authentication is an open issue needing tailored security policies and mechanisms that control what and how data is created, accessed, processed and protected.

At first glance, this challenge cannot be solved without establishing a baseline for communication security: it depends on the ability to provide confidentiality of communications and authenticate the communication partner. For example, the policies that control the access to data need to be transported to the enforcement points over communication links, which themselves need to be protected against attackers.

Users need to build trust relationships with devices, while devices might know each other. Independent thereof, this requires that security policies interoperate on all components and that those are enforced. While these not directly look like communication security problems, the ability to execute these functions depends highly on the ability to provide confidentiality of communications and authenticate the communication partner.

7.2.5 Privacy, Trust and Data Minimisation

Devices are owned by companies, governments or by individuals providing a service. It is in first instance the devices' owners that get access to the information collected. Information that allows inference of actions of their users—information sensitive to leak. Devices should only collect and store information essential for the service and delete data once no longer required, while further processing of collected data (in unintended ways) should be avoided.

So far there is only a privacy assurance—on organisational level—which is not good enough due to a potential conflict of economic interests. There is high risk that user profile data is collected and users are tracked without need. Algorithms to prevent data collection and enforce reduction are still an open issue. And once security and privacy challenges can be technically addressed, their use still needs to be understandable and manageable for human users. Data minimisation is the foundation of the Privacy-by-Design principles presented in [15] and discussed in [16].

7.3 Information Security Goals Related to IoT Security Challenges

Computer systems can become corrupted in many aspects. We need to consider at least three when developing secure systems—the so-called *CIA triad*—also known as the three 'pillars' of information security, namely *confidentiality*, data *integrity* and *availability*. Confidentiality and integrity can be addressed using basic, well-known cryptographic mechanisms. Furthermore, there are numerous additional factors, such as the *triple A*, namely *authentication*, *authorisation* and *accountability*. Excellent all-around introductions into the basic concepts of computer security are provided by [1, 2].

Gaining access to a system is an *active attack*. This might involve stealing it physically or infiltrating it to gain logical control. Let us stress that also active attacks on the communication channel, like inserting messages or downgrading the communication channel, must be considered. By contrast, to eavesdrop or record communication is a *passive attack* and is detectable directly as it involves no alteration of data.

It is of great importance to acknowledge that information security is not limited to the protection of the data that is stored and transmitted as the content of a message. We also need to take into account that other sensitive information is revealed by traffic data, so-called metadata. Due to the need to first achieve security for the messages' content and then for reasons of efficiency the handling of metadata is less widely implemented. Nevertheless it is a key consideration for the successfully establishment of privately usable services, in particular with regard to the SmartCity. We will cover private communication technologies in more detail in Sect. 7.4.

We will now briefly re-state current definitions and provide pointers to textbooks and standards. It will give you a quick glimpse, but do follow the pointers to find more background information. Additionally, we describe the most applicable security primitives/mechanisms and explain how they could be used to enhance the security and privacy in SmartCity applications. A detailed and entertaining introduction into cryptography is provided by [17]. Find with [18] a recommended and more recent mathematical textbook on cryptography.

7.3.1 Confidentiality

Confidentiality is the property that protects message content from passive attacks. It holds whenever someone not authorised fails to disclose confidential information from stored or transmitted data. The main idea is that an attacker will not learn information he is not authorised to know. On the communication side, the attacker is assumed to be able to eavesdrop on the communication. Therefore on unencrypted data at least the content of the message and the corresponding metadata are known; like the number, the size, the frequency of the messages on the communication link [19]. For data-at-rest the attacker is assumed to have physical access to a device.

7.3.1.1 Encryption to Protect Confidentiality

The cryptographic mechanism to protect confidentiality is encryption. Encryption transforms plaintext into ciphertext using an cryptographic algorithm; the reverse operation is called decryption. A cypher is a complementary pair of algorithms providing both cryptography operations. By introducing a key parameter the encryption function is extended to describe a transformation per individual key value. Classical encryption comes in two easily distinguished forms: symmetric and asymmetric.

Symmetric encryption is named for its use of either the same key for both encryption and decryption, or a key pair that can be easily computed with knowing at least one. For that reason this is specified as a one-key-scheme, since we treat them as effectively the same key. This symmetry yields cryptographic algorithms for encryption and decryption which are relatively simple, easy to implement, and are relatively secure, depending on the key length.

Further, actual symmetric encryption algorithms are often divided into two camps: stream and block cyphers. The *stream-cypher* encrypts a plaintext message bitwise using a keystream. In a perfect cypher the keystream is fully random, of same length like the message, and only used once. This is called a *one-time-pad* [18] and provides perfect confidentiality. The major drawback is that the keystream needs to be transported to the recipient via a secure channel. Therefore, implementations sacrifice unbreakability by generating the keystream using a pseudo-random number generator generated based off a shorter initial secret sequence (also referred to as seed). The cypher can be broken by predicting the secret sequence.

These days *block-cypher*s are more common, which were developed to counter shortcomings of stream cyphers. A block cypher divides the plaintext into 'blocks' of a specified size (128 Bit for AES) which are then processed resulting in an encrypted block.

Historically, the first openly available (digital) symmetric encryption scheme was the US Data-Encryption Standard (DES), which was progressively strengthened to become triple-DES (3DES). As for standards, 3DES was long superseded by the Advanced Encryption Standard (AES). Note, it is important that systems' cryptographic primitives can be updated to the most recent and thus currently considered secure algorithms. You shall not base the security of a system just on the current standard but have a process to update devices when new standards appear.

Symmetric encryption schemes compared to asymmetric ones are typically cheap to implement, use little memory and are with good performance, and are therefore suited to bulk encryption of large amounts of data. This implementation simplicity means that hardware implementations abound, requiring very little silicon real-estate, making them suitable for smart cards. Most modern CPUs include 'assist instructions' to further accelerate symmetric encryption.

The symmetry also implies that the encryption/decryption key must be kept as a secret between only the sender and the receiver, necessitating some secure key-exchange channel or mechanism. This makes it best suited for systems in which some other key-exchange exists, for example using public key encryption or a channel secured in some other way, such as physical handover; or where the encryption and decryption occur on the same system. The latter use case is particularly interesting given the speed of symmetric encryption for encrypting local files, messages prior to communication, and local memory.

As the name implies, *asymmetric encryption* requires two different keys: one each for encryption and decryption. The encryption key can be made public this mechanism is also called *public key encryption*. In detail the pair of keys differ depending on the particular encryption algorithm. Often, the key length parameter can be used to adapt to projected advances in computing power to prohibit attackers to compute the secret decryption key from the encryption key (in reasonable time). Of course, publicly distributable keys are very convenient, resolving the issue of how to exchange a secret encryption key in the first place.

The first openly available asymmetric encryption algorithm was Diffie–Hellman (DH) [20], followed soon by the highly popular Rivest, Shamir and Adleman (RSA) [21], and more recently the efficient Elliptic Curve Cryptography (ECC) [22, 23]. All three of these are based on number theoretic properties which are difficult to reverse, the key component in keeping the decryption key secret. For example, RSA uses keys based on the product of two very large prime numbers; obtaining those numbers from the encryption key is thought to be the only way to obtain the corresponding decryption key, but factoring such a large composite number is extremely costly.

In comparison to symmetric encryption algorithms, asymmetric encryption is relatively expensive to implement. Operations such as exponentiation and multiplication are required in all of the prominent algorithms, needing many more instructions/clock cycles (and therefore power) than a symmetric scheme such as 3DES or AES. An hardware implementation is also costly in terms of silicon real-estate, and is rarely done. The convenience of public keys are not always an appropriate tradeoff for the costs involved, and most real-life systems actually use asymmetric encryption at the beginning of a communication session only to exchange a (secret) symmetric encryption key which is subsequently used for the remainder of the session.

Recent advances in Elliptic Curve Cryptography (ECC) have yielded asymmetric cryptography implementations to get significant more efficient than RSA [18, 24]. They work on a different mathematical assumption: a discrete logarithm problem on elliptic curves. The result are keys which are much shorter, while giving the same cryptographic strength, and allowing for more efficient implementations. Most importantly, strong ECC can be mathematically tailored to run in very constrained devices and very small silicon footprint, even in RFIDs [25–27]. Most public key encryption schemes play a further role in signatures.

7.3.1.2 Confidentiality in SmartCities

Today's citizens may assume that data flowing between devices at their private home or in their immediate vicinity is confidential. That it is inaccessible to any other parties without consent or warrant, and that at least not everyone is able to collect and process this kind of data at will. In a SmartCity environment, this rather clear separation between private and public environment blurs.

To give an example: The city of Amsterdam in the Netherlands supports more then 40 smart city projects ranging from smart parking to the development of home energy storage for integration into the smart grid [28]. One of these projects concerns the installation of smart energy metres with incentives provided to households who plan to actively save energy. Smart metres record energy consumption in households and report these in short intervals (e.g. 2s) to the provider. The benefit for the energy provider is that frequent reports allow demand management. Consumer can then benefit from lower rates during off-peak times, whereas this also depends on people actively changing their energy use. The downside is that these frequent measurements reveal detailed information on household activities, including presence, electrical devices in use, and even what content consumers watch on television.

Critical problem is reliable and secure communication, and as well trust towards the provider processing the information. Depending on location and environment of the smart metre communication might be via powerline, WiFi, cell phone network or the Internet. In contrast to energy monitors, smart metres permit two-way low-latency communication. These networks are more or less accessible and prone to eavesdropping and remote exploitation.

Here, encryption can harden unauthorized access to collected data and helps to protect information considered confidential. For example, symmetric encryption can be hardwired into network infrastructure on a hop-by-hop basis with reasonable effort and resources, providing a basic layer of protection between devices forwarding traffic. Asymmetric encryption can help to establish secure connections between end devices. In this use case between the smart metre and a trusted energy provider. If the energy provider is not trusted, then additional steps need to be taken to ensure that high frequency readings are difficult to associate with a specific consumer or metre [29–34].

Confidentiality protects against disclosure of information to unauthorized parties. This property helps to build secure communication channels to distribute software and updates (see Sect. 7.2.2), and to ensure secure collaboration (see Sect. 7.2.4). It is as well helpful for building long-term trust relationships (see Sect. 7.2.5). It is important to note that confidentiality is an essential security property required to build private storage and communication systems.

7.3.2 Integrity and Veracity

Integrity is violated whenever data or a system is modified in an unauthorized way, without being detectable. Modification can occur due to transmission errors or due to an active attack. However, for the correct and expected behaviour of an IT system the reason for a modification does not matter. Altered data must become reliably distinguishable from unchanged data. The protection mechanisms against malicious modifications however differ from those that detect random transmission errors. Thus, a Cyclic Redundancy Check value (CRC) can be used to protect against random small transmission errors [35], but cannot provide integrity against a malicious attack, as anyone—including the attacker—can compute a new CRC valid for the changed data.

Integrity protection mechanisms can be further distinguished into detective and preventive measures. The former fulfils its duty as it 'detects the violation of internal consistency' [36]. The preventive view of integrity protection (see classic definitions in [37]) requires mechanisms that prohibit unauthorized modifications upfront. In a nutshell, data integrity is '[t]he property that data has not been changed, destroyed, or lost in an unauthorized or accidental manner' [38].

As with every security goal, let us dig a bit deeper and understand the limits of integrity. *Veracity* is the property that the data in an IT system truthfully reflects the real-world aspect it makes a statement about. It is important to acknowledge that integrity notions and protection mechanisms are not concerned with the quality of the data, specifically from an application's perspective. If information is incorrectly captured into the system data integrity protection will prohibit undetected modifications to this false data. Thus, this must be covered by mechanisms to protect the veracity, like increasing the tamper-resistance of the sensing device, making a consistency check by comparing the actual value to a prediction made, or comparing it with assertions made by several witnesses assuming the adversary being unable to corrupt a sufficient number of assertions [39].

For a better understanding of the limitation of integrity with respect to veracity let us look at a traffic estimation example: Integrity protection allows to check at the receiving application that only verifiably unmodified traffic flow information gathered by known and authenticated sensors of a road segment will be used to do the traffic estimation. However, an obstructed or broken sensor starts with a wrong observation and thus records wrong data. This data does not truthfully reflect the correct information about the road segment's current traffic flow—it does not offer Veracity, as the data is wrong with respect to the physical world—but if the integrity protection is still working this data is still verifiably integrity protected.

With this limitation in mind, let us look at technical mechanisms to protect the integrity.

7.3.2.1 Digital Signatures Schemes (DSS) and Other Protected Checksums to Protect Integrity and Gain Entity Authentication

In order to protect the integrity of a message, a simple checksum like CRC, is not sufficient against a dedicated attack. The mechanism of choice is the protected checksum. It prevents an attacker to adjust the checksum to match changes made to the data [38].

We will briefly introduce the two most common forms: digital signatures and keyed hashes, also known as Message Authentication Codes (MAC). In general, a *cryptographic hash function* is a good mathematical hash H function that additionally offers the avalanche-, the one-way property and one of the two collision-free properties:

- Avalanche property: Given two inputs s' and s'' that differ only slightly the outputs $H(s')$ and $H(s'')$ must differ in each bit with a probability of $1/2$.
- One-way property: Given H and a hash result $h = H(s)$, it is computationally infeasible[2] to find s. Of course, given H and an input s, it must be relatively easy to compute the hash result $H(s)$.
- Weakly collision-free property: Given H and an input s, it is computationally infeasible to find a different input, s', such that $H(s) = H(s')$.
- Strongly collision-free property:
  Given H, it is computationally infeasible to find any pair of inputs s and s' such that $H(s) = H(s')$.

Using a hash function over a message gives something comparable to a fingerprint of the message. It has the main advantage that it maps an arbitrarily large message into a fixed length. Thus, if applied on the input first, all subsequent algorithms work on a fixed sized data object.

[2] Computationally infeasible means that it is possible, but that doing so would require a very long time and very powerful resources.

Digital signatures are a cryptographic tool to gain integrity protection for a message. A digital signature scheme (DSS) is based on asymmetric cryptography. Like in asymmetric encryption, two related keys are involved: the secret signature creation key and a related public verification key. The key algorithms of DSS implement the two functionalities: Sign and Verify. The sign algorithm takes the secret signing key and the message and generates a signature value. The verify algorithm takes the signature value, the message and the public verification key to obtain the result. If the result is positive this means that the signature is valid. In turn this yields two things: First, the document has not been altered according to an integrity policy and second, the signature value has been created involving the secret key corresponding to public key used for verification.

If a trusted link between an entity and the public verification key exists, a valid signature on a message implies that the message originated at the entity. This gives origin authentication and can be used to build entity authentication protocols. Those protocols need to be designed very carefully [40].

The most widely known algorithms are ECDSA (based on elliptic curves) and DSA (based on RSA). Both embody the use of aforementioned cryptographic hash functions, like SHA256 or SHA512. The message is first hashed with a secure, especially collision-free, hash algorithm, and then the resulting hash is used as input for the signature generation and validation algorithms. Due to the avalanche property the hash value of two messages will be already different if a single bit is changed. Hence, the integrity policy of those schemes (by choice of the hash algorithms) is to detect any subsequent change, even a single bit, as a violation of the signed data's integrity.

Additionally, the collision freeness properties do not allow an attacker, which by definition does not have access to the secret signing key, to take an existing signature on a message and then compute (or find) a colliding message that has the same hash. If it would be feasibly to find such a collision an attacker would be able to break the unforgeability property [41] as the attacker has then a valid signature on a message that was never signed by the signer, a valid forgery.

Keyed hashes allow detecting, based on the cryptographic hash function, if the input data object is changed. However, they forbid an attacker to simply re-calculate the hash value, as the hash value can be only correctly computed with the knowledge of the secret key. It can be built by several ways, one of which is to use any normal cryptographic hash (which is keyless) and concatenate the message with a shared secret key, not forgetting padding. An example is the HMAC algorithm [42].

In terms of the integrity protection this gives a symmetric pendant to a digital signature. However, due to the shared secret it does not offer the same level of authentication of origin, nor does it allow to prove the integrity to third parties. The reason for this missing authenticity of origin is that in a symmetric key-based integrity check the verifier needs to know the secret that was used to generate the integrity check value. Hence, in theory the verifier, as well as the 'signer', can generate a valid keyed hash for any message.

Comparing keyed hashes and digital signatures the main advantage of symmetric methods are speed and simplicity in the implementation, however the missing authenticity protection the key management issues related to symmetric keys are a severe drawback. Digital signatures allow for a simpler key management as the public keys are not needed to be kept confidential. Thus, they allow presenting the public key alongside the signature and the message to any other entity which is able to verify the signature based only on public information. Only the entities (the importance here is that it can be as little as *one* single entity) that are in the possession of the secret signing key can produce a valid signature.

7.3.2.2 Integrity (and Related) in SmartCities

In SmartCities sensed data is gathered and used by algorithms to enable smarter decisions. Thus, the decisions are based on the data gathered, and bad quality data can lead to bad decisions. Imagine a SmartCity without integrity protected messages were every sensor value could have been tampered with. An faulty air pollution sensor in one city area might cause that area to be declared into a Zero Emission Zone (ZEZ). This leads to neighbourhood cars not being allowed to access the area, giving the citizens in the area far less noise, but congested roads in the neighbourhood due to through traffic. Would that not be an incentive to falsify pollution messages?

Likewise all control messages could be manipulated, so why not change the 'Access denied' message from the barrier control system into an 'Open barrier' message? The solution to that attack is that you detect any subsequent tampering with sensor data by applying integrity protection. Additionally, the origin of those integrity protected messages must be known to be a trusted source. Of course, a defect or even manipulated sensor could sent false readings. Then they would still be integrity protected, hence SmartCities need to deploy additional processing logic to detect such wrong readings and send repair personnel to investigate the actual sensor.

Integrity protection is a basic security functionality. It is needed for the authentication it aids with secure collaboration and with ownership (see Sect. 7.2.4). Finally, they can be used to authenticate software and support securing the distribution channel of software maintenance (see Sect. 7.2.2) against injected malicious code. Integrity protection detects erroneously or maliciously modified information and helps using them as input for further processing, and thus can be used as early detectors for a fail-safe behaviour (see Sect. 7.2.1). This is even more true if they are used in combination with veracity protection mechanisms.

7.3.3 Availability

Availability ensures timely and reliable access to devices and services. It goes with the assumption that a particular resource is not accessed in a non-legitimate manner, whether authorised or unauthorized. The availability property is violated if an attack succeeds by degrading a computer resource or rendering it unavailable. Typically, this is done with so-called Denial-of-Service (DoS) attacks using up all the available resources and therefore increasing the response time. If the service is unavailable when accessed by an authorised entity, then the result is as bad as the service or device does not exist at all.

To ensure availability under all circumstances is a hard to solve problem.

7.3.3.1 Availability in SmartCities

When monitoring critical environment, sensor values and alert messages need to arrive timely, otherwise detection fails and no alarm is triggered. Critical values might be related to industrial contamination with potentially hazardous consequences for example in terms of air pollution, high radiation levels or decrease of water quality.

Take for example the city of Tarragona in Spain, which is next to chemical factories. Here the constant and reliable monitoring of air pollution for critical substances is vital to protect citizens. It is essential that potential air contamination can be detected before it reaches the closest households. In order to achieve this, the detection sensor devices deployed need to send their data to a monitoring server. The city council can then detect, or even preview critical events, with automated alarms and potentially react in a timely manner. If parts of the infrastructure, like the sensor, the network, or the monitoring server, gets unavailable, this detection will fail and the population can therefore not be warned in time.

To improve availability helps to address issues related to the challenge of fault-tolerance and fail-save behaviour (see Sect. 7.2.1).

7.3.4 Authentication, Authorisation, Accountability

The three goals are often grouped together and then referred to as 'AAA' or 'triple A'. Successful authentication and authorisation allows achieving accountability for a certain action by a certain entity.

Accountability enables the detection of actions (e.g. violations) or attempted actions to be traced to the potentially responsible entity [38]. In order to achieve this the entity needs to be first authenticated and then the request for access is subject to authorisation. Depending on the level with which entities can be differentiated by the authentication mechanism, they can be held accountable. Related to accountability is the notion of *non-repudiation*, which 'is a service to generate, collect, maintain, make available and validate evidence concerning a claimed event or action in order to resolve disputes about the occurrence or non occurrence of the event or action' [43].

Initially *authentication* is required. It is the 'process of verifying a claim that a system entity or system resource has a certain attribute value' [38]. This definition was carefully chosen because it highlights that it is concerned with checking a certain attribute value. Authentication in general is not concerned with checking the identity of an entity. While the identity can be an attribute, this certain attribute should be seen as the feature of interest.

In order to restrict access you need to check if an entity is authorised to carry out a certain action: *Authorisation* is the 'process for granting approval to a system entity to access a system resource' [38].

7.3.4.1 SmartCities Authentication and Authorisation Problems

There are authentication problems on different layers of a system as complex as the SmartCity. As an example, assume the SmartCity detected a high risk of pollution and wants to restrict access to the inner city area. Assume that today only electric cars are allowed because the inner city is declared a Zero-Emission Zone (ZEZ). So the first question is: 'Who is the entity of the system that you want to authenticate?', which can be hard to answer in technical detail (see [40]). Do we want to authenticate a single car, the car's on-board-device, or its passengers? Here it becomes obvious that peer-entity authentication can happen on various layers.

For example, the city of Milan in Italy was reported in 2004 by the World Health Organisation as being one of Europe's most air polluted urban centres caused by very high downtown traffic volumes [44]. In 2008, the city introduced electronic road pricing to address traffic congestion, to promote sustainable mobility and public transport, and to decrease levels of smog. The restricted inner city so-called 'Area C' [45] and toll income is reinvested into sustainable energy projects. The area is accessible via gates monitored by video cameras equipped with automatic number plate recognition technology.

In this use case, the attribute value of the car is its license plate number. If cars can proof to the road toll system that they have a certain license plate, then this information can individually account individual cars for its road usage. Nevertheless, we acknowledge that the system design is very bad for privacy, since this data stored and correlated contributes to surveillance of citizens. There is no need for an application to have authentication based a unique identifier.

For example, to restrict inner city access to electric cars, it is sufficient if these are distinguishable from non-electronic ones. This means that the relevant attribute value is 'I am an electric car'. Of course this claim has to be proven, such that the system controlling access can be sure that it grants access for an electric car. The showing of attributes for access control should be done in a privacy preserving fashion using for example anonymous credentials [46].

Apart from having all these protocols, especially the ones that allow for privacy preserving authorisation, in all the IoT devices and in all the systems, each entity needs to have the credentials to prove its own claims to the system that asks for authorisation. This means that all participating systems need some form of keys and some form of trust that is associated with it. For example, would the city trust the car manufacturer's to vouch for the claim 'I am an electric car', or would the car need to be regularly inspected and be issued with a token issued by a state trusted institution.

However you do it, you are in need of sophisticated key management and key distribution. If not, some attacker might disguise his hybrid-car on demand as an electric car to cruise in the inner city.

The 'triple A' in general are important building blocks to tackle the challenges of secure collaboration (see Sect. 7.2.4) and to build trust relationships (see Sect. 7.2.5). They are also needed to authorise interconnectivity towards an existing infrastructure (see Sect. 7.2.2).

7.3.5 End-To-End versus Hop-To-Hop Protection

Before we move on, we want to highlight that all the above mentioned goals in general can be achieved between two parties. This means authentication could happen directly between the application in a smart phone talking with some individual barrier to access the inner city via some network infrastructure. But it could also happen between the smart phone application and an application server, and then again between the application server and the barrier control system, and then again between the control system and the individual barrier. The former is called end-to-end, while the latter is called hop-to-hop.

From a networking perspective each communication link between two individual systems is called a hop. Each system at the end of such a hop might require to know with whom they exchange data with, meaning they might need to authenticate, and be able to communicate in a secure manner. This means they might apply mechanisms to gain confidentiality, integrity and availability.

Depending on what system and what communication link you need the protection, the data transmitted in the communication system can be protected by different means. One option is to protect the transport link, the other is to protect every message separately. Both options have serious disadvantages suggesting a layered approach. See Fig. 7.1 for an example. To make the differences obvious, let us consider confidentiality protection by encryption. Hop-to-hop, or so-called link-level protection, encrypts data between neighbouring network nodes. The advantage is that encryption keys can be 'hardwired', which avoids issues with key-exchange. Major disadvantages are that all forwarding nodes have access to the unencrypted data and once the key leaked security is compromised.

In end-to-end security the confidentiality and the integrity protection is between the endpoints of the communication. As such, authentication (and finally the authorisation) can be performed between the endpoints as well. Note that for authorisation decisions it is important to understand what you want to authorise, in other words, what are the endpoints of your communication. For example, to logically authenticate a specific car you need to be sure that what you technically authenticate is affixed to that specific car, so it cannot be easily removed and placed into another car. Achieving end-to-end protection means that the need to trust the intermediate systems is removed. While this is preferable, it cannot always be achieved due to layered approaches and independent subsystems. Then a good system design shall allow identifying these hop-to-hop or system-to-system protection and highlight those trusted systems for the risk assessment.

Implementing end-to-end security is essential for secure collaboration (see Sect. 7.2.4), building trust and to implement private communication (see Sect. 7.2.5).

7.4 Technical Communication Mechanisms Towards Privacy

The key concepts of information security are confidentiality, integrity and availability. Confidentiality and integrity, we learned that these can be addressed with end-to-end encryption and signatures using public key cryptography. We also learned that the use of encryption should be enabled as the default. But irrespective from the successful use of modern cryptographic techniques, attackers may eavesdrop communication and analyse network traffic.

The SmartCities wide area networks are impossible to physically secure against unauthorized access. In the IoT domain the local network access is predominantly wireless and is therefore prone to eavesdropping. To protect citizens from any kind of hidden loss of personal information when accessing public resources the communication also needs to be protected against traffic analysis. For example SmartCity applications do heavily depend on users location information to provide location-based services. Nevertheless there is no public interest to leave citizens traceable and facilitate continuous surveillance. The traditional information security goals are unable to protect location information, since these leak from metadata and can be extracted by traffic analysis. Here we need new means of protection.

Traffic analysis [19, 47] focuses solely on communication patterns and the extraction of information out of metadata, irrespectively from the use of content data encryption. In traffic analysis, network traffic is captured with the aim to gather information about the network, its devices, and its users. This discloses not only the communication partners and their frequency of communication, but also reveals information about the area of network alignment. Information extracted can be further processed and analysed by combining techniques from data mining and machine learning [48]. Further extracted information can be individual usage patterns, but as well identifiers used for future tracking.

Traffic analysis works without any knowledge of messages' contents, and is therefore applicable to encrypted messages. Large numbers of captured messages make traffic analysis more effective. It might reveal even more information if the traffic flow or the messages themselves can be modified or tampered with by other means.

Find an excellent introduction into privacy and data protection by design in [49], with more details covered in [50]. The Privacy-by-Design principles are presented in [15] and discussed in [16].

7.4.1 Network Properties for Private Communication

Even if the protection of user data is addressed by means of end-to-end encryption, we still need to look into information loss caused by leaking protocol metadata. This leakage can go up to the point, which may render end-to-end encryption obsolete. Traffic analysis attacks were successfully run on SSH [51] and Skype [52]. To limit success of these kind of attacks, at least the following properties [47] shall be taken into consideration by the network of devices:

- Coding—All messages with the same encoding can be traced.
- Size—Messages with the same size can be correlated.
- Timing—By observing the duration of a communication and considering average round-trip times between the communication partners patterns of network participation can be extracted.
- Counting—The number of messages exchanged between the communicating parties can be observed.
- Volume—Volume combines information gained from message size and count. The volume of data transmitted can be observed.
- Pattern—By observing communication activity, patterns of sending and receiving can be observed.

Finally, the observer can also perform a long-term intersection/disclosure analysis of the network by observing devices and the network for long time and reducing the set of possible communication paths and recipients by analysing online and offline periods. Characteristic usage patterns, such as an IoT device connecting every minute, may appear and can be used to further reduce the number of possible paths.

The following Table summarises the message properties and how they can be addressed (Table 7.1).

To counter traffic analysis we need to minimise any kind of information leakage. Therefore the network property we ideally aim for is unobservability, or at least anonymity. The *unobservability* property ensures that messages and random noise are indistinguishable from each another. In terms of network nodes it insures that their activity goes unnoticeable and that messages cannot be correlated. It is a very powerful property combining anonymity and *dummy traffic*. Anonymity breaks down into the unlinkability and unidentifiability property. The *unlinkability* property ensures that neither messages nor network nodes system can be correlated. Whereas *unidentifiability* ensures that these are indistinguishable, building a so-called anonymity set. Given that the anonymity set is always greater one, the system provides the *anonymity* property.

The word anonymity is derived from the Greek word 'anonymia', meaning nameless, and is interpreted to mean *the right not to be identified*. These aforementioned terms are defined in detail in [53]. In sum they are related as follows:

Unobservability = Anonymity + Dummy Traffic with
Anonymity = Unidentifiability + Unlinkability

Table 7.1 Protection against passive attacks

Attacks based on	Proposed solutions
Message coding	Change coding during transmission e.g. with k-nested encryption
Message timing	(1) batched forwarding of messages
	(2) random delay of messages (delaymin ≥ latencymax)
Message size	Use a predefined message size and padding small messages
Message counting	Receive and forward a standard number of messages and use dummy traffic
Communication volume	Protect message size and communication volume
Communication pattern	Continuous network participation
Message frequency	Use a standardized message exchange pattern
Brute force	No clear protection, dummy traffic helps
Long-term intersection	No clear protection, continuous connectivity and dummy traffic help

7.4.2 Proxies, VPNs and Dummy Traffic

A certain degree of anonymity can be achieved by using a proxy or a VPN, whereas both solutions route traffic via a relay. Proxies were initially developed for surfing the web, but SOCKs proxies can also forward TCP streams. Unfortunately, an observer with access to the traffic entering and leaving the proxy over extended periods of time, can reveal the communication relation. VPN networks on the other hand are slightly more robust, since they provide a layer of encryption to the incoming traffic. Therefore, incoming and outgoing traffic cannot be mapped easily.

Message frequencies and flow can still be analysed. The message flow between parties includes both the traffic volume and communication pattern. Communication partners have a unique distinguished behaviour that can be fingerprinted. An observer can perform a brute force analysis of the network by observing all possible paths of communication and generating a list of all possible recipients. Dummy traffic inserts additional messages with meaningless content into the network during times of less communication. The sender inserts them into the network in a form not distinguishable from real messages. This is done to pretend a constant high traffic that makes traffic analysis to require significantly more resources. Most efficient is the concept of dummy traffic used in conjunction with mix or ring networks with an implicit addressing scheme to prevent mapping.

7.4.3 Anonymity: Mix Networks and Onion Routing

Leakage of metadata can be further reduced by providing increased protection against network traffic analysis. This includes hiding network endpoints, timing and location information. Traffic analysis is resisted by ensuring networks support the anonymity property as implemented by anonymising networks, most commonly using proxy chains. Anonymising proxy networks have started with the implementation of Chaums Mix in 1981 [54]. The system tunnels encrypted traffic through a number of low-latency proxies.

Initially, interest in this field was primarily theoretical, but in the last 30 years a lot of research in this field has looked at developing practical and usable systems for preserving anonymity [19, 55].

Onion Routing [56] was primarily developed to allow anonymous web browsing in close to real-time, but the concept is applicable to prevent traffic analysis in any network. Here the traffic is forwarded by multiply relays in ways that it is hard to tell which actually carry the traffic. This is achieved by using dummy traffic and nested point-to-point encryption. Once traffic leaves the onion routing network it can be observed, and therefore end-to-end encryption is needed which remains the responsibility of the end nodes. A well-known implementation of Onion Routing is 'The Onion Router' (TOR) [57].

Mix networks provide a unidirectional communication channel only, but provide stronger protection in comparison to onion routing. They additionally enforce a uniform message format and introduce extended delays. Mixing possesses the following attributes. They can

- hide the relationship between sender and receiver of a message,
- guarantee anonymity of the sender, the receiver, and both to all third parties and
- protect against revelation of signalisation relations, location updates, time of communication, and kind of service.

Mix networks and onion routing offer a high degree of end node controlled protection. The overhead of techniques based on Mixing is moderate. With Onion Routing there is a sufficiently mature solution available for SmartCity environments.

7.4.3.1 Anonymity in SmartCities

Smart metres provide automatic metre readings in high frequency for intervals defined by the electricity provider. In Sect. 7.3.1.2, we discussed the need that readings need to be confidential, since the frequent measurement reveal detailed information on household activities. Reference [29] discusses the need to further anonymise metre readings stored at the provider, so that it is hard to map the measurements with a particular metre or customer.

Nevertheless without protection, metre traffic can be mapped to communication partners, traced to a specific metre, and remains prone to traffic analysis and interception. Traffic can be modified, packets can be injected, and replayed. Triggered or natural changes in the communication pattern and volume can reveal as well sensitive household information, to the point of encryption getting obsolete. Classic traffic anonymisation techniques, like proxy chains and mixing, can support to address this thread with moderate costs.

7.4.4 Unobservability: Broadcast and DC-Networks

Broadcast and multicast based concepts offer full receiver anonymity. They protect the receiver of a message by sending it to all, or a set of, recipients of a network. An implicit destination address is utilised for enabling only the recipient to recognise the message. This can be done by public key encryption, which also provides authenticity, integrity and confidentiality. Here every recipient attempts to decrypt the message, whereas only the intended will succeed by using the correct private key.

With DC-Net a different mechanism was introduced [58]. The DC-Net is a round-based broadcast protocol where members can unobservable publish a one bit message per round. This is called 'superposed sending' and is very secure but prone to Denial-of-Service attacks. To address these protections against disrupting nodes were proposed [59, 60]. By 'superposed receiving' [60] DC-Net was extended to support anonymous receiving of messages. See [47] for a detailed description of the basic DC-Net-algorithm. Only few implementations exist [61, 62], probably due to DC-Nets sensitivity to disruption.

It is possible to categorise concepts into re-routeing-based and non re-routeing-based concepts [63]. Broadcast or multicast and DC-Net are the only non-re-routeing-based systems. These come expensive as the network grows since all members of the network need to at least send or receive one message. The situations changes dramatically whenever a shared medium becomes available [64], like access networks based on WirelessLAN and IPv6LowPAN.

We note that the overhead of broadcast-based solutions in todays switched networks comes with significant costs and, even worse, do not scale. Currently there are no mature solutions to obtain the unobservability property at reasonable costs. It is an open issue, which requires additional research.

7.4.4.1 Unobservability in SmartCities

One of the most sensitive data of citizens more recently collected and stored by governments in a SmartCity context is medical data. For example, the city of Johannesburg in South Africa aims to go paperless by 2016. This includes deploying a digital media health record to improve record keeping and as well patient care.[3] There is high risk that sensitive medical information will leak at some point and be processed in unintended ways, most probably not for the benefit of citizens. We conclude that the protection of stored data, location information and usage patterns might not be sufficient for medical data. Here, we may consider to protect as well the very existence of stored information and the access to it. Communication and storage systems that provide the unobservability property can further help to protect overly sensitive information, that cannot be stored by other means (e.g. paper).

[3]http://ehealthnews.co.za/joburg-invests-in-ehealth-to-benefit-patients.

7.5 Summary and Conclusion

There are many security challenges for Information and Communication Technologies (ICT) in the SmartCity domain. Some of them are well-known challenges in computer systems, like fault-tolerance, monitoring and software maintenance. Some are rooted from the networking domain, like traffic analysis, availability especially of wireless communication, and the plethora of problems when it comes to establishing secure communication links. These are magnified due to limited resources of devices in the Internet of Things, like cryptography must be especially efficient (see Sect. 7.2.3). Luckily, for a lot of these, technical and cryptographic solutions exist and are pushed towards general use and are currently pushed towards standardization.

This means that any active new development must reconsider if it has enabled all the latest security functionality, all people involved must stay vigilant and up-to-date with security enhancing tools. In general, no system as big as a SmartCity can be build securely without considering security from the design phase and always enable it as default in each and every subsystem. Still, it needs security professionals to gain the oversight to constantly judge the overall's system. This needs organisational strategies that also allow to update and react on technological changes. If some algorithm or system is known to be insecure it needs to be phased out quickly. Unfortunately, note that for secure system's engineering, the equation *secure subsystem + secure subsystem = secure system* does *not* hold. Adding new or combining existing systems is very likely introducing new security problems.

A good start to address these are the traditional information security goals. There are various models. We covered the 'three pillars' of information security, namely confidentiality, integrity and availability. We as well covered the 'triple A': authentication, authorisation and accountability. With the provided corresponding examples for the SmartCity domain you get a good starting point to discuss these topics with security experts. Table 7.2 provides an overview which challenges discussed in Sect. 7.2 can be addressed by which security and privacy functionality.

Finally, there are very hard technical challenges like identifying and managing the ownership over data. This is also known under the term of data provenance. Please consider that under EU privacy laws it is not the one who gathered or processed the data who legally has a say in what can be done with it; it is the data subject, the citizen, who the data is about or connected to. This will require a usable set of interfaces allowing interactions of ordinary citizens with data collecting and processing backends not yet technically foreseen.

Finally, technically supporting privacy of communications still remains to be adequately addressed. Privacy will come with a performance hit. The potential impact especially for achieving communication privacy is far more drastic than switching on encryption. And privacy is not achieved with just switching on encryption. We showed that the issues of information leakage can render even the best encryption by-passable given sufficient metadata and computing resources. Here, anonymous and unobservable communication helps to minimise leaking information. We outlined Proxies, Onion Routing and DC-Net as potential solutions, whereas the latter

Table 7.2 Mapping which challenges can be addressed by which security and privacy functionality

	Confidentiality	Integrity	Availability	'Triple A'	'End-to end'	Anonymity and unobservability
	Sect. 7.3.1	Sect. 7.3.2	Sect. 7.3.3	Sect. 7.3.4	Sect. 7.3.5	Sect. 7.4
Sect. 7.2.1 Fault-tolerance and fail-safe behaviour		✓	✓			
Sect. 7.2.2 Infrastructure integration, monitoring, updates	✓	✓		✓		
Sect. 7.2.4 Ownership and secure collaboration	✓	✓		✓	✓	
Sect. 7.2.5 Privacy, trust and data minimisation	✓			✓	✓	✓

lacks maturity and requires significant more research. Privacy is also expensive when just used for storing the data in a secure and privacy-preserving fashion, e.g. cryptographic well established methods like Shamir's secret splitting would require at least three replications [65].

In SmartCity environments, we need to deal with all the well-known security challenges. Most of them can be addressed with cryptography. The leakage of metadata can be addressed with anonymity systems.

Be prepared that technically adequate security always comes at a cost. However the hidden costs and dangers to the society as a whole that come from **any** insecurity or privacy-breach of an ICT-supported nerve system of a SmartCity are far greater. Once that the technical systems enabling a SmartCity become (or are just perceived as) an enemy, the general public will start fighting them; this will stall adoption and kill active participation of citizens in their SmartCity. In the end, any system has to offer incentives for collaboration; so you better build secure and privacy-preserving SmartCities that can gain and maintain their citizen's trust.

Acknowledgements H.C. Pöhls and R.C. Staudemeyer were supported by the European Unions 7th Framework Programme (FP7) under grant agreement n° 609094 (RERUM). H. C. Pöhls was also partly supported by the European Unions Horizon 2020 Programme under grant agreement n° 644962 (PRISMACLOUD).

References

1. Gollmann D (2011) Computer security, 3rd edn. John Wiley & Sons
2. Stallings W, Brown L (2014) Computer security: principles and practice, 3rd edn. Pearson Education

3. ISO/IEC (2014) ISO/IEC 27001: Information technology—Security techniques—Information security management systems—Overview and vocabulary. Technical report
4. Mitnick KD, Simon WL (2003) The art of deception: controlling the human element of security. John Wiley & Sons
5. Slay J, Koronios A (2005) Information technology, security and risk management. John Wiley & Sons, Australia Ltd
6. Paul M (2012) The 7 qualities of highly secure software. CRC Press
7. McGraw G (2006) Software security: building security, vol 1. Addison-Wesley
8. Viega J, McGraw G (2001) Building secure software: how to avoid security problems the right way. Addison Wesley
9. Tragos EZ, Pöhls HC, Staudemeyer RC, Slamanig D, Kapovits A, Suppan S, Fragkiadakis A, Baldini G, Neisse R, Langendörfer P, Dyka Z, Wittke C (2015) Securing the internet of things— security and privacy in a hyperconnected world. In: Vermesan O, Friess P (eds) Building the hyperconnected society- internet of things research and innovation value chains, ecosystems and markets. River Publishers Series of Communications. pp 189–219
10. Issarny V, Georgantas N, Hachem S, Zarras A, Vassiliadist P, Autili M, Gerosa MA, Hamida AB (2011) Service-oriented middleware for the future internet: state of the art and research directions. J Internet Serv Appl 2(1):23–45
11. Tragos EZ, Bernabe JB, Staudemeyer RC, Luis J, Ramos H, Fragkiadakis A, Skarmeta A, Nati M, Gluhak A (2016) Trusted IoT in the complex landscape of governance, security, privacy, availability and savety. In: Digitising the industry - internet of things connecting the physical, digital and virtual worlds. River Publishers Series of Communications. pp 210–239
12. Heer T, Garcia-Morchon O, Hummen R, Keoh SL, Kumar SS, Wehrle K (2011) Security challenges in the IP-based internet of things. Wireless Pers Commun 61(3):527–542
13. Weber RH (2010) Internet of things new security and privacy challenges. Comput Law Secur Rev 26(1):23–30
14. Lamport L, Shostak R, Pease M (1982) The Byzantine generals problem. ACM Trans Program Lang Syst 4(3):382–401
15. Cavoukian A (2009) Privacy by design ... take the challenge
16. Gürses S, Troncoso C, Diaz C (2011) Engineering privacy by design. Comput Priv Data Prot 14:25
17. Schneier B (1996) Applied cryptography: protocols, algorithms, and source code in C, 2nd edn. John Wiley & Sons, New York
18. Katz J, Lindell Y (2014) Introduction to modern cryptography, 2nd edn. Chapman & Hall/CRC
19. Danezis G, Clayton R (2007) Introducing traffic analysis. In: Digital privacy: theory, technologies, and practices, pp 1–24
20. Diffie W, Hellman ME, Diffie W, Hellman ME (1976) New directions in cryptography. IEEE Trans Inf Theory 22(6):644–654
21. Rivest RL, Shamir A, Adleman L (1978) A method for obtaining digital signatures and public-key cryptosystems. Commun ACM 21(2):120–126
22. Koblitz N (1987) Elliptic curve cryptosystems. Math Comput 48(177):203–203
23. Miller V (1986) Use of elliptic curves in cryptography. In: Proceedings of advances in cryptology (CRYPTO85). Springer, pp 417–426
24. Hankerson D, Menezes AJ, Vanstone S (2006) Guide to elliptic curve cryptography. Springer Science & Business Media
25. Bock H, Braun M, Dichtl M, Hess E, Heyszl J, Kargl W, Koroschetz H, Meyer B, Seuschek H (2008) A milestone towards RFID products offering asymmetric authentication based on elliptic curve cryptography. Invited talk at RFIDsec
26. Braun M, Hess E, Meyer B (2008) Using elliptic curves on RFID tags. Int J Comput Sci Netw Secur 2:1–9
27. Hein D, Wolkerstorfer J, Felber N (2009) ECC is ready for RFID a proof in silicon. In: Avanzi RM, Keliher L, Sica F (eds) Selected areas in cryptography. Lecture notes in computer science, vol 5381, pp 401–413
28. Municipality of Amsterdam. Amsterdam—SmartCity

29. Efthymiou C, Kalogridis G (2010) Smart grid privacy via anonymization of smart metering data. In: 1st IEEE international conference on smart grid communications, Oct 2010, pp 238–243
30. Jawurek M (2013) Privacy in smart grids. Ph.D. thesis, Friedrich-Alexander-University Erlangen-Nuernberg
31. Lahoti G, Mashima D, Chen W-P (2013) Customer-centric energy usage data management and sharing in smart grid systems. In: Proceedings of the first ACM workshop on smart energy grid security, SEGS '13. ACM, New York, NY, USA, pp 53–64
32. Danezis G, Jawurek M, Kerschbaum F (2011) Sok: privacy technologies for smart grids—a survey of options
33. Mashima D, Roy A (2014) Privacy preserving disclosure of authenticated energy usage data. In: 2014 IEEE international conference on smart grid communications (SmartGridComm), Nov 2014, pp 866–871
34. Pöhls, HC, Karwe M (2014) Redactable signatures to control the maximum noise for differential privacy in the smart grid. In: Cuellar J (ed) Proceedings of the 2nd workshop on smart grid security (SmartGridSec 2014). Lecture notes in computer science (LNCS), vol 8448. Springer International Publishing
35. Peterson W, Brown D (1961) Cyclic codes for error detection. Proc IRE 49(1):228–235
36. Michiels EF (1996) ISO/IEC 10181-6: 1996 Information technology—Open systems interconnection—Security frameworks for open systems: integrity framework. ISO Geneve, Switzerland
37. Clark DD, Wilson DR (1987) A comparison of commercial and military computer security policies. In: 1987 IEEE symposium on security and privacy. Los Alamitos, CA, USA, Apr 1987, pp 184–184
38. Shirey R (2007) RFC 4949–Internet Security Glossary
39. Gollmann D (2012) Veracity, plausibility, and reputation. In: Information security theory and practice. Security, privacy and trust in computing systems and ambient intelligent ecosystems, pp 20–28
40. Gollmann D (1996) What do we mean by entity authentication? In: Proceedings of 1996 IEEE symposium on security and privacy, pp 46–54
41. Goldwasser S, Micali S, Rivest RL (1988) A digital signature scheme secure against adaptive chosen-message attacks. SIAM J Comput 17(2):281–308
42. Turner S, Chen L (2007) RFC 6151–updated security considerations for the MD5 message-digest and the HMAC-MD5 algorithms
43. ISO/IEC (1997) ISO/IEC 13888-1: Information technology—security techniques—non-repudiation, Part 1: General. ISO Geneve, Switzerland
44. World Health Organisation Europe (WHO/E) (2013) Health impact assessment of air pollution in the eight major italian cities, p 65
45. Municipality of Milan. Milan—Area C
46. Camenisch J, Dubovitskaya M, Haralambiev K, Kohlweiss M (2015) Composable and modular anonymous credentials: definitions and practical constructions. In: Lecture notes in computer science (including subseries lecture notes in artificial intelligence and lecture notes in bioinformatics), vol 9453. Springer Verlag, pp 262–288
47. Raymond J-F (2001) Traffic analysis: protocols, attacks, design issues, and open problems. In: Designing privacy enhancing technologies, pp 10–29
48. Fawcett T, Provost F (1996) Combining data mining and machine learning for effective user profiling. Sci Technol 42:8–13
49. Danezis G, Domingo-Ferrer J, Hansen M, Hoepman J-H, Métayer DL, Tirtea R, Schiffner S, Agency (2014) Privacy and data protection by design—from policy to engineering. Technical report, European Union Agency for Network and Information Security, Dec 2014
50. Danezis G, Diaz C (2008) A survey of anonymous communication channels 1–61
51. Song DX, Wagner D, Tian X (2001) Timing analysis of keystrokes and timing attacks on SSH. In: 10th USENIX security symposium 28913:25

52. Dupasquier B, Burschka S, McLaughlin K, Sezer S (2010) Analysis of information leakage from encrypted Skype conversations. Int J Inf Secur 9(5):313–325 Jul
53. Pfitzmann A, Hansen M (2010) A terminology for talking about privacy by data minimization: anonymity, unlinkability, undetectability, unobservability, pseudonymity, and identity management. Technical report
54. Chaum DL (1981) Untraceable electronic mail, return addresses, and digital pseudonyms, Feb 1981
55. Ruiz-Martínez A (2012) A survey on solutions and main free tools for privacy enhancing web communications. J Netw Comput Appl 35(5):1473–1492
56. Goldschlag D, Reed M, Syverson P (1999) Onion routing. Commun ACM 42(2):39–41
57. Dingledine R, Mathewson N, Syverson P (2004) Tor: the second-generation onion router. In: Proceedings of the 13th USENIX security symposium, vol 13. USENIX Association, pp 303–320
58. Chaum D (1988) The dining cryptographers problem: unconditional sender and recipient untraceability. J Cryptology 1(1):65–75
59. Golle P, Juels A (2004) Dining cryptographers revisited. In: Proceedings of advances in cryptology (EUROCRYPT 2004), pp 456–473
60. Waidner M, Pfitzmann B (1990) The dining cryptographers in the disco: unconditional sender and recipient untraceability with computationally secure serviceability. In: Proceedings of the workshop on the theory and application of cryptographic techniques on advances in cryptology (EUROCRYPT '89) 89:690
61. Corrigan-Gibbs H, Ford B (2010) Dissent: accountable anonymous group messaging, p 12
62. Goel S, Robson M, Polte M, Sirer E (2003) Herbivore: a scalable and efficient protocol for anonymous communication. Technical report, Cornell University
63. Guan Y, Fu X, Bettati R, Zhao W (2002) An optimal strategy for anonymous communication protocols. In: Proceedings of the 22nd international conference on distributed computing systems 2002, pp 257–266
64. Stajano F, Anderson R (2000) The cocaine auction protocol: on the power of anonymous broadcast. Inf Hiding 1768:434–447
65. Shamir A (1979) How to share a secret. Commun ACM 22(11):612–613

Chapter 8
IoT Communication Technologies for Smart Cities

Matteo Cesana and Alessandro E.C. Redondi

8.1 Introduction

According to the latest studies, by 2050 70 % of the world population will be living in towns and cities which are responsible of 75 % of GreenHouse Gas (GHG) emissions even if they only cover 2 % of the Earth surface [6, 45]. In this context, the vision of Smart City entails the development of methodologies, solutions, and procedures to improve the efficiency of urban environments and facilitate their sustainable development. Realizing such a vision calls for the active participation of different stakeholders which naturally share/use the urban ecosystem, including city governing bodies, law-makers, utilities, Information and Communication service providers/producers and citizens.

In particular, the capillary use of Information and Communication Technologies (ICT) will provide the backbone for improving the efficiency of existing services and for fostering the creation of new ones in the urban environment. Among the ICT solutions which can make our cities smarter, the Internet of Things (IoT) paradigm is one of the most promising ones [17]. The IoT envisions scenarios where everyday-life objects equipped with sensing peripherals, processing/storage units and communication technologies have a "presence" on the Internet, that is, they can be reachable from the Internet and they can further deliver data up to the Internet on the surrounding environment they are immersed in.

The IoT paradigm finds application in many different domains which are relevant to the vision of Smart Cities including home automation, industrial automation, med-

M. Cesana (✉) · A.E.C. Redondi
Dipartimento di Elettronica, Informazione e Bioingegneria,
Politecnico di Milano, Piazza L. da Vinci 32, 20133 Milan, Italy
e-mail: matteo.cesana@polimi.it

A.E.C. Redondi
e-mail: alessandroenrico.redondi@polimi.it

V. Angelakis et al. (eds.), *Designing, Developing, and Facilitating Smart Cities*,
DOI 10.1007/978-3-319-44924-1_8

ical aids, mobile health care, elderly assistance, intelligent energy management and smart grids, automotive, traffic management, and many others [15].

The concepts of IoT and Smart Cities have become more and more coupled in the last few years. On the one hand, such connection has been stimulated by the strong push from local and national governments to adopt ICT solutions oriented to the urban administration. The possibility of connecting urban objects, resources and services to the Internet in order to facilitate their management and utilization is a great plus for both the citizens and the governments, as it allows for better quality of services for the one and lower administrative costs for the others. On the other hand, Smart Cities offer a perfect application scenario for many IoT solutions: therefore, many technological advancements in different areas related to the IoT paradigm have been motivated, designed, and tested expressly for such a scenario.

This chapter provides a general overview of the main communication technologies in the field of the Internet of Things which can have a beneficial impact in the realization of Smart Cities. We focus here on the available solutions to provide connectivity to/from smart objects, and propose a classification of the different technologies based on the reference network architecture used to "cover" the urban environment; the alternative solutions are critically categorized on the basis of quantitative/qualitative key performance indicators including supported data rate, communication latency, coverage width, cost, flexibility, robustness and maturity/availability/diffusion.

We start off by analyzing the most promising application domains for Smart Cities in Sect. 8.2; Sect. 8.3 provides the reference classification of the most common alternatives of IoT architectures for Smart Cities, which are then described and evaluated in Sects. 8.4–8.6. Finally, Sect. 8.7 reports a discussion on the open challenges of the presented technologies, together with our concluding remarks.

8.2 IoT-Based Services for Smart Cities

We organize the plethora of IoT services/applications for Smart Cities which are envisioned to be implemented in the near future in three main categories, namely (i) smart urban mobility, (ii) services for urban sustainability and (iii) services aimed at enhancing the quality of life of citizens. In the following, we provide details and give examples for each one of these macro-areas.

8.2.1 Smart Urban Mobility

Management and optimization of urban mobility is one of the main challenges that any municipal administration has to face. It includes all the activities related to the management of vehicular traffic within the urban boundaries, with the ultimate goal of allowing easy and smooth mobility to anyone, anywhere and at any time. This requires not only a careful planning of urban spaces devoted to vehicular traffic (i.e., offline management), but also the capacity to quickly operate when needed, in an "online" fashion. Having such a capacity is clearly connected to the availability of real-time data of various types from the urban vehicular environment, and this is where the IoT plays a key role. In the following, we list several IoT applications and services that will be or have already been implemented to support smart urban mobility:

- **Traffic monitoring**: the ability to monitor traffic congestion and detect traffic incidents in real time is crucial for obtaining safer roads and smoother traffic flows. Such capability may be achieved either with the use of statically deployed cameras or other sensors [23], or using real-time measurements coming from the vehicles themselves [25, 33].
- **Smart parking**: cameras or other sensors [27] may be used to monitor the availability of vacant parking lots in the city, in order to direct drivers along the best path for parking. Such a service may produce many benefits, such as lesser traffic congestion, fewer emissions and less stressed citizens.
- **Smart traffic lights**: communications between traffic lights and vehicles may be established to inform the latter of the optimal speed in order to e.g., hit a green light or other important information [44]. Also, specific traffic lights in the city may be controlled in real time to facilitate the mobility of emergency vehicles.

8.2.2 Services for Urban Sustainability

Having "greener" cities have nowadays become not only a good intention, but a global goal regulated by international agreements. While part of the transition to environmentally aware cities will be pursued through fairly easy technological improvements (e.g., switching to energy-efficient LED lighting) and administrative regulations (e.g., creating low-emission zones), the role of information and communication technologies, and the IoT in particular, is of key importance. Several examples falls within the area of environmental-aware IoT services:

- **Smart lighting**: adapting the intensity of public lights to movement of pedestrians and cars may allow for notable energy saving, reduction of light pollution and increased safety. Also, specific sensor for detecting malfunctioning may be installed in order to reduce the maintenance costs. The same ideas may be also applied in indoor scenarios [46].
- **Waste management**: capacity sensors may be used to disseminate the status of each trash bin in the city, and such information may be used to optimize routing and scheduling of vehicles deputed to waste collection [14].
- **Energy consumption monitoring and optimization**: future Smart Cities electricity services will be based on the concept of Smart Grid, where smart meters and devices will operate to control and optimize the production and distribution of

electricity. In this context, the IoT paradigm is expected to play a key role, especially for the integration of customers' premises and appliances with the smart grid owned by power distributors [42].

8.2.3 Services Aimed at Enhancing the Quality of Life of Citizens

Cities are made of citizens, whose quality of life (QoL) is critical for the success of the city itself. The IoT will play a major role in the development of services and applications to enhance the quality of life of citizens. Besides improvements in urban mobility and a cleaner environment, the IoT may enable additional services, such as

- **Noise monitoring**: exposure to excessive noise levels is known to negatively impact the quality of life, producing annoyance, sleep disruption, anxiety and other disturbances. Noise data coming from several sound sensors dislocated in the city may help municipalities in monitoring the level of noise [26].
- **Air quality monitoring**: sensors for monitoring the quality of air and the level of pollution may be deployed in public spaces and such data distributed publicly to citizens [28].
- **Automation of public buildings**: the IoT paradigm may also be employed to implement building automation systems, supporting applications such as electronic devices management and maintenance, energy monitoring, smart rooms and many others [29].

Table 8.1 Qualitative comparison of Smart City services requirements

Application	Coverage	Range [m]	# of devices	Tolerated delay	Rate
Traffic monitoring	Full	~1000	~1000	Minutes	Low
Smart parking	Hotspot	~100	~100	Seconds	Med-high
Smart traffic lights	Full	~10	~1000	Seconds	Med-high
Smart lighting	Full	~1000	~1000	Seconds	Low
Waste management	Full	~1000	~1000	Minutes	Low
Smart grid	Full	~10	~100	Seconds	Med-high
Noise monitoring	Full	~1000	~1000	Minutes	Low
Air quality monitoring	Full	~1000	~1000	Minutes	Low
Home automation	Hotspot	~10	~10	Seconds	Low

Table 8.1 briefly summarizes the requirements of each of the aforementioned services in terms of degree of coverage (full or hotspot), transmission range, number of devices, tolerated delay and produced traffic rate. As one can see, such requirements may vary a lot from case to case, justifying the adoption of different communication technologies, tailored to the particular application scenario. Such communication technologies are detailed in the following sections.

8.3 Interconnecting Objects in Smart Cities: Working Architectures

Urban environments are extremely complex and heterogeneous in terms of available communication technologies and architectures. The plethora of IoT services and applications envisioned for Smart Cities and described in Sect. 8.2 requires the interplay of different communication technologies and different system's architectures. Regardless of the specific application/service, Smart Cities generally require some type of ICT infrastructure to support the exchange of information among the different agents in the urban environment.

As far as the communication is concerned, data must travel from devices which are immersed in the urban environment toward information sinks, and vice versa. Generally speaking, there are three most commonly used ways to realize such communication patterns: (i) through **Cellular Mobile Networks**, (ii) through **IoT-Dedicated Cellular Networks**, (iii) through **Multi-Tier Networks**.

Figure 8.1 reports the main layout for the three different architectures.

In the case of **Cellular Mobile Networks**, the reference architecture is the one of "legacy" mobile radio networks (2G/3G/4G) with a Radio Access Network (RAN) in the front end and a Core Network (CN) at the backhand. The RAN often works over licensed spectrum bands and the CN includes several entities to manage users mobility, registration, etc. As an example, Fig. 8.1a reports some of the entities in the CN of the Long Term Evolution (LTE) system including the Serving Gateway (SGW), the Packet Data Network Gateway (PGW), the Home Subscriber Server (HSS), the Mobility Management Entity (MME) and the Policy and Charging Rules Function (PCRF) server.

Whilst cellular mobile networks are designed to serve primarily human-to-human and human-to-machine traffic, **IoT-Dedicated Cellular Networks** are stand-alone networks dedicated to service only data traffic to/from unmanned field devices. The "last-mile" connection to the field devices is implemented via long-range transmission technologies over unlicensed spectrum bands, and the backhand infrastructure is much simpler than the CN of mobile radio networks.

Multi-Tier Networks feature traffic concentrators or gateways which, on one side, collect the traffic from the field devices through short/medium-range wireless technologies, and on the other side deliver the collected traffic to the backhand via long-range backhauling communication technologies.

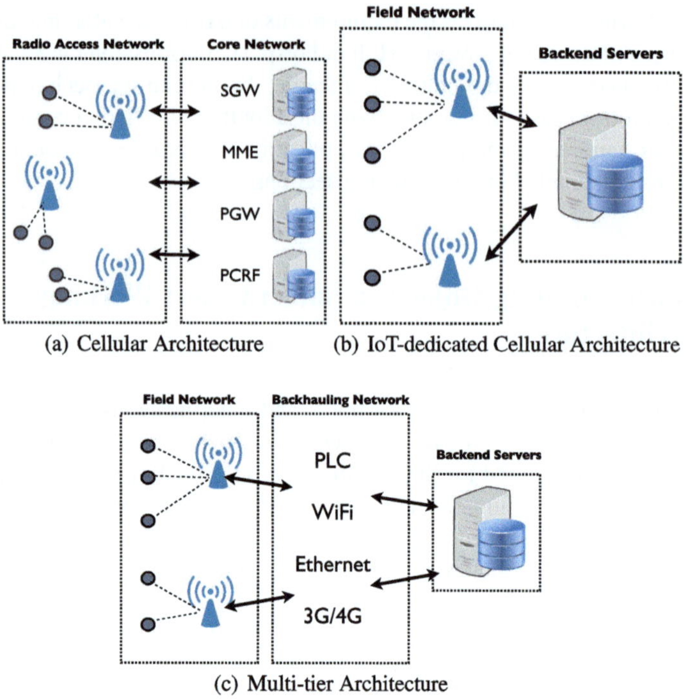

Fig. 8.1 Architectures to support M2M communications in Smart Cities

The following sections describe the main technologies which are available in all the three architectural classes.

8.4 Cellular Mobile Networks

Current cellular mobile networks were mainly designed for human-to-human and human-to-machine interactions targeting specific applications/services like telephony, SMS/MMS exchange, multimedia download and streaming. The device ecosystem of Smart Cities further includes unmanned devices which are immersed in the environment for monitoring and reaction functionalities, thus requiring data exchange capabilities to/from the backend. This de facto defines a new communication paradigm which involve little or no human interaction, and thus it is often referred to as Machine-To-Machine (M2M) communications or Machine-Type Communications (MTCs).

M2M communications are characterized by distinctive features with respect to "legacy" human-to-human communications including larger number of devices, periodic or intermittent network access and small amount of data per device [13].

Although most of M2M communications is currently serviced by legacy 2G cellular technologies (GSM, GPRS), the massive growth of the M2M traffic poses specific challenges both in the RAN and CN of cellular mobile networks [38]. To this extent, efforts are in place to improve the cellular architectures to effectively accommodate M2M communications.

According to the Third Generation Partnership Project (3GPP) which is standardizing the future generation mobile cellular networks, the main distinctive features of MTCs with respect to human-based communication include [11, 24]:

- different market scenarios; MTC can be actually used to support diverse applications in different market fields. Sample use cases include the support for smart metering/smart grid applications, environmental monitoring and crowdsensing applications;
- lower costs and effort: the user equipment must be much cheaper than the "legacy" devices with extreme capabilities in terms of energy efficiency;
- a potentially very large number of communicating terminals;
- to a large extent, little traffic per terminal: MTC mainly required the exchange of very small and intermittent data from the field devices to the network.

On the user equipment's side, the main open issues deal with the cost reduction and the definition of network-assisted power saving functionalities to prolong the device lifetime; on the network's side, the major issues include coverage enhancement, the definition of lightweight signaling procedure for M2M devices to avoid problems of overload and congestion at the radio and core network levels [16, 48] and the study of effective radio resource allocation techniques to manage the interplay between M2M communication and human-to-human ones [50]. To this extent, the technical specification groups of 3GPP has launched several initiatives to define specific modifications to support MTCs in the Global System for Mobile communications (GSM) and the Long-Term Evolution (LTE) standards. Table 8.2 reports an overview of the main features of the upcoming standardization efforts in the field of cellular IoT.

8.4.1 GSM Evolutions

The working groups dealing with the management of the GSM/GPRS and Edge Radio Access Networks (GERAN) are focusing on two complementary approaches to make GSM more efficient for M2M [24]: an *evolutionary* approach and a *clean slate* one; the evolutionary approach targets the modification of the legacy GERAN architecture to increase uplink capacity, extend downlink coverage for both control and data channels, and reduce power consumption/complexity of M2M devices while maintaining full compliancy with the current GERAN structure; two proposals are currently competing within the evolutionary approach, the most promising one, according to the latest plenary meeting of GERAN working groups [36], being the so-called Extended Coverage GSM (EC-GSM). In EC-GSM, the uplink uses Frequency Division Multiple Access overlaid with Code Division Multiple Access, that

Table 8.2 Evolution in the cellular technologies to accommodate M2M/MTC

	Release 8	Release 8	Release 12	Release 12/13	Release 13		
	Cat-4	Cat-1	Cat-0	LTE-M	NB LTE-M	EC-GSM	CS IoT
Spectrum (MHz)	700–900	700–900	700–900	700–900	700–900	800–900	700–900
Channel width	20 MHz	20 MHz	20 MHz	1.4 MHz	200 kHz	200 kHz	5 kHz (UL) 3.75 kHz (DL)
TX Rate DL	150 Mb/s	10 Mb/s	1 Mb/s	200 kb/s	200 kb/s	~300 kb/s[a]	200 kb/s
TX Rate UL	50 Mb/s	5 Mb/s	1 Mb/s	200 kb/s	144 kb/s	≤10 kb/s	~48 kb/s[b]
Duplexing	Full duplex	Full duplex	Half duplex	Half duplex	Half duplex	Half duplex	Half duplex
TX power UL (dBm)	23	23	23	20	23	23–33	≤23
Cost[c]	1.4	1	0.4	0.2	<0.15	<0.15	<0.15
Availability	Available	Available	Available	2016	2016	2016	2016

[a]Peak rate of the EC-PDTCH when base station is transmitting at 43 dBm [24]
[b]Peak rate of the PUSHC with a bonding factor of 8 [24]
[c]scaling factor w.r.t. Release 8 Cat-1

is, in order to allow more devices to transmit at the same time in the same frequency multiplexing based on overlaid code division multiple access technique is proposed to separate the users simultaneously transmitting in the same time slot. Coverage extension for all the transport channels is essentially achieved through blind repetition, that is, the same data block is repeated several times by the transmitter, thus allowing higher receiving gains; different repetition levels are defined based on the coverage class the device belongs to. Other enhancements include definition of new control messages with smaller payload sizes and introduction of a new lower power class.

The *clean-slate* approach targets the re-farming of the GSM spectrum to support a brand new narrowband air interface compatible with GSM channelization of 200 kHz. Four proposals are under investigation, even if the one which seems to be reaching the largest consensus is called NarrowBand Cellular IoT (NB-CIoT) and is based on asymmetric narrow band channels in the downlink and in the uplink; in the downlink, each chunk of 200 kHz is subdivided into 48 narrowband sub-channels of 3.75 kHz width, whereas the uplink defines 36 sub-channels of 5 kHz width. The downlink adopts Orthogonal Frequency Domain Multiple Access modulation, whereas the uplink sub-channels are "assigned" according to a Frequency Division Multiple Access scheme. Sub-channel bonding is further allowed in the uplink to increase the nominal uplink throughput. The reference spectrum bands include the GSM spectrum and the guard bands of the LTE. The base station operates in RF full duplex mode in order to maximize network capacity while the devices operate in half duplex mode to reduce the RF cost.

8.4.2 LTE Evolutions

As far as the evolution of LTE is concerned, LTE Rel-11 has focused on RAN overload functionalities to handle the access of large numbers of M2M devices, and on device power differentiation. The Release 12 of LTE introduces low-cost M2M devices with reduced capability, Category 0 devices, whose cost is approximately 40–50 % of regular LTE Release 8 Cat1 devices. Cost/complexity reduction is mainly achieved by reducing the number of radio transceiver (1 receiving antenna versus 2 receiving antennas of legacy LTE devices), by limiting the maximum transport block sizes (up to 1000 bits per sub-frame) and by further allowing an optional FDD half duplex operation mode. Moreover, to improve the lifetime of Cat 0 devices, a device power saving mode is introduced, which is mainly intended for user equipment with infrequent uplink (mobile-originated) traffic. Devices in power save mode remain registered to the network but are not reachable as they do not check for paging. The device remains in power saving mode until it needs to initiate a "session" toward the network (e.g., issue a tracking area update or start a new uplink transmissions). In addition, scheduling prioritization and service differentiation solutions have been introduced to minimize the impact of MTC data on human-based traffic.

The Release 13 is in the works for better response to M2M requirements leading to the so-called LTE for M2M (LTE-M) [9, 10]. The main improvements at the physical layer include the definition of narrowband channels for transmission of 1.4 MHz and 200 KHz which allow the use of less expensive (and more energy-efficient) hardware at the UE side while improving on the coverage; moreover, features which are already available in the Release 12 are being further improved and extended including an Enhanced Power Saving mode (EPS), and an Extended Discontinuous Reception (DRX) functionality.

8.5 IoT-Dedicated Cellular Networks

IoT-dedicated cellular networks are taking pace to fill in the need of designing low-cost, low-energy M2M applications with limited traffic requirements. IoT-dedicated cellular operators often share the same proposition value which includes reduced energy consumption and Total Cost of Ownership (TCO) with respect to classical cellular operators, global reach and plug-and-play connectivity.

As far as the architecture is concerned, IoT-dedicated cellular networks share a common star topology with base stations serving wide areas and large numbers of unmanned field devices, mostly targeting new uses (smoke alarms, parking sensors, maintenance alerts, environmental monitoring) that have not been viable with GPRS/GSMs higher silicon costs, subscription prices, and power consumption [22]. The different IoT-dedicated cellular technologies mainly differ in the used spectrum band, in the capability to supporting bidirectional traffic and in the

Table 8.3 Comparison of different short-range communication standards for multi-tier IoT architectures

	SigFOX	LoRaWAN		Weightless			Ingenu
		EU	US	-W	-N	-P	
Spectrum	868–902 MHz	863–870 MHz 433 MHz	902–928 MHz	470–790 MHz TV white spaces	Sub GHz (ISM)	Sub GHz (ISM)	2450 MHz
Channel width	100 Hz	125–250 kHz	125–500 kHz	6–8 MHz	200 Hz	12.5 kHz	1 MHz
TX Rate UL	≤100 b/s	250–50 b/s	980 b/s–kb/s	250 b/s–50 kb/s	250 b/s	200 b/s–100 kb/s	624 kb/s
TX Rate DL	256 b/day	250 b/s–50 kb/s	980 b/s–21.9 kb/s	2.5 kb/s–16 Mb/s	None	200 bytes –100 kb/s	156 [kb/s]
Packet size	≤12 bytes	≤222 bytes	≤222 bytes	≥10 bytes	≤20 bytes	≥10 bytes	6 bytes–10 kbytes
Max range (km)	10–50	2–15		5	3	2	100
TX power UL	10 μW–100 mW	14 dBm	20 dBm	17 dBm	17 dBm	17 dBm	20 dBm
Standard (if any)	Proprietary	Standard available	Standard available	Standard available	Standard available	Standard in the works	Proprietary

maturity/availability of the proposed solutions. In the following, we briefly overview the major technologies and commercial solutions which are compared at glance in Table 8.3.

8.5.1 SigFOX

SigFOX uses ultra-narrowband (UNB) radios which are built in the field devices and talk directly to a SigFOX base station according to a star-like network topology [40]. Reliability is enhanced by making each device reachable by multiple base stations. The communication protocols at the Physical and MAC layers are proprietary and leverage 100 Hz channels out of a 200 kHz spectrum around 868 or 902 MHz, depending on the region of use. The communication pattern is mostly uplink (from the field devices to the base stations) with the possibility of activating a tiny downlink channel for control purposes. The data exchange protocol is based on messages with payload up to 12 bytes. The very same message is repeated multiple times over different frequency channels to make reception more robust. The message rate can be customized on the specific application needs with a maximum number of per day, per device messages equal to 140 which leads to a maximum uplink throughput of 100 bps. As for the coverage characteristics, SigFOX transceiver generally feature a maximum output power of 15 dBm with a receiver sensitivity of −126 dBm. The claimed coverage is up to 10 km in urban areas and up to 50 km in rural ones.

SigFOX operates by providing the reference technology including base station development/upgrade, and methods/tools for deployment to SigFOX Network Operators (SNOs) in return of a monthly/yearly fee which depends on the specific traffic and coverage requirements of the reference market segment. The SNOs are usually responsible for the upfront investments to build up, plan and maintain the low power network, as well as the business development in the reference market sector [4]. SigFOX is, at the moment, the market leader in the provision of low-power, low-cost IoT connectivity with partnerships of several SNOs across Europe and the US.

8.5.2 *Weightless*

Weightless Special Interest Group (SIG) [5] is a nonprofit standard organization created to manage the standardization activities of low-power, wide-area technologies. The Weightless system is represented in Fig. 8.2 [7]. Going bottom-up, uplink and downlink transmissions are distinguished at the physical layer through time division duplexing transmissions; downlink transmissions are multiplexed on a time division manner and leverage a Direct Sequence Spread Spectrum (DSSS) approach. The uplink is operated according to a Frequency Division Multiple Access (FDMA), that is, multiple uplink concurrent transmissions may be operated over different noninterference frequency channels. Besides FDMA, the concurrent access of multiple uplink transmissions is mostly managed in a time-scheduled way with the base station notifying the field devices the proper time slots for transmission.

Two modulation/physical layer approaches are further introduced for the uplink: Narrow Band FDMA (NB-FDMA) with a reference channel bandwidth of 200 kHz and Wide Band FDMA (WB-FDMA) with reference channel bandwidth of 6 or 8 MHz.

The data link layer includes different sub-layers

- the *Baseband* which is responsible for multiplexing and de-multiplexing, further managing the send/receive data for the field devices
- the *Lower Link Layer* which is responsible for acknowledgements/retransmissions and data fragmentation/de-fragmentation.

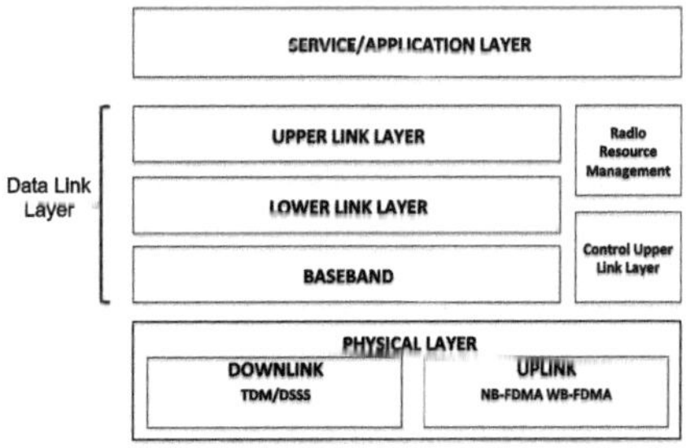

Fig. 8.2 Weightless reference architecture

- the *Upper Link Layer*, responsible for encryption and sequencing and delivery of data to and from the service layer.
- the *Radio Resource Manager* is responsible for managing the Radio resources of the MAC layer, including network configuration.

Differently than SigFox, Weightless system is richer in functionalities at the data link layer as it currently supports acknowledged transmission, data fragmentation/de-fragmentation, multicast transmissions from the base stations and interrupt capabilities which allow devices to raise alarms for specific events such as power outage.

Weightless includes, at the moment of writing, three different solutions for low-power wide-area networks. The **Weightless-W** is designed primarily to operate in unlicensed spectrum including the white space spectrum frequencies between 470 and 790 MHz previously allocated solely for TV broadcast and wireless microphone applications. Unlike 3G and LTE spectrum, these frequencies are not being auctioned by government communications regulators and are being offered license- and cost-free for use.

The **Weightless-N** is the standard version targeting low-cost applications needing only unidirectional data transmission. It operates in sub-GHz spectrum using the NB-FDMA physical layer described above. Pilot networks operated with the Weightless-N standard have been deployed in the cities of London, UK , Copenhagen and Esbjerg, Denmark.[1]

The **Weightless-P** version proposes itself as a solution targeting reliability and performances similar to cellular systems at a fraction of the cost. The standard uses the narrow band modulation scheme offering bidirectional communications capabilities with fully acknowledged two-way communications. The standard is currently in the works; base stations, endpoints and development kits are expected to be available in the second half of 2016.

8.5.3 LoRAWAN

The LoRa Alliance [1] has been created with similar objectives as Weightless SIG, that is, standardizing low-power, wide area communication technologies for the Internet of Things. The reference architecture of LoRa communication protocol stack, *LoRaWAN*, is reported in Fig. 8.3. The standard is frequency-agnostic in the sense that it can operate in different ISM band portions depending on the specific regional rules. The key technology at the physical layer is a proprietary Chirp Spread Spectrum (CSS) modulation scheme which allows to set up bidirectional connections between the end devices and the base stations/gateways [41].

[1]See Weightless SIG web site, press release section, http://www.weightless.org/news/type/press-releases.

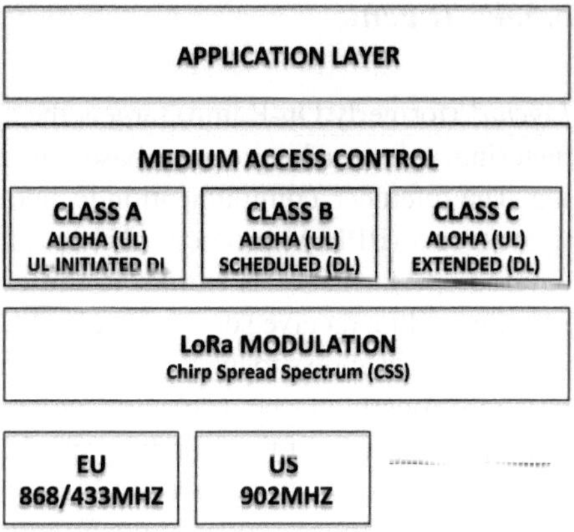

Fig. 8.3 LoraWAN reference architecture

At the Medium Access Control layer, LoRaWAN defines three operation modes which entail different medium access control modes and different balance between uplink and downlink transmission capabilities;

- the Class A operation is the standard baseline meant for the lowest power end device requiring limited downlink communications from the base stations; devices of Class A may initiate uplink transmissions according to an ALOHA-like access protocol; conversely, downlink transmissions from the base station are allowed only in two short receiving time-windows which follow each uplink transmission. Class A devices can optionally require an acknowledgement to their uplink transmissions.
- Class B devices share the same ALOHA-like access protocol for the uplink, but they have additional receiving time windows with respect to class A devices. In Class B operation mode, the base station periodically broadcast a beacon message to synchronize the field devices so that they can schedule in time the required additional receiving time-windows. Devices of Class B can further support downlink multicast transmissions.
- Class C devices share the same ALOHA-like access protocol for the uplink, but they have almost continuous receiving time windows. Like devices of Class B, Class C devices can support multicast transmissions.

Network pilots using LoRa technology are already active in Europe[2] and US.[3]

[2]See press release at http://www.semtech.com/Press-Releases/2015/Semtech-LoRa-based-Internet-of-Things-Wide-Area-Network-to-Deploy-with-Telecom-Operator-Orange.html.

[3]See press release at https://www.semtech.com/Press-Releases/2015/Senet-Deploys-First-Low-Power-Wide-Area-Network-in-North-America-for-IoT-Applications-Based-on-Semtech-LoRaT-RF-Platform.html.

8.5.4 Ingenu

Ingenu[4] (formerly On-Ramp) targets the application segments of smart grids/smart metering, asset tracking, usage-based insurance and critical infrastructure monitoring. The reference communication technology is based on the Random Phase Multiple Access (RPMA) protocol [35] operating in the unlicensed 2.4 GHz band. The coverage performance is similar to those of the other IoT-Dedicated cellular technologies with a receive sensitivity of -142 dBm and a maximum transmit power of 20 dBm which allows to have up to 400 square miles of coverage with a single basestation if properly placed. Capacity can be tens of thousands of devices per basestation. Ingenu is one of the founding members of the IEEE 802.15.4k standardization working group which is currently working to extend the IEEE 802.15.4 PHY and MAC layer for the support of Low Energy Critical Infrastructures (LECIM) [8].

8.6 Multi-tier Architectures

Differently from the cellular scenario, multi-tier architectures are characterized by a layered design in which Things are used both to sense data and to form the network infrastructure, in a multi-hop/mesh fashion. Data collected from such devices is then generally forwarded to a central collection point (gateway, concentrator), which then conveys such data to the Internet through other technologies. Multi-hop transmission is generally needed to compensate for the limited communication range achievable by the radio technologies used in such scenarios. On the one hand this is a consequence of the extremely low power consumption exhibited by such solutions, which is key for certain applications. On the other hand, the limited radio range may cause to use more devices than what is actually needed, just for ensuring connectivity. In the following, we give details on the main technologic solutions proposed so far in the field of short-range, multi-tier architectures for supporting IoT applications in Smart City scenarios. The main features of each solution are compared in Table 8.4.

8.6.1 Solutions Based on IEEE 802.15.4

The IEEE 802.15.4 standard, specified for wireless personal area networks, offers the fundamental lower network layers for low-cost, low-rate, and low-power communication. The standard specifies only the PHY and MAC layers of the protocol stack: at the physical layer, three unlicensed frequency bands may be used (868/915/2450 MHz). Originally, the direct sequence spread spectrum (DSSS) modulation scheme was specified, allowing a data rate of 20, 40 and 250 kbps for the three bands, respectively. The 2006 revision improved the data rates in the 868/915 bands to 100 and

[4]http://www.ingenu.com/.

Table 8.4 Comparison of different short-range communication standards for multi-tier IoT architectures

Standard	Frequency bands	Max Tx rate	Max range (m)	TX power	Application
ZigBee (802.15.4)	868/915/2450 MHz	250 kbps	100	1–100 mW	Home automation Backhaul for WSN
WI-SUN (802.15.4g)	sub-1 GHz, 2.4 GHz	1 Mpbs	200	1–100 mW	Home automation Backhaul for WSN
ULP (802.15.4q)	868/915/2450 MHz	100 kbps	100	5–15 mW	Ultra low power applications
Wireless M-Bus	169/433/868 MHz	100 kbps	300	1–100 mW	Metering
Z-Wave	908 MHz	100 kbps	100	1–100 mW	Home automation
Bluetooth Low Energy (BLE)	2450 MHz	1 Mbps	30	1–100 mW	e-Health, Sport, Multimedia
WiFi Low Power (802.11ah)	Sub-1 GHz	7.8 Mbps	1000	10 mW–1 W	Long range WSN Backhaul for WSN

250 kbps, respectively. Other amendments were made to the standard in the following years, all targeted to expand the available PHYs with several additions. The MAC layer employs the CSMA-CA mechanism for channel access and is responsible for maintaining the connectivity (beacons transmission and synchronization, PAN association/disassociation, etc.). The frame size is generally 127 bytes. On top of the PHY and MAC layers defined by the IEEE 802.15.4 standard, several solutions have been proposed to enable communication between smart devices, which are briefly addressed in the following.

8.6.1.1 ZigBee

Zigbee [2] is probably the most known high-level communication protocol based on IEEE 802.15.4. It supports star, tree and mesh topologies, and two types of devices. The coordinator (full-function device, FFD) is responsible for maintaining the network, composed by routers and end devices (reduced-function devices, RFD). ZigBee supports both non-beacon and beacon-enabled networks. In the former type, medium access is achieved through the IEEE 802.15.4 CSMA-CA mechanism. In the latter, beacons are used to schedule the transmissions of network nodes, thus lowering their duty cycle and consequently extending their battery. At the application layer, ZigBee also includes methods for secure communication, such as key establishment and transport and frame protection.

8.6.1.2 6LoWPAN

6LoWPAN (IPv6 over Low Power PAN) specifies a set of rules to apply the IP protocol to low-power devices for the Internet-of-Things. Clearly, such integration allows for easy interoperability with other types of IP-enabled devices (e.g., WiFi based) and the Internet. Mapping the IP network layer to the 802.15.4 lowest layers requires several functionalities, all provided by 6LoWPAN: packet size adaptation, header compression, address resolution and management, routing and security.

8.6.1.3 802.15.4 Amendments and Other Protocols

The IEEE 802.15.4 standard has been used as starting point for several working solutions, and it is still being refined in order to support full interoperability.

Examples include WirelessHART and MiWi. The former uses the 802.15.4 PHY layer and redefines the upper layer. In particular, it is based on a TDMA protocol and allows to create self-organizing and self-healing mesh networks [20]. The latter is a trimmed-down, economical version of ZigBee, proprietary of MicroChip, which uses low data rates and very short communication distances [21].

It is also worth mentioning two amendments of the IEEE 802.15.4 protocol, namely WI-SUN and Ultra Low Power (ULP). WI-SUN is under study by the 802.15.4g Task Group, and focuses on Smart Utility Networks (SUN) with the objective to provide a standard that facilitates very large-scale process control applications. In particular, 802.15.4g includes operation in ISM bands (700 MHz–1 GHz and the 2.4 GHz band), data rates from 40 kbps to 1 Mbps and a PHY frame size up to a minimum of 1500 octets to support IP packets without fragmentation. Different multi-rate and multi-regional (MR) PHYs are specified, in order to ensure interoperability with existing systems [19]. The Ultra Low Power version (ULP, 802.15.4q Task Group) explicitly focuses on ultra low power applications, with a target peak power consumption for the PHY layer of maximum 15 mW.

8.6.2 Z-Wave

Developed by the Z-Wave alliance [3], this protocol defines all layers of the protocol stack and targets mainly home automation applications. Z-Wave operates at 908 MHz and uses GFSK encoding as modulation scheme. Different data rates are available (9.6/40/100 kbps) and the communication range is comparable to 802.15.

4-based solution (tens of meters). Similarly to ZigBee, Z-Wave utilizes a mesh network architecture and provides basic routing and security functionalities. Differently from the "open" 802.15.4, Z-Wave is a proprietary system made and licensed by one single company (Sigma Designs): this is not necessarily a drawback, since the tight control on how devices should communicate may facilitate interoperability between products from different vendors.

8.6.3 Wireless M-Bus

The Metering Bus (M-Bus) is a field bus specialized for transmission of metering data from gas, electricity, heat, and other meters to a data collector. The Wireless M-Bus is a radio variant of M-Bus: it can work within three bands (169/433/868 MHz) and allows the creation of star and mesh topologies with the help of a time synchronized TDMA source routing protocol. Data transmission rate can be as high as 100 kbps for a communication range up to 300 m. Many off-the-shelf commercial products based on such protocol are already available on the market with a claimed lifespan of more than 10 years with a single battery.

8.6.4 Bluetooth Low Energy

Stimulated by the popularity Bluetooth recently enjoyed in the field of audio streaming, Bluetooth Low Energy (BLE, also called Bluetooth Smart) was introduced in 2010 to be suitable for M2M and IoT applications. As its name says, the main focus is on the reduction of the power consumption so that such protocol may be used in battery-powered devices for a long period of time. BLE uses GFSK modulation with rate data rate of 1 Mbps in the 2.4 GHz ISM band. 40 different channels are available, divided in 3 advertising channels (carefully chosen in order to minimize interference with WiFi) and 37 data channels. The BLE protocol stack is tailored to easy integration with IPv6, supporting packet fragmentation and providing basic security primitives. The biggest drawback of BLE is that it supports only star topology (not mesh networks), therefore limiting its application to real-life scenarios. Recently, the Bluetooth SIG launched a study group to define an industry standard BLE mesh protocol. This should close the gap between BLE and mesh capable protocols such as IEEE 802.15.4 and Z-Wave. [18]

8.6.5 WiFi Low Power

IEEE 802.11ah operates in the sub 1 GHz band (900 MHz) and provides extended range WiFi networks with an eye on reducing power consumption. Therefore, it is particularly tailored to IoT and M2M applications. At the PHY layer, 802.11ah uses OFDM-based waveforms and supports BPSK, QPSK and 16 to 256-QAM modulations. This allows to have data rates from 150 kbps to nearly 8 Mbps. The MAC layer is designed to maximize the number of connected devices (up to 8191) and includes power saving modes to reduce the energy consumption by deactivating the radio module during non-traffic periods. The protocol also includes optimizations for small data transmission and long sleeping periods [12, 32]. Due to their large coverage, IEEE 802.11ah networks may be also used as backhaul, acting as an intermediate step between device (e.g., 802.15.4 nodes) and data collectors.

8.6.6 Gateway-to-Internet

As mentioned before, multi-tier architectures generally deliver data from Things (sensing domain) to a central collection point which bridges such data to the Internet (network domain). Such gateway should have specific features, such as support for multiple sensing domain protocols (e.g., ZigBee, Z-Wave, BLE, etc.), protocol translation and conversion and easy manageability. Connection to the Internet may be provided using different technologies, namely (i) classic access through an ISP, (ii) access through cellular architecture, or (iii) access through Power Line Communication (PLC).

8.6.7 Multi-tier Network Testbeds and Realizations

In parallel with the development and standardization of the different IoT communication technologies mentioned in the previous sections, in the last few years there has been an increasing interest for the realization of demonstrators and testbeds of IoT solutions for the Smart City scenario.

Probably, the most interesting example is given by the *SmartSantander* project [39], which propose a unique city scale experimental research facility. The testbed, deployed in the city of Santander, is composed of around 3000 IEEE 802.15.4-compliant devices and 200 devices (mostly mobile) with GPRS communication capabilities. Such devices are used to test different use cases developed within the project, including static and mobile environmental monitoring, parking management, traffic monitoring and irrigation management. These applications share a three-tiered architecture in which 802.15.4-compliant nodes transmit the sensed information to gateways equipped with several communication interfaces (IEEE 802.15.4, WiFi, GPRS, Bluetooth and Ethernet). Such gateways have either a local database accessible remotely or transmit all the data to a central server using Internet connection. Such a testbed has been developed not only for demonstrating the benefits of IoT solutions in a smart city scenario, but also for giving researchers the possibility of testing experiments (e.g., routing protocols) with the deployed nodes. The project envisions the deployment of a total of 20,000 sensors in Santander, Belgrade, Guildford and Lbeck, exploiting a large variety of technologies.

The *Padova Smart City* project [47] uses IEEE 802.15.4-compliant nodes placed on streetlight poles and connected to the network of the city municipality by means of a gateway. Each node is equipped with different sensors, including photometer, temperature, humidity, and benzene sensors for monitoring the environment. Data is delivered to the gateway using 6LoWPAN and the RPL routing protocol. Several considerations and inferences were possible from the analysis of the collected data.

The *Smart Berlin Testbed* [30] is composed by nearly 300 IEEE 802.11-compliant nodes organized in a mesh topology and support WSNs operated by 6LoWPAN. The testbed is remotely manageable and has been used to perform white space detection in the area of deployment.

Finally, an interesting example is given by the work in [31], where a city scale mobile sensing infrastructure that relies on bicycles is proposed. The *NITOS BikesNet* architecture consists of fixed gateways (WiFi access points and custom-made ZigBee gateways). Bikes in the city are equipped with both WiFI and Zigbee interfaces and several sensors (GPS, temperature humidity and light-intensity). All data is stored locally on the bike memory and delivered to the gateways when in range. The testbed has been implemented in the city of Volos (Greece) and used to populate a database of the available WiFi networks in the city.

8.7 Discussion and Concluding Remarks

Smart Cities are complex environments with diverse applications, stakeholders, and governing bodies which lead to the coexistence of different business propositions and value chains for different services. Such complexity and heterogeneity is reflected also in the technological offer to support wide are coverage and connectivity in urban environment, which is vast and diverse. In the following, we summarize the high-level features and discuss the main challenges of the three architectural alternatives explained in the previous chapters:

- **Cellular Mobile Networks** are particularly fit for Smart City applications requiring high coverage and flexibility in terms of supported data rate. Moreover, cellular architectures can leverage the embedded support for worldwide mobility and security. On the other hand, the integration of MTC into cellular architectures opens up new technical challenges including the differentiation of traffic at the RAN and CR and the management of massive access loads in the RAN and the CN. In terms of nontechnical challenges, the standardization activities to support massive MTC in mobile cellular networks are relatively "young", due to the unavoidable inertia the mobile operators have in enhancing their complex network architectures and technologies to accommodate tiny M2M traffic. Such inertia has opened up business opportunities for different technological solutions to support M2M communication which have quickly spread out in the last few year and will be described in the next section.

- **IoT-dedicated Cellular Network** operators have a clear time advantage over cellular operators in offering IoT-specialized connectivity solutions. Cellular IoT architectures are in general characterized by lower costs, both in the network equipment and in the network devices, compared to classical cellular IoT architectures, which, at the moment, allows them to be extremely aggressive in their business models. On the other side, IoT-dedicated architectures generally target low-rate applications with highly customized network deployment with scarce flexibility. Moreover, IoT-dedicated architectures are often asymmetric in the supported channel rate at the air interface, with limited downlink channels.
- **Multi-tier Architectures** are particularly tailored to those applications characterized by limited number of nodes and low communication range. Such requirements are typically encountered in indoor scenarios like home automation or industrial control/metering, where multi-tier architectures are generally preferred to cellular-based solutions. On the one hand, multi-tier architectures allow for very flexible setups, easily customizable based on the customer's needs (in terms of transmission rate, delay and power consumption). On the other hand, the low transmission range sometimes constitutes a drawback, and more devices than needed have to be installed just to provide the required communication coverage. The standards presented in the previous sections all constitute a possible solution for implementing personal area networks (PAN), which form the basis of IoT applications for Smart Buildings and Smart Homes. Solutions based on IEEE 802.15.4 (e.g., ZigBee) and Z-Wave, which were specifically designed to overcome the power and range limitations of traditional WiFi and Bluetooth solutions, are still struggling to find their way on the market and to become a widely used standard. At the time of writing, no clear winner is emerging in such a battle of standards. On the one hand Z-Wave, being controlled by a single company, allows for easy interoperability between different products and it is thus very attractive to manufacturers. Also, working in the 900 MHz band allows to reduce the number of collisions and transmission retries compared to the 2.4 GHz band and that may translate in lower power consumption. On the other hand, the IEEE 802.15.4 open-standard has clear advantages (e.g., global standardization, products can be made by a variety of manufactures, etc.) but still lacks full interoperability, although several efforts are being made in this direction (e.g., ZigBee 3.0, will provide seamless interoperability among products from different manufacturers). Between the two dogs striving for the bone, WiFi and Bluetooth are trying to close the gap with their low-power versions. Interestingly, the solutions proposed by mobile giants such as Samsung, Apple and Google do not give any hint on the final outcome of such battle: Samsung's Artik chip supports WiFi, Bluetooth Low Energy, Zigbee but not Z-Wave; Google's Thread standard is based on 802.15.4 and 6LoWPAN, while Apple's HomeKit proposes a completely different solution based on either WiFi or Bluetooth Low Energy, so that all smart devices can be controlled directly from a smartphone without the need of installing a hub or an additional radio interface. In the long run, it is unclear which standard will emerge as a clear winner, and it is possible that they will coexist for a long period, making product developers continually reevaluate which wireless standard is the best for their needs.

As it emerges from our previous discussion, it is likely that diverse communication technologies and architectures for IoT in Smart Cities will forcedly coexist in the same environment serving different subsets of applications. In such ecosystem, besides the challenges related to the improvement of the specific communication technologies which have been already discussed in the previous sections, the additional challenge will be to exploit such coexistence to make smart cities even smarter. In this view, three factors will likely play a key role:

- the definition of **unifying architectures** to orchestrate the interplay among different communication technologies through the definition of proper abstraction layers that can be readily embedded within various hardware and software, and relied upon to connect the myriad of devices in the field with Smart Cities. Efforts in this respect are already in place in the research community and in standardization bodies [37, 43];
- the **interconnectedness** between applications and services operated by diverse stakeholders through different communication technologies/architectures; related to the previous item, data coming from diverse Smart City applications and services should be exposed and made available to foster the creation of novel composed value added services through proper **programming interfaces** [34];
- the availability of **easy-to-use management platforms** to build-up novel applications for Smart Cities [49]; the final users of smart city applications are heterogeneous and diverse (citizens, group of citizens, governing bodies, law enforcement bodies) which call for different types of interaction with the application/service itself; as an example, citizen-oriented applications may require simple but effective data visualization plug-ins, whereas, services targeting urban efficiency and sustainability may require, besides data visualization, advanced data analytics and business intelligence tools. To this extent, the design and availability of management platforms for Smart Cities will be central.

Acknowledgements This work has been partially supported by the Italian Ministry for Education, University and Research (MIUR) through the national cluster project SHELL, Smart Living technologies (grant number: CTN01 00128 111357).

References

1. https://www.lora-alliance.org/
2. https://www.zigbee.org/
3. http://www.z-wave.com/
4. http://www.sigfox.org
5. www.weightless.org
6. United Nations Secretary-Generals High-Level Panel on Global Sustainability (2012) Resilient people. A future worth choosing, Resilient Planet
7. Weightless System Specifications, v0.9 (2012)
8. IEEE Standard for Local and Metropolitan Area Networks Part 15.4 (2013) Low-rate wireless personal area networks (lr-wpans)amendment 5: physical layer specifications for low energy, critical infrastructure monitoring networks. IEEE P802.15.4k/D5, Apr 2013, pp 1–152

9. Study on Provision of Low-Cost Machine-Type Communications (MTC) User Equipments (UES) based on lTE. 3GPP TR 36.888 (2013)
10. Technical Specification Group Services and System Aspects; Service Requirements for Machine-Type Communications (MTC); stage 1. 3GPP TS 22.368 (2014)
11. Technical Specification Group Services and System Aspects; Architecture Enhancements to Facilitate Communications with Packet Data Networks and Applications. 3GPP TS 23.682 (2015)
12. Adame T, Bel A, Bellalta B, Barcelo J, Oliver M (2014) IEEE 802.11ah: the wifi approach for M2M communications. IEEE Wirel Commun 21(6):144–152. doi:10.1109/MWC.2014.7000982
13. Akyildiz IF, Gutierrez-Estevez DM, Balakrishnan R, Chavarria-Reyes E (2014) LTE-advanced and the evolution to beyond 4G (B4G) systems. Phys Commun 10:31–60. doi:10.1016/j.phycom.2013.11.009. http://www.sciencedirect.com/science/article/pii/S1874490713000864
14. Anagnostopoulos T, Zaslavsky A, Medvedev A (2015) Robust waste collection exploiting cost efficiency of iot potentiality in smart cities. In: 2015 International conference on recent advances in internet of things (RIoT), pp 1–6. doi:10.1109/RIOT.2015.7104901
15. Atzori L, Iera A, Morabito G (2010) The internet of things: a survey. Comput Netw 54(15):2787–2805. doi:10.1016/j.comnet.2010.05.010. http://dx.doi.org/10.1016/j.comnet.2010.05.010
16. Biral A, Centenaro M, Zanella A, Vangelista L, Zorzi M (2015) The challenges of M2M massive access in wireless cellular networks. Digit Commun Netw. doi:10.1016/j.dcan.2015.02.001. http://www.sciencedirect.com/science/article/pii/S235286481500005X
17. Borgia E (2014) The internet of things vision: key features, applications and open issues. Comput Commun 54, 1–31. doi:10.1016/j.comcom.2014.09.008. http://www.sciencedirect.com/science/article/pii/S0140366414003168
18. Chang KH (2014) Bluetooth: a viable solution for IoT? [industry perspectives]. IEEE Wirel Commun 21(6):6–7. doi:10.1109/MWC.2014.7000963
19. Chang KH, Mason B (2012) The IEEE 802.15.4G standard for smart metering utility networks. In: 2012 IEEE Third international conference on smart grid communications (SmartGridComm), pp 476–480. doi:10.1109/SmartGridComm.2012.6486030
20. Chen D, Nixon M, Han S, Mok A, Zhu X (2014) Wirelesshart and IEEE 802.15.4E. In: 2014 IEEE International conference on industrial technology (ICIT), pp 760–765. doi:10.1109/ICIT.2014.6895027
21. Chhajed S, Sabir M, Singh K (2014) Wireless sensor network implementation using miwi wireless protocol stack. In: 2014 IEEE International advance computing conference (IACC), pp 239–244. doi:10.1109/IAdCC.2014.6779327
22. Evans-Pughe C (2013) The M2M connection. Eng Technol 8(11):39–43. doi:10.1049/et.2013.1102
23. Foschini L, Taleb T, Corradi A, Bottazzi D (2011) M2M-based metropolitan platform for IMS-enabled road traffic management in IoT. IEEE Commun Mag 49(11):50–57. doi:10.1109/MCOM.2011.6069709
24. 3rd Generation Partnership Project (3GPP) T.S.G.G.R.A.N. (2015) Cellular system support for ultra-low complexity and low throughput internet of things (CIoT) (release 13)
25. Gerla M, Lee EK, Pau G, Lee U (2014) Internet of vehicles: from intelligent grid to autonomous cars and vehicular clouds. In: 2014 IEEE World forum on internet of things (WF-IoT), pp 241–246. doi:10.1109/WF-IoT.2014.6803166
26. Gubbi J, Marusic S, Rao A, Law YW, Palaniswami M (2013) A pilot study of urban noise monitoring architecture using wireless sensor networks. In: 2013 International conference on advances in computing, communications and informatics (ICACCI), pp 1047–1052. doi:10.1109/ICACCI.2013.6637321
27. He W, Yan G, Xu LD (2014) Developing vehicular data cloud services in the IoT environment. IEEE Trans Ind Inf 10(2):1587–1595. doi:10.1109/TII.2014.2299233
28. Hromic H, Le Phuoc D, Serrano M, Antonic A, Zarko IP, Hayes C, Decker S (2015) Real time analysis of sensor data for the internet of things by means of clustering and event processing.

In: 2015 IEEE International conference on communications (ICC), pp 685–691. doi:10.1109/ICC.2015.7248401
29. Jung M, Reinisch C, Kastner W (2012) Integrating building automation systems and IPv6 in the internet of things. In: 2012 Sixth international conference on innovative mobile and internet services in ubiquitous computing (IMIS), pp 683–688. doi:10.1109/IMIS.2012.134
30. Juraschek F, Zubow A, Hahm O, Scheidgen M, Blywis B, Sombrutzki R, Gunes M, Fischer J (2012) Towards smart berlin—an experimental facility for heterogeneous Smart City infrastructures. In: 2012 IEEE 37th Conference on local computer networks workshops (LCN Workshops), pp 886–892. doi:10.1109/LCNW.2012.6424078
31. Kazdaridis G, Stavropoulos D, Maglogiannis V, Korakis T, Lalis S, Tassiulas L (2014) NITOS BikesNet: enabling mobile sensing experiments through the OMF framework in a city-wide environment. In: 2014 IEEE 15th International conference on mobile data management (MDM), vol 1, pp 89–98. doi:10.1109/MDM.2014.17
32. Khorov E, Lyakhov A, Krotov A, Guschin A (2015) A survey on IEEE802.11ah: an enabling networking technology for smart cities. Comput Commun 58:53–69. doi:10.1016/j.comcom.2014.08.008. http://www.sciencedirect.com/science/article/pii/S0140366414002989. Special Issue on Networking and Communications for Smart Cities
33. Li X, Shu W, Li M, Huang HY, Luo PE, Wu MY (2009) Performance evaluation of vehicle-based mobile sensor networks for traffic monitoring. IEEE Trans Veh Technol 58(4):1647–1653. doi:10.1109/TVT.2008.2005775
34. Liu J, Li Y, Chen M, Dong W, Jin D (2015) Software-defined internet of things for smart urban sensing. IEEE Commun Mag 53(9):55–63. doi:10.1109/MCOM.2015.7263373
35. Myers TJ (2010) Random phase multiple access system with meshing
36. Network T.S.G.G.R.A. (2015) Draft report of tsg geran meeting 67, version 0.0.1
37. ONEM2M E (2015) Functional architecture
38. Ratasuk R, Prasad A, Li Z, Ghosh A, Uusitalo M (2015) Recent advancements in M2M communications in 4G networks and evolution towards 5G. In: 2015 18th International conference on intelligence in next generation networks (ICIN), pp 52–57. doi:10.1109/ICIN.2015.7073806
39. Sanchez L, Muoz L, Galache JA, Sotres P, Santana JR, Gutierrez V, Ramdhany R, Gluhak A, Krco S, Theodoridis E, Pfisterer D (2014) Smartsantander: IoT experimentation over a smart city testbed. Comput Netw 61:217–238. doi:10.1016/j.bjp.2013.12.020. http://www.sciencedirect.com/science/article/pii/S1389128613004337. Special issue on Future Internet Testbeds Part I
40. SigFOX. M2M and IoT redefined through cost effective and energy optimized connectivity
41. Sornin N, Luis M, Eirich T, Kramp T, Hersent O (2015) Lorawan specification
42. Spano E, Niccolini L, Di Pascoli S, Iannacconeluca G (2015) Last-meter smart grid embedded in an internet-of-things platform. IEEE Trans Smart Grid 6(1):468–476. doi:10.1109/TSG.2014.2342796
43. Swetina J, Lu G, Jacobs P, Ennesser F, Song J (2014) Toward a standardized common M2M service layer platform: introduction to oneM2M. IEEE Wirel Commun 21(3):20–26. doi:10.1109/MWC.2014.6845045
44. Tielert T, Killat M, Hartenstein H, Luz R, Hausberger S, Benz T (2010) The impact of traffic-light-to-vehicle communication on fuel consumption and emissions. Internet of Things (IoT) 2010:1–8. doi:10.1109/IOT.2010.5678454
45. (WHO) W.H.O. (2010) Urbanization and health. Bull World Health Organ 88(4):245–246
46. Zaidi S, Imran A, McLernon D, Ghogho M (2014) Enabling IoT empowered smart lighting solutions: a communication theoretic perspective. In: 2014 IEEE Wireless communications and networking conference workshops (WCNCW), pp 140–144. doi:10.1109/WCNCW.2014.6934875
47. Zanella A, Bui N, Castellani A, Vangelista L, Zorzi M (2014) Internet of things for smart cities. IEEE Internet of Things J 1(1):22–32. doi:10.1109/JIOT.2014.2306328
48. Zanella A, Zorzi M, dos Santos A, Popovski P, Pratas N, Stefanovic C, Dekorsy A, Bockelmann C, Busropan B, Norp T (2013) M2M massive wireless access: challenges, research issues, and ways forward. In: 2013 IEEE Globecom workshops (GC Wkshps), pp 151–156. doi:10.1109/GLOCOMW.2013.6824978

49. Zaslavsky A, Georgakopoulos D (2015) Internet of things: challenges and state-of-the-art solutions in internet-scale sensor information management and mobile analytics. In: 2015 16th IEEE International conference on mobile data management (MDM), vol 2, pp 3–6. doi:10.1109/MDM.2015.72
50. Zheng K, Hu F, Wang W, Xiang W, Dohler M (2012) Radio resource allocation in LTE-advanced cellular networks with M2M communications. IEEE Commun Mag 50(7):184–192. doi:10.1109/MCOM.2012.6231296

Chapter 9
Cloud Internet of Things Framework for Enabling Services in Smart Cities

Dimitrios Kelaidonis, Panagiotis Vlacheas, Vera Stavroulaki, Stylianos Georgoulas, Klaus Moessner, Yuichi Hashi, Kazuo Hashimoto, Yutaka Miyake, Keiji Yamada and Panagiotis Demestichas

9.1 Introduction

The contemporary era of the future internet has already been deployed, in the form of innovative Internet of Things (IoT) infrastructures, in different application domains that range from smart health and smart buildings to smart cities. Focusing on the smart cities domain, by 2030 it is anticipated that more than 27 large cities with more than ten million of population will exist, while sixty percent (60 %) of the world population will live in an urban environment [1]. Major challenges will be presented due to the increase of cities' population and some of the most important of them will refer to energy and waste management, citizens' health status monitoring, traffic management and air pollution monitoring and control. Such challenges will require the combination of existing heterogeneous Information and Communication

D. Kelaidonis (✉) · V. Stavroulaki · P. Demestichas
Department of Digital Systems, University of Piraeus, Piraeus, Greece
e-mail: dkelaid@unipi.gr

V. Stavroulaki
e-mail: veras@unipi.gr

P. Demestichas
e-mail: pdemest@unipi.gr

P. Vlacheas
WINGS ICT Solutions, Athens, Greece
e-mail: panvlah@wings-ict-solutions.eu

S. Georgoulas · K. Moessner
University of Surrey, Guildford, UK
e-mail: s.georgoulas@surrey.ac.uk

K. Moessner
e-mail: k.moessner@surrey.ac.uk

V. Angelakis et al. (eds.), *Designing, Developing, and Facilitating Smart Cities*,
DOI 10.1007/978-3-319-44924-1_9

Technologies (ICT) solutions, so as to provide innovative and more efficient composite services to the people that will live in these cities [2, 3].

The application of the IoT paradigm in the context of smart cities implies a wide range of heterogeneous devices such as sensors, actuators, smartphones, cars, computers, home appliances, buildings, smart city infrastructure elements, that will be connected to the internet, as well as with each other via heterogeneous access networks, with the aim of providing smart, personalized applications and services any-time, anywhere [4]. A vast variety of IoT applications is envisaged; however most current developments have mainly focused on specific applications, leading to domain-centric silos that lack flexibility and interoperability. Given the size and functional needs of the true IoT vision and especially, of large-scale IoT applications such as those deployed/required in smart cities, there are key challenges to be solved to achieve ubiquity, scalability, dependability and sustainability of real-time IoT service provision. The vast amount of objects and devices that have to be handled and the variety of networking and communication technologies, as well as the administrative boundaries that should be supported, require a different management approach. The added value of IoT consideration consists not only in providing the objects with the infrastructure for physical communication, but also in exchanging derived data and knowledge, which can be (re-)used by other objects and applications and, therefore, making possible the development of totally new or already existing applications but with more enhanced features/functionalities [5, 6].

The last decade, various fundamental technologies have been introduced and used in various domains so as to realize the IoT [7], including technologies such as the Radio Frequency Identification (RFID) [8, 9] and Near Field Communications (NFC) [10, 11], Wireless Sensor Networks (WSN) [12], etc. As it has been highlighted in previous research activities [13–17], the IoT paradigm of the "7 trillion devices for 7 billion people" requires cognitive self-x (configuration, optimization, healing) solutions that will effectively contribute to the overcoming of the various issues that arise. Moreover, whereas various research efforts, such as [17–27] have significantly contributed to the definition of IoT architectures and cognitive communication technologies to ensure interactions and facilitate information exchange, there has been less focus on the distribution of the solutions that will enable greater

Y. Hashi
Hitachi Solutions East Japan, Ltd., Sendai, Japan
e-mail: yuichi.hashi.wg@hitachi-solutions.com

K. Hashimoto
Waseda University Tokyo, Tokyo, Japan
e-mail: kazuo.hashimoto@aoni.waseda.jp

Y. Miyake
KDDI R&D Laboratories Inc., Fujimino, Japan
e-mail: miyake@kddilabs.jp

K. Hashimoto · K. Yamada
Kokusai Kogyo Co., Ltd., Tokyo, Japan
e-mail: keiji_yamada@kk-grp.jp

scalability and extensibility of the capabilities of IoT infrastructures and relevant data processing. Furthermore, there is a lack of distributed management functionality to overcome the technological heterogeneity of the capillary networks, to enhance the context awareness, as aspects that can be resolved though the design and the deployment of cloud-based applications from the IoT field.

This work focuses on the current trends in the context of cloud infrastructure services combined with IoT integrated systems/platforms and/or frameworks towards the provision of large-scale cloud-based distributed IoT services in the context of the smart cities of the future. These trends include [28] the integration of special devices in the whole computing continuum, from high-performance computing ones to mobile devices, and the design of decentralized service-oriented systems. Moreover, the main contemporary trends include the improvement of the virtualization technologies, the achievement of portability, and interoperability requirements, or the automation of the organization and management of the back-end resources. Consequently, cloud-based applications from the fields of the IoT and Big Data are expected to guide the new services.

The rest of this work is structured as follows: Sect. 9.2 presents the identified challenges for the cloud-based IoT architectures, and Sect. 9.3 presents an overview of the work in progress in this field. Furthermore, Sect. 9.4 analyzes in detail the case of one specific promising multi-Cloud-IoT architectural solution, while the work concludes in Sect. 9.5 with the lessons learnt through the study and the elaboration of the Cloud-IoT concepts.

9.2 Challenges

The Smart City environment includes a vast amount of devices, services, applications and ICT-related solutions, which can be ideally manipulated over IoT infrastructures. The attributes needed by IoT services and the characteristics of Cloud systems clearly motivate the merging of the Cloud and IoT worlds. A wide variety of Cloud features can ideally contribute to the enhancement of IoT solutions, by building innovative integrated smart cities Cloud-IoT environments with features such as on demand service provision, ubiquitous access, resource pooling, as well as elasticity, which actually are essential for the IoT world. Towards this direction a set of different challenges have been identified [28, 29] that should be taken into account for the design and the development of innovative inter-Cloud-IoT solutions. Inter-Cloud (aka multi-Cloud) is defined as the approach that facilitates scalable resource provisioning across multiple cloud infrastructures [30, 31]. In the rest of this section some of the major identified challenges are described in more details.

The heterogeneity of the entities that are involved in the context of the smart cities, including devices, services, and applications, requires the deployment of innovative mechanisms that will enable the production, the discovery, the mix and the reuse of different service components and the creation of new added value public services through pooling and sharing of resources, data, content and tools, even across national borders. In this way, they will facilitate the collaboration between different stakeholders, end-users and public administrations. The "*automated discovery and composition of services*" [24, 32, 33] will enable the establishment of federated environments that will support publicly available cloud services. The last requires interoperable, reusable modules for public service functionalities, which ideally address the smart cities requirements.

The "*federated environments*" concept [34] is one of the core concepts for the design, the deployment and the management of decentralized cloud infrastructures. Such a concept may be applicable to "*software-defined data centers*" *(SDDC)* [35, 36], '*software-defined networks*" *(SDN)* [37, 38] and additional appropriate cloud-based mechanisms to enable incorporation of resources and services independent of their location across distributed computing and data storage infrastructures.

Through the federation of distributed environments, different approaches towards the design and the development of standards should be elaborated. These approaches will lead to increased interoperability between cloud services and infrastructure providers. The "*interoperability and standardization*" [39, 40] aspects, in turn, will enable efficient migration of services, applications and data. Having overcome potential borders with respect to interoperability aspects, the design and the development of distributed, federated and heterogeneous cloud computing architectural models [41, 42] will ensure the harmonization of the heterogeneous cloud-based systems interworking in an efficient and high-performance way.

In this direction, "*services deployment and management*" [43, 44] in a decentralized and autonomous way should be supported. This will be achieved through the introduction of tools for automatic and dynamic deployment, configuration and management of services to enhance availability, flexibility, and elasticity to meet targeted performance constraints (e.g., resources management based on "minimum requirements vs. available resources"). In addition, the introduction of innovative software and hardware solutions should be elaborated, so as to facilitate the coherent deployment of distributed applications over heterogeneous infrastructures and platforms from multiple providers, as well as the design and the deployment of mechanisms to offload computation and storage tasks from mobile devices onto the cloud at both design and execution time.

"*Security and privacy*" [45, 46] constitute challenging aspects for cloud systems by including mechanisms, tools and techniques ensuring the security and transparency of cloud infrastructures and services, including data integrity, localization, and confidentiality, also when using third-party cloud resources.

The "*trust*" [47] aspect focuses on data and services from different cloud providers that refers to the collaborative development, adaptation, and testing of open-source software for innovative and trusted cloud-based services. In this way, "*trust*" will allow the collaboration across different platforms and different technical environments for the deployment of integrated systems.

Table 9.1 summarizes the identified topics and challenges in the context of the research field that focuses on the design and the development of Cloud-IoT architectures for smart cities environments. Bearing in mind the identified challenges, the next section presents the latest research activities towards the realization of the Cloud-IoT architectures.

Table 9.1 Identified topics and challenges for Cloud-IoT smart city environments

Challenge	Description
Development of distributed, federated, and heterogeneous cloud computing models	This challenge refers to the design and the development of multi-Cloud architectural models that will allow the distribution of the resources, the services and applications in multi-IaaS (Infrastructure-as-a-Service) environments
Deployment and management of resources: in a decentralized, autonomous way	This challenge is related to the issues with respect to the introduction of tools for autonomic and dynamic deployment, configuration and management of services to enhance availability, flexibility, and elasticity to meet targeted performance constraints (e.g., resources management based on "minimum requirements vs. available resources")
Security, privacy	Cloud systems with built-in security mechanisms and tools that will exploit techniques ensuring the security and transparency of cloud infrastructures and services, including data integrity, localization and confidentiality
Trust: data and services from different cloud providers	Definition of specification and standards for the provision of Cloud mechanisms that will address multi-Cloud concerns with respect to data manipulation and exploitation, as well as will take into account the trust issues to define the suitability of data generating devices
Federated environments	The federated cloud environments deal with the deployment and the management of cloud-based mechanisms to enable incorporation of resources and services independent of their location across distributed computing and data storage infrastructures. It can be ideally combined with the outcomes of the modeling activities with respect to the design and the development of multi-Cloud architectures, as they are investigated in the context of the "development of distributed, federated and heterogeneous cloud computing model" challenge

9.3 Cloud Internet of Things Architectures for Smart Cities

An important number of research initiatives in the context of the identified challenges for the Cloud-IoT architectures have been kicked off from different European Research and Innovation projects. The outcomes of these research activities will contribute to advanced, innovative architectural solutions to enable Cloud-IoT capabilities in the context of diverse application domains, with particular focus on large-scale experimentation environments in smart cities. In the following sub-sections some indicative research projects that are currently being performed are introduced and reviewed.

9.3.1 ClouT: Cloud of Things

ClouT [48], which stands for "cloud of things", is providing infrastructure, services, tools and applications that will be used in the context of smart cities (e.g., by municipalities, citizens, service developers, etc.) integrating advances from the IoT and Cloud domains. The main features of the proposed architecture include: (a) the support of the capability for acquisition of Smart City data leveraging IoT and Internet of People, (b) the functionality for the city data provision so as to allow the easy development of scalable, dependable, and semantic services, and (c) the deployment of innovative Smart City applications in pilot cities. ClouT projects proposed reference architecture for the integration of IoT enabled Cities deployments with the Cloud computing. The architecture includes IoT related technologies in terms of devices (e.g., sensors, actuators, etc.) and appropriate software for the device management, data gathering, and management (Sensorisation and Actuatorisation, IoT Kernel and Interoperability and City Resource Virtualization building blocks). Moreover, the architecture includes components from the Cloud field, for the provision of computing and storage capabilities (Computing and Storage building block). The Cloud-IoT integration is realized through the ClouT architecture that is structured in the City Platform-as-a-Service (CPaaS) and the City Infrastructure-as-a-Service (CIaaS) view, complemented with security aspects in a cross architecture view.

9.3.2 MUSA: Multi-Cloud Secure Applications

In the context of heterogeneous cloud ecosystems a key factor refers to the deployment of applications that are able to maximize the benefits of the combination of the cloud resources in use in the involved clouds, therefore to the creation of multi-Cloud applications. The MUSA project [49, 50] aims to the design and the development of a framework that will include security-by-design mechanisms to allow application's self-protection at runtime, as well as methods and tools for the integrated security assurance in both the engineering and operation of multi-Cloud applications. Aiming to the deployment and the support of a multi-Cloud environment (that among other integrates IoT deployments) the proposed framework is composed by four different architectural building blocks. First introduces an integrated development environment (IDE) for creating the multi-Cloud application taking into account its security requirements together with functional and business requirements. Advanced security mechanisms designed and embedded in the application components so as to allow the self-protection, ensuring in the same time the dynamic, automated and secure collaboration among different applications, such as distributed IoT deployments. In large-scale IoT deployment and Cloud applications, framework's intelligent decision support system will realize an automated deployment environment that will allow the dynamic distribution of the components according to security needs. Finally, a security assurance platform in form of Software-as-a-Service (SaaS) that will support multi-Cloud application's runtime security control and transparency to increase user trust, that constitutes a critical factor for the integration of distributed Cloud-based IoT environments.

9.3.3 RAPID: Heterogeneous Secure Multi-level Remote Acceleration Service for Low-Power Integrated Systems and Devices

The RAPID project [51] focuses on the development of efficient and powerful computing infrastructures that among others will be exploited so as to accelerate cloud-based applications for the IoT, by improving significantly their performance. In particular, the project focuses on the introduction of innovative Cloud computing processing capabilities, enable by an efficient heterogeneous CPU-GPU (Central Processing Unit—Graphics Processing Unit) cloud computing infrastructure. Such infrastructure can be used to seamlessly offload CPU-based and GPU-based (using CUDA API [52]) tasks of applications running on low-power devices such as Wireless Sensor Networks (WSN) nodes, wearable devices, etc., to more powerful devices over a heterogeneous network (HetNet). This will benefit the existing IoT infrastructures in terms of energy management, distributed processing and load balancing on IoT deployments. The proposed solution by the project, consists of three different core modules: (a) the accelerator client that take over the decision-making with respect to the execution host of particular tasks either locally or remotely on accelerator servers, (b) the accelerator server that takes over the execution of particular tasks and returns the results and (c) the directory server that keeps status and resource information of the RAPID accelerators. Through the introduction of RAPID enablers (complemented with appropriate APIs) it will be achieved the realization of evolved Cloud-based IoT deployments and software solutions that require high performance and resource availability so as to achieve the best possible result.

9.3.4 INPUT: In-Network Programmability for Next-Generation Personal Cloud Service Support

The INPUT project [53] aims realize a distributed Cloud-IoT architecture for the deployment of Cloud-based IoT environments, through the virtualization of physical infrastructures in the smart infrastructures, including smart city concepts such as Smart Homes. In particular, through the replacement of the Smart Devices (SDs) with virtualized images that will provide their functionalities, the INPUT system will provide to the end-users integrated virtualized devices functionalities as services. The project aims to overcome current limitations on the cloud service design by introducing computing and storage capabilities to edge network devices in order to allow users, operators and service providers to create/manage private clouds "in the network". Essentially, through the introduction of Cloud capabilities such as cloud-images and virtualization it is realized a Cloud-IoT environment for the future internet, focusing on the end-users. Specifically, particular interest has been positioned on moving Cloud services closer to the end-user. Moving the cloud services closer to end-users and smart devices, the project aims to avoid pointless network infrastructure and data-center overloading, and to provide lower latency to services. The project technologies allow the integration among the Cloud and IoT, and enable next-generation cloud applications to go beyond classical service models, namely Infrastructure-as-a-Service (IaaS), Platform-as-a-Service (PaaS), and Software-as-a-Service (SaaS). The project enablers aims to a new Cloud infrastructure that replace physical Smart Devices (SD), which are usually placed in users' homes (e.g., network-attached storage servers, set-top-boxes, video recorders, home automation control units, etc.) or deployed around for monitoring purposes (e.g., sensors), with their "virtual images", providing them to users "as a Service" (SD as a Service—SDaaS). For the validation of the project, different proof of concept implementations are planned, such as Intelligent Transportation Systems (ITS) in large cities, including traffic status monitoring [54].

9.3.5 SPECS: Secure Provisioning of Cloud Services Based on placeSLA Management

The SPECS project aims [55] at developing and implementing an open source framework to offer Security-as-a-Service based on particular Service Level Agreements (SLAs), by providing comprehensible and enforceable security assurance by Cloud Service Providers (CSP), which is a critical factor to deploy trustworthy Cloud ecosystems, such as those related to smart cities infrastructures. The SPEC framework architecture includes tools and mechanisms that are distributed in building blocks for negotiation among multi-Cloud environment modules, monitoring of decentralized, cloud deployments, and the enforcement of real-time on-demand Cloud environment (re-)configuration. The implementation of a multi-Cloud environment, supporting among other the IoT, is based on the particular application of the above three architectural principles. Specifically, for the composition and use of Cloud services the project introduces a user-centric negotiation of security parameters in Cloud SLA, along with a trade-off evaluation process among users and CSPs. The Cloud SLAs are evaluated fulfilling a minimum required security level that is called Quality of Security (QoSec). Moreover, the framework, through the introduction of advanced tools and techniques for the enforcing agreed Cloud SLAs, aims to keep a sustained QoSec that fulfills the specified security parameters. SPECS' enforcement framework will also "react and adapt" in real-time to fluctuations in the QoSec by advising/applying the correct countermeasures (e.g., triggering a two-factor authentication mechanism). Consequently, the project aims to realize multi-Cloud environments focusing on security aspects for the secure and reliable collaboration among distributed modules.

9.3.6 SOCIOTAL: Creating a Socially Aware and Citizen-Centric Internet of Things

The SOCIOTAL Project [56] works towards an IoT ecosystem that has trust, user control and transparency at its heart in order to gain the confidence of everyday users and citizens. By providing adequate socially aware tools and mechanisms that simplify complexity and lower the barriers of entry this encourages citizen participation in the Internet of Things. The project designs and provides key enablers for a reliable, secure, and trusted IoT environment that enable creation of a socially aware citizen-centric Internet of Things by encouraging people to contribute their IoT devices and information flows. The project proposes an architecture combined by three core parts that allow the integration among heterogeneous distributed IoT deployments with Cloud services. The architectural building modules are: (a) the application level components, (b) the core components and (c) the communication layer. The application levels includes all the related components to applications, services, management interfaces and Cloud clients for the interconnection among the platform components with distributed Cloud infrastructures for the service provisioning. The core components include different technological aspects, with main focus on security modules and IoT services orchestration and management. Finally, the communication layer provides the modules for the bridging between the platform and the external entities, such as IoT devices. Essentially, this project realizes the Cloud-IoT architectural approach, not by hosting technological solutions over the Cloud, but through the integration of IoT services with Cloud services for the provision of secure IoT applications.

9.3.7 COSMOS: Cultivate Resilient Smart Objects for Sustainable City Applications

The COSMOS project [57] aims at enhancing the sustainability of Smart City applications by allowing IoT based systems to reach their full potential, by enabling in the same time things to evolve and act in a more autonomous way, becoming more reliable and smarter. The project combines multiple technological aspects from the

Cloud computing field and the IoT area, aiming to create a combined solution that will enable the things of the IoT to be able to learn based on others experiences, being in the same, aware about their context time through the acquisition of situational knowledge and analysis. The introduction of the Complex Event Processing (CEP) building block, combined with appropriate mechanisms and/or APIs that enable the interaction with the social media, leads to the deployment of a Cloud-based IoT platform. This platform enables innovative feature and capabilities on the data delivering and processing, as well as the information management with appropriate mechanisms for the handling of the exponentially increasing "born digital" data. Essentially, the project aims to realize a Cloud-based IoT ecosystem, realized by smart object, innovative data gathering mechanisms connected to social media, and advanced data processing mechanism including complex event processing of multi-dimensional datasets.

9.3.8 CITY PULSE: Real-Time IoT Stream Processing and Large-Scale Data Analytics for Smart City Applications

The CITY PULSE project [58] focuses on the smart city applications challenges, such as integration of heterogeneous data sources and the challenge of extracting up-to-date information in real-time from large-scale dynamic data. The project's main aim is to provide a scalable, adaptive and robust framework that provides Virtualization capabilities (virtualization of Internet of People and IoT), supports large-scale data analytics (cloud-based infrastructure for the distributed data processing), integrates semantic technologies for machine-interpretable descriptions of data and knowledge and allows the easy creation of real-time smart city applications by reusable intelligent components. As general purpose, the project aims to develop, build and test a framework for semantic discovery and processing of large-scale real-time IoT and relevant social data streams for reliable knowledge extraction in a real city environment. The project introduces a distributed architecture, enriched with Cloud computing capabilities and features that realize the distributed IoT large scale data analysis and processing, by producing data for the end-user applications. The project provides a set of different tools and datasets that work as enablers for the Cloud-IoT architectural aspects. Such tools are the ontologies such as the Quality Ontology and the Stream Annotation Ontology, tools for the sensor data streaming, annotation and managements, knowledge building tools using the sensor data and application management tools. Consequently, the project integrates IoT and Cloud in the level of the sensors streaming data processing over distributed cloud-based facilities, as well as through the provision and support of innovative IoT applications that cooperate with existing Cloud services and CITY PULSE project enablers' tools.

9.3.9 FIESTA: Federated Interoperable Semantic IoT/Cloud Testbeds and Applications

The FIESTA project [59] investigates the aspects with respect to the production of a "first-of-a-kind" blueprint experimental infrastructure (tools, techniques, and best practices) enabling testbed operators to interconnect their facilities in an interoperable way, while at the same time facilitating researchers in deploying integrated experiments, which seamlessly transcend the boundaries of multiple IoT platforms. Essentially the project is providing a Cloud-IoT testbed infrastructure, for experimentation purposes, that is realized over the interconnected/interoperable underlying testbeds. The main idea is the provision a single entry point to all FIESTA Experimentation-as-a-Service (EaaS) services using a single set of credentials. They will be able to design and execute experiments across a virtualized infrastructure i.e., access the data and resources from multiple testbeds and IoT platforms using a common approach. FIESTA offers various Cloud-based tools for enabling capabilities such as (a) to design and execute experimental work-flows, (b) dynamically discover IoT resources, and (c) access data in a testbed agnostic manner. FIESTA enables Cloud-IoT architectural infrastructure that lies on the provision of an IoT EaaS atop a middle-ware infrastructure that adapts and federates existing IoT platforms and testbeds. This entails the adaptation of the data of those testbeds to a common FIESTA ontology (i.e., compliance to common semantics), as well as the provision of a common standards based API for accessing the IoT services of the testbeds. FIESTA will be validated and evaluated based on the interconnection of different testbeds, as well as based on the execution of novel experiments in the areas of mobile crowd-sensing, IoT applications portability, and dynamic intelligent discovery of IoT resources.

9.3.10 iKaaS: Intelligent Knowledge-as-a-Service

The work in the context of the iKaaS project [60] relies on the definition of the multi-Cloud architectures for the IoT, and proposes the design and the deployment of an "intelligent Knowledge-as-a-Service" (iKaaS) platform for the IoT environments. The iKaaS project is developing an intelligent and secure multi-Cloud-IoT Smart City platform based on Semantics and Data Models, Big Data resources, and heterogeneous Cloud environments, with data collected from a variety of sensors from IoT environments deployed as mobile terminals, smart devices, and smart homes. The platform features will be applied by means of Smart City applications promoting self-management of health and safety of citizens, as well as an information system improving data analysis for a smarter life in the city. A well-defined approach has been structured so as to address the whole life-cycle of such a platform. This approach focuses on the definition of a universal data model based on the Semantic Web for data collected from IoT and stored in cloud platforms, complemented with security-by-design features for the data. Furthermore, the project's approach focuses on the development of a decentralized heterogeneous secure multi-Cloud environment, building an integrated knowledge-as-a-service platform concept. A multi-dimensional Cloud architecture with multiple Local Clouds and a Global Cloud will be deployed, validating its operation with multiple applications, as well as determining the utility of the distributed knowledge-base through experiments in model smart cities and Smart Homes. The separation between 'Local' and 'Global' cloud, in the context of iKaaS, is determined as follows. The 'Local Cloud' is a cloud infrastructure that covers a specific geographical area in a specific time period, by providing sufficient computing, storage and networking capabilities, responsible for the provisioning of particular requested services. The 'Global Cloud' is considered as a traditional cloud infrastructure that provides on-demand/elastic (illusion of infinite) processing power and storage capability, ensuring at the same time the increase of the business opportunities for service providers, and the ubiquity, reliability, performance, efficiency, scalability of service provision.

Having identified the different challenges (Sect. 9.2), as well as having described the different research projects, initiatives and related research activities, with respect to the design and the development of Cloud-IoT architectures for smart cities, Table 9.2 presents, a correlation between the challenges and the projects. The correlation is performed in terms of the challenges that are covered in the context of the corresponding project.

Table 9.2 Identified challenges versus projects for smart cities Cloud-IoT architectures

Challenges	Development of distributed, federated, and heterogeneous cloud computing model	Deployment and management of resources	Security, privacy	Trust	Federated environments
Projects					
ClouT		X			X
MUSA	X		X		
RAPID	X	X	X		
INPUT		X			
SPECS		X	X	X	X
SOCIOTAL			X	X	X
COSMOS		X			X
CITY PULSE		X			X
FIESTA	X	X			X
iKaaS	X	X	X	X	X

9.4 The iKaaS Case in Detail

This section intends to present a detailed case study based on a particular multi-Cloud-IoT architecture, by analyzing the architectural capabilities and features, as well as their application to Smart City related concepts. In particular, the case study is based on the iKaaS Platform architecture, and in the next sub-sections an overview of the proposed architecture will be provided, complemented with three different scenarios/approaches with respect to the application of the platform in a Smart City environment.

9.4.1 Multi-Cloud-IoT Architecture Overview

The iKaaS architecture (Fig. 9.1) introduces the approach of the Local Cloud environments and a Global Cloud environment, having the ability to interconnect with each other. The introduction of the Local Clouds and the Global Cloud focuses to the development of an intelligent, privacy preserving and secure Big Data analytics engine build atop a multi-Cloud infrastructure, that will be fed with large-scale ubiquitous data collected from heterogeneous sensing networks and data sources, including cyber-physical systems, wearable sensors, and social or crowd-sources.

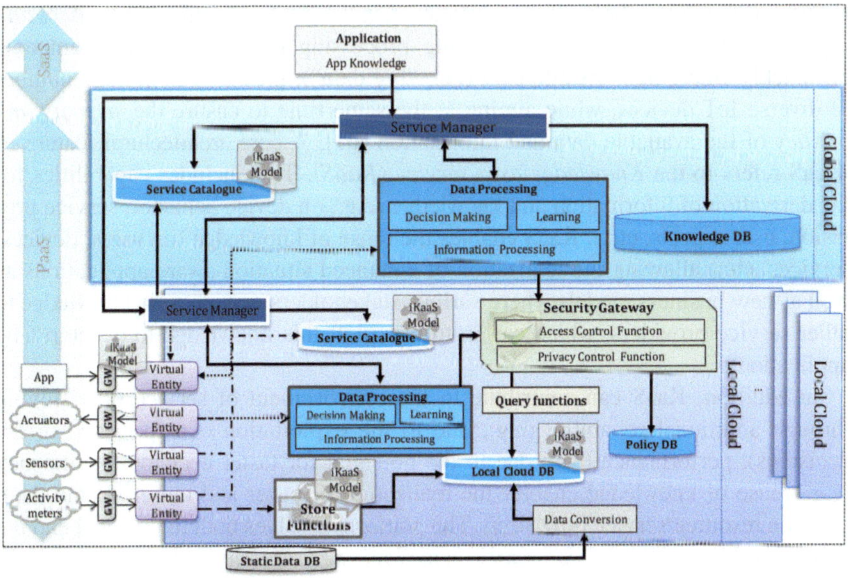

Fig. 9.1 iKaaS multi-Cloud architecture [60]

The Local Cloud environment comprises the sufficient computing, storage and networking capabilities, and provides requested services to users in a certain geographical area and time period, as well as it offers additional processing and storage capability to services. The Local Clouds can involve an arbitrary large number of nodes (sensors, actuators, smartphones, etc). The aggregation of resources comprises sufficient processing power and storage space and networking can rely on heterogeneous network technologies. The goal is mainly to serve users of a certain area. In this respect, a Local Cloud is the virtualized processing, storage and networking environment, which comprises IoT devices in the vicinity of the users; users will exploit the various services composed of the Local Cloud devices capabilities.

The Global Cloud is seen in the "traditional" sense, as a construct with on-demand/elastic (illusion of infinite) processing power and storage capability. It is a "backbone infrastructure", which increases the business opportunities for service providers, and the ubiquity, reliability, performance, efficiency, scalability of service provision. In this way, the Global Cloud offers more opportunities for offering services, more options on which to base service features in case of context changes, more resources for contributing to the decisions, elastic provision of resources on-demand, etc. Further to that, the Global Cloud can enable, as a special (yet important) case, the existence of IoT service providers capable of providing larger scale services without owning actual IoT infrastructure.

The iKaaS platform encompasses a wide range of innovative functionalities that range from semantic data storage, management and protection to autonomic service management and knowledge building on various concepts including the service provision. In particular, the platform provides *distributed data storage and processing* capabilities for efficiently communicating, processing and storing massive amounts of, quickly-emerging, versatile data (i.e., "big data"), produced by a huge number of diverse IoT devices, while aiming at the same time to ensure the *security and privacy* of the available data and information [61]. A core architectural feature of iKaaS refers to the *Knowledge-as-a-service (KaaS)*. This includes capabilities for the derivation of information and knowledge (e.g., on device behavior, service provision, user aspects, etc.). KaaS enables the reuse of knowledge (on users, devices, services, etc.) allowing the realization of enhanced situation aware applications as well as new business models where various stakeholders may provide knowledge to other service providers, which can in turn exploit this knowledge to develop new applications.

In addition, KaaS can contribute to the improvement of service provisioning through additional reliability (e.g., due to the exploitation of experience in the decisions), performance (e.g., due to the potential for faster decisions), efficiency (e.g., reuse of knowledge enables the realization of faster and more reliable decisions on resource/service provision). The platform involves both in the local and the global level, *consolidated data and service logic descriptions and semantic storage repositories*, referred as catalogues. These enable the reuse of data/data sources, services and service logic, enabling capabilities for *autonomic service management*. The latter includes (a) dynamically understanding the requirements, decomposing the service (finding the components that are needed), (b) finding the best service configuration and migration (service component deployment) pattern, and (c) during the service execution, reconfiguring the service, (i.e., conducting dynamic additions, cessations, substitutions of components).

As it can be observed, part of the iKaaS functionality determines the optimal way to offer a service. For instance, service components may need to be migrated as close as possible to the required (IoT) data sources. IoT services may need generic service support functionality that is offered within the Global Cloud, and at the same time, they do rely on local information (e.g., streams of data collected by sensors in a given geographic area), therefore, the migration of components close to the data sources (i.e., in Local Clouds) will help in the reduction of the data traffic. The next sub-sections provide the short description of the core iKaaS architectural components in Global and Local Cloud respectively. Moreover, Sects. 9.4.2 provides a storyline summary and corresponding proof of concepts, that showcase the collaboration among the architectural components so as to realize the interaction between the Global Cloud and the Local Clouds environments.

9.4.1.1 Global Cloud Components

Service Manager

The Service Manager, in the Global Cloud, includes a set of different functionalities. Such functionalities allow the dynamic/on-demand creation of complex services through the analysis of the application request(s), the interconnection among the applications with the rest of the system components, as well as the communication among the architectural components (in local and global level). Regarding the complex services creation, the application request(s) can, ideally, be fulfilled by particular existing services that will be combined, by the Service Manager, so as to create a complex service with integrated functionality. A composite/complex service refers to the semantic based association between different distributed services with the main aim to provide an integrated more complex functionality. The application requests are analyzed based on particular semantic data models (e.g., Service Model, Virtual Entity Model, Knowledge Model, etc.) that are included in the iKaaS Data Model [62]. The Service Manager supports also the flexible/autonomic selection of more appropriate cloud resources that allow the deployment and execution of services in the Global Cloud. In addition, it supports the service migration from 'limited-resources' cloud environments (e.g., a smart home Local Cloud) to infrastructures with more resource power (e.g., another more powerful Local Cloud, such as the smart city, or directly to the Global Cloud). Moreover, this component supports the orchestration of service sub-components and delivery to an Application, as well as the end-to-end monitoring of the executed processes. Further to the above, the Service Manager supports the communication among third-party entities and systems' components such as the Knowledge database (DB). More specifically, for the knowledge DB the Service Manager supports credential-based authorization, for the access control of the database's clients.

Service Catalogue

The Global Service Catalogue stores and provides information on existing, deployed services. Services can be linked to functions of devices, e.g., sensors, actuators, smartphones, etc. (e.g., temperature monitoring) or they can be more complex services composed by several other elementary ones (e.g., monitor traffic conditions on A25). Information from the Service catalogue can be exploited by the Service Manager in order to find the optimal composition of services that fulfills application requirements, user preferences, policies and Cloud resources.

The Global Service Catalogue works as a federation of semantic data stores, that aggregate the semantic data from the distributed local service catalogues with respect to the available IoT services in each local cloud. It supports the capability for performing federated queries against the distributed stored data over a particular dedicated cloud network infrastructure that constitutes the Storage Network of the iKaaS cloud platform infrastructure. It embeds a software module that is called Federation Manager and is responsible for the data manipulation of distributed federation clients that have been registered to its (Domain Name System) DNS-like indexing system, as available semantic data storages over the internet. In addition to the federated semantic data, the global service catalogue stores, natively, semantic data that corresponds to iKaaS Service Model instantiations for the description of the complex services. The complex service description is based on the service data model and the semantic data that is related to the complex service, is stored in the Service Catalogue.

Data Processing

The Data processing comprises functionality for big data analytics, learning, knowledge based inference and reasoning to support autonomous behavior and self-adaptation of applications. In other words, the data processing component encompasses methods for data analytics, knowledge building and inference so as to realize (a) enhanced awareness with respect to external situation(s), operational context and human/social aspects, (b) learning capabilities for building knowledge and experience related to situation and past application behavior adaptation, so as to enable faster processing of data, more efficient and reliable behavior adaptation and control, etc. and (c) reasoning capabilities to support the optimal autonomous application/system behavior adaptation, taking into account current context, knowledge and policies. The Data processing mechanisms continuously run, retrieving data from the local data processing components running services and applications, and update the knowledge data stored in the and Knowledge DB, through the Global Service Manger.

Knowledge Database (DB)

This component maintains knowledge related to users and situations in the real world. For example, part of the information includes user profile, data and devices repository, which can comprise "static" user information such as age, gender as well as information on current user status based on analyzed data acquired from user devices, information on user devices (wearables) being used by the particular user, e.g., type, state (active/nonactive), communication details, etc. The data stored in this DB instantiates the iKaaS Knowledge Model that constructs the abstract semantic representation of different identified knowledge concepts. Such data constitutes the input and/or the output for innovative machine learning algorithms inspired by Time-series and Bayesian Networks techniques. The Knowledge DB combined with the Data processing mechanisms, are essential towards the provision of "Knowledge–as-a-Service", and it can be distributed across the Global Cloud infrastructure. The access on the stored knowledge data is restricted by particular authentication based access control security functionality. This functionality is based on the authentication of the knowledge DB clients using credentials. This is to ensure, for example, that a service request (and its corresponding associated Global Data Processing functionalities) that may simply need a generic (anonymized) mobility pattern from the Global Knowledge DB cannot access information about a particular user's age and state, also stored in the Global Knowledge DB.

9.4.1.2 Local Cloud Components

Service Manager

A key role of the Local Service Manager in the Local Cloud is to provide a contact point with local services, offering the Local Cloud resources "as a service" to the Global Cloud. The Local Service Manager may retrieve information on local required services to serve an application from the local service catalogue and may also retrieve additional required information from the Local Data Processing. The Local Service Manager can provide information/actions to an Application. Furthermore, this component provides support for application request(s) locally, in case no intervention of the Global Cloud is required and executes optimally the migrated by the Global Cloud subservices using the appropriate Local Cloud resources.

Service Catalogue

The role of the Local Service Catalogue is similar to the Global Service Catalogue with the difference that this component stores semantic data that corresponds to the instantiations of the iKaaS Service Model for the Local Services in each Local Cloud. In addition, the Local Service Catalogue supports the federation of data to the Global Service Catalogue, through the Local Service Manager, over the Storage Network.

In the case of the semantic data federation, the Local Service Catalogue works as federation client to the federation manager that is hosted on the Global Service Catalogue.

Data Processing

The Local Data Processing is complementary to the Global Data Processing. In fact some functionality for learning and knowledge derivation may be distributed over the Global and Local cloud data processing to allow for greater efficiency. In general local data processing comprises mechanisms for the processing of data and the generation of knowledge close to the data sources. The processed data and/or knowledge may be further distributed to the Global Cloud or may be stored in the Local Cloud database (DB) if being only relevant to local services. In the scope of IoT, that is characterized by huge amount of data streaming from heterogeneous sources, such as sensors, smart devices, etc. it is important to have the ability of processing this information in real-time, not just in the sense of fulfilling a fixed set of time constraints, but having the required data when it is needed.

Security Gateway

The Security Gateway [61] constitutes the module that ensures the security and privacy of the data across the Local Cloud instantiated infrastructures. Further to that, the security gateway controls the access of third-party entities coming from the global cloud level to the particular local level. The security gateway bases its functionality on the policy-based access control and it is collaborating with the Policy database (DB) so as to ensure the security of the data and the processes. The incoming requests from Global Cloud are handled by the corresponding Local Service Manager and they are filtered by the local security gateway before they are forwarded to the corresponding entities either for the data fetching or for data modification. In this way, the local cloud can control the traffic from external entities, ensuring that only authorized entities will have access on the data, following in addition particular security and privacy rules and access policies.

Policy Database (DB)

The policy DB is a database for storing security and privacy policies that are related with the corresponding local cloud instance. The security policy is used by the Security Gateway to interpret the rules of the country where the local cloud is set up when the security gateway grants access permissions to the service components linked to an Application. The security policy maintains two tables: the expiry periods of access permissions and the definitions related to data privacy. On the privacy policy the status of the consent on the transfer of data to third parties for each data owner is listed.

The privacy policy makes it possible for the Security Gateway to provide personal data to the service components linked to an Application while preserving the privacy of the data owner. The policy DB is used only in the scope of the local cloud, aiming to ensure proper and secure functionality in this, serving incoming requests forwarded from the global cloud service manager and/or direct application requests.

Local Cloud Database (DB)

The Local Cloud DB stores and manages data from the IoT devices that are associated with the corresponding Local Cloud platform instance. It works as a type of scalable broker for the real-time collected measured data provided for the devices and it is exploited by higher level cloud mechanisms such as the Data Processing. It offers a kind of abstraction for the data in terms of their retrieval and manipulation since a single data management Application Programming Interface (API) complements this mechanism, offering in this way a homogenous IoT data management across the heterogeneous Cloud platforms. Furthermore, an appropriate abstract data model for the structuring of the Local DB data is provided on the context of iKaaS architecture and defines the core principles for the data structuring of the collected measurements from the available IoT devices.

Data Conversion

The role of this component is the homogenization of raw static data towards a common format and/or data model. Knowledge in iKaaS will most likely be an RDF DB, since the semantic web technology is assumed for the scalable treatment of knowledge over the Internet. Geospatial data stored in a Static Data DB are defined in GML format, which is an iKaaS extension of CityGML. The data conversion process is to convert data from an iKaaS GML format to a RDF format. The contents of iKaaS extension of CityGML shall be categorized to buildings, tunnels, bridges, water body, transportation, vegetation, city furniture land use, environment sensors, and underground pipes. The contents shall also be classified into either open or non-open.

Virtual Entity

The Virtual Entity (VE) is the virtual abstract representation of the IoT devices [63, 64], that is comprised by software for the virtualization of the real device capabilities, complemented with the corresponding semantic description that constitutes instantiation of the VE Model. Each instantiation of the VE Model is associated with a related iKaaS Model instantiation for a Local/Simple service, and this description data is stored into the Service Catalogues. Moreover, the VEs support both the real-time data streaming from the real IoT devices to the iKaaS platform mechanisms, as well as the data storing and management into the Local Cloud DBs. Essentially, a VE bridges the distributed IoT devices and/or IoT platforms with the heterogeneous Cloud environments.

9.4.2 Storyline Summary

The storyline of this case study includes Smart City concepts with particular focus on the citizens' health care, as well as on the improvement of their daily life in urban environments. In particular, it is considered that Mr. T. uses a ubiquitous Ambient Assisted Living (AAL) application in order to monitor his health status, as well as to monitor and/or control/adjust the conditions of the surrounding environment (Fig. 9.2). Moreover, when the user is moving in the Smart City he can be informed, through the application, about the suitability of the conditions in accordance with his health status, as well as to navigate in the city, avoiding congested and "low-quality air" areas. When the user is located in the home, the Smart Home Local Cloud takes over to host and provide the AAL service, whereas when the user is outside, the AAL service is dynamically reconfigured, so as the Smart City Local Cloud can take over the support of its capabilities.

The storyline introduces three different scenarios that the iKaaS platform capabilities are applied to including: (a) the introduction of IoT enablers in Smart City environments, (b) the Cloud-IoT services creation on-demand, and (c) the service migration on inter-Cloud environments. The next sub-sections describe the three different scenarios, as well as present the corresponding Message Sequence Charts

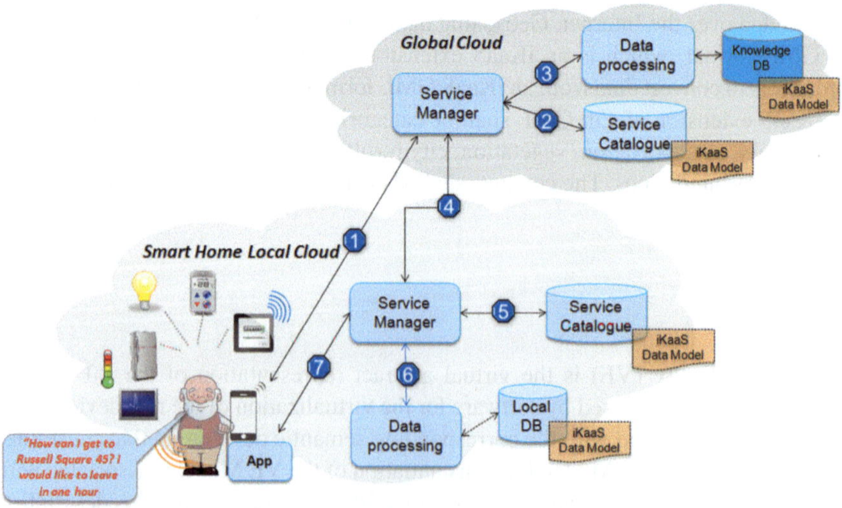

Fig. 9.2 AAL application in a Smart City environment

(MSCs) for the representation of the iKaaS architectural components interaction in each different case.

9.4.2.1 Introduction of IoT Enablers in Smart City Environments

The core component that bridges the real-world IoT devices with the iKaaS platform Local Clouds is the VE, while each VE is directly associated with a local/simple service that is deployed in the Local Cloud. For instance, a sensor that measures the air pollution is associated with a "pollution monitoring" service, described by semantic data that correspond to an instance of the iKaaS Service Model. Consequently, the introduction of IoT enablers in the Smart City environment, through the iKaaS platform, corresponds to the registration of local services (Fig. 9.3).

This process can be executed either in an automated way by the VEs due to their activation in the iKaaS platform, or manually by the corresponding local cloud administrator/manufacturer, etc. The information on the newly registered local services is federated to the Global Service Catalogue that interconnects, through the Local Service Manager over the storage network, with the Local Service Catalogue. Essentially, the Global Service Catalogue supports the creation of federated requests

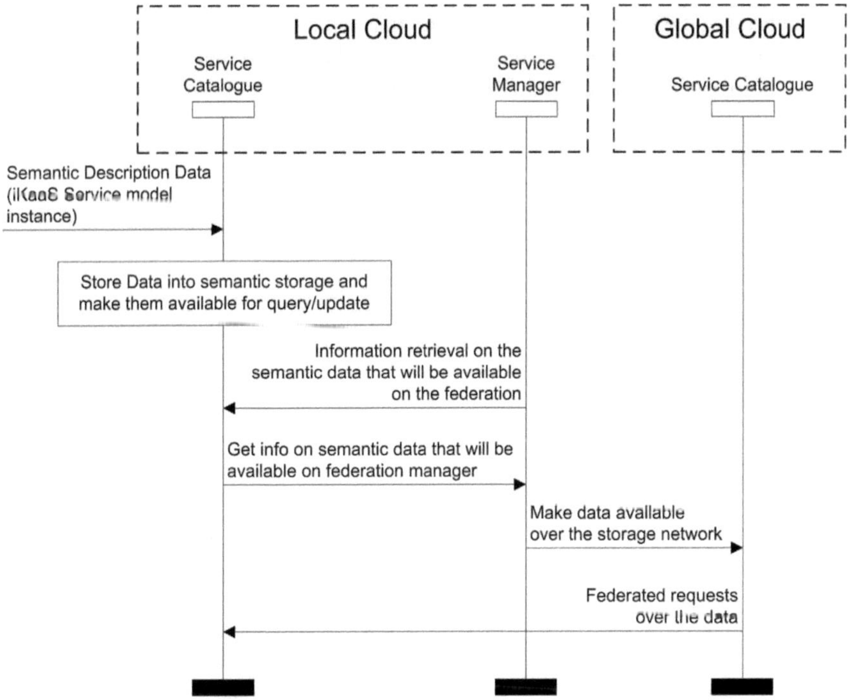

Fig. 9.3 IoT enablers, introduction process

(queries/updates) towards the semantic data that is stored into the distributed Local Service Catalogues. Thus, the Global Service Catalogue works as the architectural component that allows the centralized management of the distributed data across heterogeneous Cloud environments that can be deployed in a Smart City, including heterogeneous IoT devices.

9.4.2.2 Cloud-IoT Services Creation and Deployment on-Demand

Considering that numerous local services have been registered in the distributed local clouds of a Smart City environment (Sect. 9.4.2.1), these services can be used for the creation of composite multi-Cloud services, based on the end-user requirements. Based on the storyline, the composite service will support the AAL service that is comprised by local services from the Smart Home and the Smart City domain (Fig. 9.4).

The end-user uses a mobile application so as to perform the application request to the iKaaS platform, by sending the request to the Global Service Manager. The Global Service Manager retrieves from the Global Service Catalogue the information on relevant local services, taking into account the domain, geographical location and requested functions. Thanks to federation, the data from all available local clouds, which relate to the given location, are available on the Global Service Catalogue, through the Local Service Managers that actually work as federation manager clients to a federation manager module that is embedded in the Global Service Catalogue component. The Global Service Manager runs an optimization process for

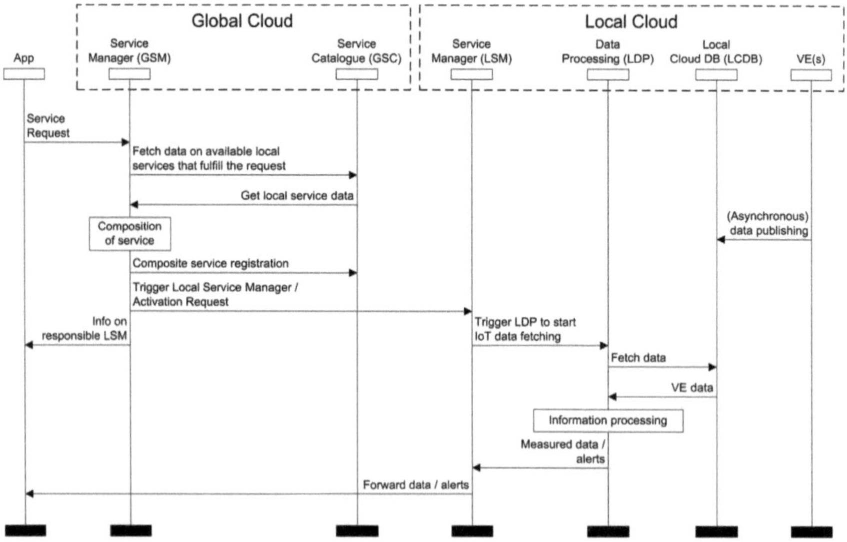

Fig. 9.4 Service creation over multi-Cloud-IoT environments

the selection of the most appropriate local services/VEs and composes the requested service. In turn, the Global Service Manager registers the Complex Service in the Global Cloud Service Catalogue and triggers all selected Local Service Managers. The Local Service Manager triggers the corresponding Local Cloud Data Processing components that in their turn start to fetch data either from the Local Cloud DB or directly from relevant VEs that support data streaming. The Local Data Processing sends to the Local Service Manager the retrieved measurements, alerts, as well as any decision for device/VE configuration (e.g.,: on/off). Essentially, the Local Data Processing performs an initial data processing in terms of real-data and defined thresholds comparison and based on the comparison results it creates decision on devices configuration and produces alerts. The data (measurement and/or configuration commands, alerts) are forwarded to the Application (given that the Service Manager is the main contact point of the application). The Local Service Manager forwards to the Application the produced alerts (including decisions on device configuration) and the retrieved measurements to be displayed to the user.

9.4.2.3 Service Migration on Multi-Cloud-IoT Environments

It is assumed that this scenario (Fig. 9.5) occurs as a continuation of the previous one (Sect. 9.4.2.2). The user is originally located within the home and when his location changes, this trigger a "handover" to a different but more appropriate local cloud (as part of the composite service generated in the previous scenario). This second local cloud in this case is the Smart City local cloud. Initially, a VE (wearable/smartphone providing GeoLocation) sends the new location of the user to the Local Cloud DB. The Local Cloud Data Processing retrieves this context change (GeoLocation change) from the Local Cloud DB or directly from VE data streaming. The Local Data Processing identifies a context change. The Local Data Processing sends a notification to the Application via the Local Service Manager (again this is done as it is assumed that the contact point for the Application is always the (Global or Local) Service Manager). The Local Service Manager informs the Global Service Manager that a cloud switch is required. The Global Service Manager retrieves information on the composite service (complex service) from the Global Service Catalogue to check in which new local cloud the Application should be switched to. The Global Service Manager triggers the appropriate Local Service Manager (in this case it is assumed that there is a switch to the Smart City Local Service Manager). The Application is informed on the switch of Local Service Manager and starts directly the communication with this component.

The Application then sends to the Global Service Manager a request to get the best route for a particular destination (part of the smart mobility service). Having completed the Cloud switch process, it starts the execution of the knowledge exploitation processes that actually in this particular case, uses the generated knowledge so as to detect and propose the best route in the smart city so as the user can transit from one point to another. Since the Global Service Manager receives the application request, it triggers the Global Data Processing. The Global Data Processing retrieves relevant

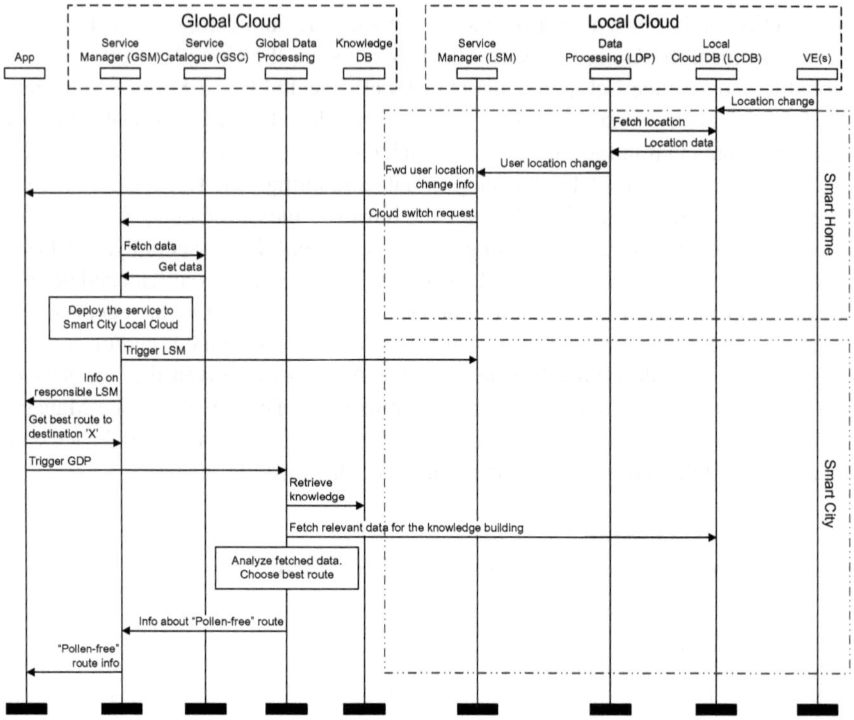

Fig. 9.5 Service migration on inter-Cloud environments

knowledge from the Global Knowledge DB. Indicatively, the knowledge may correspond to data with respect to previous user preferences for the route selection based on traffic data and/or air-quality indicators (e.g., amount of pollution or amount of pollen in the air).

The Global Data Processing interacts with the Smart City Local Cloud DB, via the Security Gateway, in order to obtain knowledge that will enable optimal service provision. Environmental conditions are retrieved, such as transportation conditions, city events (e.g., protests, concerts), weather, pollution, pollen, etc. The Global Cloud Data Processing analyzes the retrieved data. The time to destination is predicted regarding various routes. The best route is chosen, based on multiple criteria (time, pollution, etc.). The user's health status is taken into account for the optimal route selection (e.g., depending on the user sensitivity to pollen, it may be best to select a "pollen-free" route even if it requires more time. The Data Processing forwards the result to the Global Service Manager that in its turn forwards the result to the mobile application.

9.5 Concluding Remarks

The recent years the Internet of Things (IoT) has been actively introduced in the smart cities ecosystems, bringing on the forefront new challenges and requirements with respect to ubiquitous applications, seamless interoperability among heterogeneous systems, Big Data management and processing, as well as to the provision of high performance and reliable IT solutions. At the same time, Cloud Computing participates the technological innovation board, by offering enormous storage, computing facilities and data sharing opportunities.

The convergence of IoT and Cloud can lead to the deployment of Cloud-IoT architectures with innovative and important benefits. In particular, the use of the Cloud for the IoT and the deployment of Cloud-IoT architectures, realize ubiquitous sensing, allow the interconnection of devices, enables the service sharing and provisioning, and provides automated decision making in real time. Moreover, the IoT environments can be benefited by Cloud-IoT convergence in terms of the IoT Big Data storage and processing capabilities beyond the capability of individual "things", as well as through the increase of capacity for the serving of the on-demand requirements for the third-party entities, such as the end-users and applications. Thus, such convergence can enable the development of new innovative applications in various emerging areas such as smart cities, smart grids, smart healthcare and others to improve all aspects of life.

The introduction of the local and global clouds leads to the deployment of a multi-Cloud-IoT architecture that brings added value on the support and the realization of cross-domain applications that are comprised by distributed IoT services. Specifically, taking into account the smart cities paradigm, there are many different IoT platforms and infrastructures that provide services in the context of the smart city, such as traffic monitoring, air-quality monitoring, smart transportation, etc. The application and/or service migration over multiple local cloud environments, coordinated by the global cloud allows the ubiquitous IoT for the end-users, with the provision of applications that combine services in seamless and interoperable way, hiding, in the same time, the technological heterogeneity by the end-user. Moreover, the local and global cloud design approach brings important benefits on the IoT Big Data processing distribution, the storage management and the service performance in terms of accuracy, low-latency and improvements on response time. Specifically, the semantic data federation across the local clouds on the global cloud, as well as the distribution of data processing mechanisms in isolated cloud deployments enhances and evolves the IoT data management minimizing the requirements for: (a) centralized high-capacity storage warehouses, (b) centralized infrastructures with high-capacity and processing power, and (c) high-performance networking infrastructures to support distributed applications and/or services interaction, data exchange and communication. Further to the above, through the design of a multi-Cloud-IoT architecture with local and global cloud deployments, it is achieved the migration of the services near to the end-user. In particular, having local cloud deployments, such as smart home, smart city, smart car, etc., the applications that combine distributed (in local clouds) services, can interconnect directly to the local clouds establishing direct links with the services. Thus, the applications should not interact with a centralized point that will work as proxy to different service providers, but it can interact directly with the local cloud that is required each time in a dynamic and reconfigurable way, coordinated by global cloud.

For the realization of such architecture, with local and global cloud, it should be considered various requirements that are related to: (a) security and privacy aspects, (b) the data management aggregation and exploitation, (c) the real-time on-demand provisioning of services, and (d) the dynamic resource provisioning so as to support applications and services processing and storage demands. The security and privacy aspects should be supported in the local clouds by introducing particular rules and restrictions based on their terms and conditions agreements. In the global cloud level it should be realized a first stage filtering and access control (e.g., the credential-based authentication) of the third-party entities coming into the system. Having managed the security and privacy issues then it should be designed and deployed appropriate cloud-based mechanisms that will support the IoT data management in terms of storage, processing and modification. Moreover, it should be allowed the provision of services based on real-time demands of an application, in order to avoid the provisioning of services that commit Cloud resources and actually stay in an idle and/or nonproductive state. Finally, appropriate mechanisms should be designed and deployed in local and global level so as to manage the resource provisioning (in terms of hardware capabilities), the application offloading, the migration and/or dynamic reconfiguration of services, enabling in this way efficient, high performance and reliable Cloud-IoT environments.

Acknowledgements The work is supported by the collaborative European Union and Ministry of Internal Affairs and Communication, place country-region Japan, Research and Innovation action: iKaaS. EU Grant number 643262.

References

1. Naphade M, Banavar G, Harrison C, Paraszczak J, Morris R (2011) Smarter cities and their innovation challenges. Computer 44(6):32–39
2. Jalali R, El-khatib K, McGregor C (2015) Smart city architecture for community level services through the internet of things. In: 2015 18th international conference on intelligence in next generation networks (ICIN). IEEE, pp 108–113
3. European smart cities 4.0 (2015). http://www.smart-cities.eu/?cid=01&ver=4
4. Li S, Da Xu L, Zhao S (2014) The internet of things: a survey. Inf Syst Front 17(2):243–259
5. Heo N (2015) Deployment issues for the internet of things: a survey. Int Inf Inst (Tokyo). Inf 18(4):1313
6. Al-Fuqaha A, Guizani M, Mohammadi M, Aledhari M, Ayyash M (2015) Internet of things: a survey on enabling technologies, protocols and applications
7. Want R, Schilit BN, Jenson S (2015) Enabling the internet of things. Computer 1:28–35
8. Bonde MJ, Rane KP (2014) Radio frequency identification (RFID) misplaced objects. Int J Electron Electr Eng 7(7):657–662
9. RFID Journal, What is RFID. http://www.rfidjournal.com/articles/view?1339

10. Katiyar K, Gupta H, Gupta A (2014) Integrating contactless near field communication and context-aware systems: improved internet-of-things and cyberphysical systems. In: 2014 5th international conference confluence the next generation information technology summit (confluence). IEEE, pp 365–372
11. Near Field Communication (NFC). http://www.nearfieldcommunication.org/
12. Yang K (2014) Wireless sensor networks: principles, design and applications
13. Vlacheas P et al (2013) Enabling smart cities through a cognitive management framework for the internet of things. IEEE Commun Mag J Papers 51(6):
14. Foteinos V (2014) A cognitive management framework for enabling autonomous applications in the internet of things. IEEE Veh Technol Mag
15. Zanella A, Bui N, Castellani A, Vangelista L, Zorzi M (2014) Internet of things for smart cities. IEEE Internet Things J 1(1):22–32
16. Jin J, Gubbi J, Marusic S, Palaniswami M (2014) An information framework for creating a smart city through Internet of things. IEEE Internet of Things J 1(2):112–121
17. Misra P, Rajaraman V, Dhotrad K, Warrior J, Simmhan Y (2015) An interoperable realization of smart cities with plug and play based device management. arXiv:1503.00923
18. IERC—European Research Cluster on the Internet of Things. http://www.internet-of-things-research.eu/
19. IoT-A FP7 EU Project (2013) Duration: 2010–2013. Grant Agreement No. 257521. http://www.iot-a.eu/public/
20. VITAL FP7 EU Project (2013) Duration: 2007–2013. Grant Agreement No. 608682. http://vital-iot.eu/
21. EBBITS FP7 EU Project (2014) Duration: 2010–2014. Grant Agreement No. 257852. http://www.ebbits-project.eu/
22. iSURF EU FP7 Project (2013) Duration: 2007–2013. Grant Agreement No. 213031. http://www.isurfproject.eu/
23. IoT.est EU FP7 Project (2014) Duration: 2011–2014. Grant Agreement No. 288385. http://ict-iotest.eu/iotest/
24. iCore FP7 EU Project (2014) Duration: 2011–2014. Grant Agreement No. 287708. http://www.iot-icore.eu/
25. BUTLER FP7 EU Project (2014) Duration: 2011–2014. Grant Agreement No. 287901. http://www.iot-butler.eu/
26. Rabbachin A, Quek TQS, Shin H (2011) Cognitive network interference. IEEE J Sel Areas Commun 29(2):480–493
27. Howard PN (2015) Pax technica: how the internet of things may set us free or lock us up. Yale University Press
28. Clusters of European projects on cloud, new approaches for infrastructure services. https://eucloudclusters.wordpress.com/new-approaches-for-infrastructure-services/
29. Community Research and Development Information Service—CORDIS. ICT-07-2014 - Advanced Cloud Infrastructures and Services (2014). http://cordis.europa.eu/programme/rcn/664792_en.html
30. Bernstein D, Ludvigson E, Sankar K, Diamond S, Morrow M (2009) Blueprint for the intercloud - protocols and formats for cloud computing interoperability. IEEE Comput Soc:328–336. doi:10.1109/ICIW.2009.55
31. Sotiriadis S, Bessis N, Anjum A, Buyya R (2015) An inter cloud meta-scheduling (ICMS) simulation framework: architecture and evaluation
32. Ngan LD, Kanagasabai R (2013) Semantic web service discovery: state-of-the-art and research challenges. Pers Ubiquit Comput 17(8):1741–1752
33. Wang P, Ding Z, Jiang C, Zhou M (2014) Automated web service composition supporting conditional branch structures. Enterp Inf Syst 8(1):121–146
34. Pawar PS, Sajjad A, Dimitrakos T, Chadwick DW (2015) Security-as-a-Service in multi-cloud and federated cloud environments. In: Trust management IX. Springer, pp 251–261
35. Darabseh A, Al-Ayyoub M, Jararweh Y, Benkhelifa E, Vouk M, Rindos A (2015) Sdstorage: a software defined storage experimental framework. In: 2015 IEEE international conference on cloud engineering (IC2E). IEEE, pp 341–346

36. Alba A, Alatorre G, Bolik C, Corrao A, Clark T, Gopisetty S, Traeger A (2014) Efficient and agile storage management in software defined environments. IBM J Res Dev 58(2/3):5–1
37. Nunes B, Mendonca M, Nguyen XN, Obraczka K, Turletti T (2014) A survey of software-defined networking: past, present, and future of programmable networks. IEEE Commun Surv Tutor 16(3):1617–1634
38. Kreutz D, Ramos FM, Esteves Verissimo P, Esteve Rothenberg C, Azodolmolky S, Uhlig S (2015) Software-defined networking: a comprehensive survey. Proc IEEE 103(1):14–76
39. Zhou H (2012) The internet of things in the cloud: a middleware perspective. CRC Press
40. Biswas AR, Giaffreda R (2014) IoT and cloud convergence: opportunities and challenges. In: 2014 IEEE world forum on internet of things (WF-IoT). IEEE, pp 375–376
41. Botta A, de Donato W, Persico V, Pescapé A (2014) On the Integration of cloud computing and internet of things. In: 2014 international conference on future internet of things and cloud (FiCloud). IEEE, pp 23–30
42. Demchenko Y, Ngo C, De Laat C, Rodriguez J, Contreras LM, Garcia-Espin JA, Ciulli N (2013) Intercloud architecture framework for heterogeneous cloud based infrastructure services provisioning on-demand. In: 2013 27th international conference on advanced information networking and applications workshops (WAINA). IEEE, pp 777–784
43. Bruneo D, Fritz T, Keidar-Barner S, Leitner P, Longo F, Marquezan C, Woods C (2014) Cloudwave: where adaptive cloud management meets devops. In: 2014 IEEE symposium on computers and communication (ISCC). IEEE, pp 1–6
44. Hiray S, Ingle R (2013) Context-aware middleware in cyber physical cloud (CAMCPC). In: 2013 international conference on cloud & ubiquitous computing & emerging technologies (CUBE). IEEE, pp 42–47
45. Suciu G, Vulpe A, Halunga S, Fratu O, Todoran G, Suciu V (2013) Smart cities built on resilient cloud computing and secure internet of things. In: 2013 19th international conference on control systems and computer science (CSCS). IEEE, pp 513–518
46. Roman R, Zhou J, Lopez J (2013) On the features and challenges of security and privacy in distributed internet of things. Comput Netw 57(10):2266–2279
47. Villari M, Celesti A, Fazio M, Puliafito A (2014) A secure self-identification mechanism for enabling iot devices to join cloud computing. In: Internet of things. IoT infrastructures. Springer, pp 306–311
48. Cloud-of-Things—(ClouT) Project. Call: FP7-ICT-2013- EU-Japan. Type: STREP. Duration: 3 years (01 Apr 2015). http://clout-project.eu
49. MUlti-cloud Secure Applications (MUSA) Project. Call: H2020-ICT-2014-1. Topic: ICT-07-2014, Type: RIA, Duration: 3 years (1 Jan 2015). http://www.musa-project.eu
50. Rios E, Iturbe E, Orue-Echevarria L, Rak M, Casola V (2015) Towards self-protective multi-cloud applications—MUSA-a holistic framework to support the security-intelligent lifecycle management of multi-cloud applications. In: CLOSER 2015—proceedings of the 5th international conference on cloud computing and services science, Lisbon, Portugal, 20–22 May 2015. SciTePress, pp 551–558. ISBN: 978-989-758-104-5
51. Heterogeneous Secure multi-level remote acceleration service for low-power integrated systems and devices—(RAPID) Project. Call: H2020-ICT-2014-1, Topic: ICT-07-2014, Type: RIA, Duration: 3 years (1 Jan 2015). http://www.rapid-project.eu
52. CUDA Toolkit Documentation. http://docs.nvidia.com/cuda/cuda-runtime-api/#axzz4237B8rWd
53. In-network programmability for next-generation personal cloud service support (INPUT) Project. Call: H2020-ICT-2014-1, Topic: ICT-07-2014, Type: RIA, Duration: 3 years (01 Jan 2015). http://input-project.eu
54. Lyberopoulos G, Theodoropoulou E, Mesogiti I, Filis K, Bruschi R, Lago P, Lombardo C, Atzori L, Lera A, Morabito G (2015) INPUT: a distributed cloud infrastructure for intelligent transport systems. In: 19th international conference on circuits, systems, communications and computers (CSCC 2015), Zakynthos Island, Greece, 16–20 July 2015
55. SPECS: secure provisioning of cloud services based on SLA management. Call: FP7-ICT-2013-1, Topic: Trustworthy ICT, Type: STREP, Duration: 2, 5 years (1 Nov 2013). http://www.specs-project.eu

56. SOCIOTAL: creating a socially aware and citizen-centric internet of things. Project reference: 609112. Funded under: FP7-ICT. From 01 Sept 2013 to 31 Sept 2016. http://sociotal.eu/
57. COSMOS: cultivate resilient smart objects for sustainable city applications. Project reference: 609043. Funded under: FP7-ICT. From 01 Sept 2013 to 31 Sept 2016. http://iot-cosmos.eu/
58. CITY PULSE: real-time IoT stream processing and large-scale data analytics for Smart City applications. Project reference: 609035. Funded under: FP7-ICT. From 01 Sept 2013 to 31 Sept 2016. http://www.ict-citypulse.eu/page/
59. FIESTA: federated interoperable semantic IoT/cloud testbeds and applications. Project reference: 643943. Funded under: H2020-EU.2.1.1.3. From 01 Feb 2015 to 01 Feb 2018, ongoing project. http://www.fiesta-iot.eu/
60. Intelligent Knowledge-as-a-Service (iKaaS) Project. Call: H2020-EUJ-2014. Topic: EUJ-1-2014. Type: R&I. Duration: 3 years (1 Oct 2014). http://www.ikaas.com
61. Hidano S, Kiyomoto S, Murakami Y, Vlacheas P, Moessner K (2015) Design of a security gateway for iKaaS platform. In: Proceedings of the 6th EAI international conference on cloud computing, Daejeon, South Korea, 28–29 Oct 2015
62. Hashimoto K, Yamada K, Tabata K, Oda M, Suganuma T, Biswas AR, Vlacheas P, Stavroulaki V, Kelaidonis D, Georgakopoulos A (2015) iKaaS data modelling: a data model for community services and environment monitoring in Smart City. In: Proceedings of the 3rd international workshop on self-aware internet of things (Self-IoT2015), in conjunction with ICAC 2015, The 12th IEEE international conference on autonomic computing
63. Kelaidonis D, Somov A, Poulios G, Foteinos V, Stavroulaki V, Vlacheas P, Demestichas P (2013) A cognitive management framework for smart objects and applications in the internet of things. Mobile networks and management. Springer, Berlin, Heidelberg, pp 196–206
64. Kelaidonis D, Somov A, Foteinos V, Poulios G, Stavroulaki V, Vlacheas P, Demestichas P, Baranov A, Biswas AR, Giaffreda R (2012) Virtualization and cognitive management of real world objects in the internet of things. In: 2012 IEEE international conference on green computing and communications (GreenCom). IEEE, pp 187–194

Chapter 10
Future Internet Systems Design and Implementation: Cloud and IoT Services Based on IoT-A and FIWARE

Stelios Sotiriadis, Kostantinos Stravoskoufos and Euripides G.M. Petrakis

10.1 Introduction

The Cloud and the Internet of Things (IoT) are the main technologies enabling transformation of the living or work environment (e.g., houses, cities, factories) into one which is characterized by a new model of services for automated and smarter management of infrastructures and machines and of their interaction with people. The new environment (e.g., a house at the finer scale or building, neighborhood, smart city at a larger scale) becomes adaptable to the everyday needs of people enhancing their quality of life, work and productivity while taking into account environmental issues (e.g., low energy consumption, better management of resources, waste management, etc.).

A future public cloud platform will offer services that will be available wherever the end user might be located. This approach enables easy access to information and accommodates the needs of users in different time zones and geographic locations [1]. Moreover, cloud computing will offer quick deployment and ease of integration as software integration occurs automatically and organically in cloud installations. We expect that future smart city platforms will utilize a robust cloud architecture thus providing resiliency and redundancy to its users.

S. Sotiriadis (✉) · K. Stravoskoufos · E.G.M. Petrakis
Department of Electronic and Computer Engineering, Technical University of Crete, Chania, Greece
e-mail: s.sotiriadis@intelligence.tuc.gr

K. Stravoskoufos
e-mail: kgstravo@intelligence.tuc.gr

E.G.M. Petrakis
e-mail: petrakis@intelligence.tuc.gr

V. Angelakis et al. (eds.), *Designing, Developing, and Facilitating Smart Cities*,
DOI 10.1007/978-3-319-44924-1_10

Existing IoT-based smart city architectures have been conceptualized and implemented to address certain challenges based on use case-oriented requirements, thus not considering issues of openness, scalability, interoperability, and use case independence. As a result, they are less principled, lacking standards, and are vendor or domain specific. Most importantly, they are hardly replicable in the sense that the same architecture cannot be used in more than one use cases, creating the IoT silos. Cloud-based smart city deployments aim to go beyond that, by considering an open and dynamic configurable IoT platform. This goes beyond the current solutions' aims that are mainly vertically closed, so forming many "Intranets of Things" rather than an "Internet of Things" [2].

The lack of standardization in the IoT domain has resulted in the fragmentation of the approaches in IoT systems design and implementation. To address the fragmentation of existing IoT solutions, the IoT-A project [3] proposes an architecture reference model that defines the principles and standards for generating IoT architectures and promoting the interoperation of IoT solutions. IoT-A compliant architectures assure that generated knowledge will be modular and reusable across domain or use case specific boundaries. However, IoT-A addresses the architecture design problem, and does not focus on whether existing cloud platforms can offer the tools and services to support the implementation of IoT-A compliant IoT systems.

Currently, cloud computing offers a variety of services including hardware and software. FIWARE[1] moves future Internet application design a step forward by offering a large scale of cloud services to be built upon the concept of scalability and elasticity as discussed in [4]. FIWARE platform comprises a set of Generic Enablers (GEs) that are considered general purpose and common to several usage areas. We expect that cloud computing will offer flexibility so to meet variations on demand. Further, it will characterize a multi-consumer to provider model by allowing seamless interoperability between its features as discussed in [5]; this in conjunction with FIWARE services offers a new method of designing novel tools for the emerging needs of future IoT. Other benefits include, cost-effective infrastructure for IoT use cases as cloud service, as cloud services are in general available at much cheaper rates than traditional approaches and can significantly lower the overall IT expenses. Cloud computing delivers a better cash flow by eliminating the capital expense (CAPEX) associated with developing and maintaining the server infrastructure.

Leveraging on results from EU-funded research, we propose to design IoT systems deriving from IoT-A based on FIWARE, the EU initiative for cloud services provision and one the candidate reference cloud implementations that can be exploited in the Future Internet. This work extends the work of [6] by providing a comparison of the functionality of FIWARE GEs against IoT-A requirements and guidelines and point out weaknesses and missing points along with suggestions.

[1]http://www.fiware.org.

10.2 Background

Latest years, the concept of virtualized service infrastructures (also known as cloud service providers) has emerged towards developing Future Internet (FI) application tools and services for next generation systems that can be easily reused for smart city applications. Various cloud providers have developed and promoted their services and gained significant support from EU FP7 programme to allow Use Case (UC) trials projects with novel FI applications as discussed by [7]. In particular, different FI-PPP[2] projects have been formed to validate and test their features in various domains (e.g., healthcare, industry, cities) to build systems for the health care (FI-STAR[3]) media, network creativity (FI-CONTENT[4]), business collaboration services (FISPACE[5]) smart energy (FINENCE[6]), and smart virtual factories (FITMAN[7]) domains.

In this perspective, various users, including public administration, users and other stakeholders, e.g., Small Medium Enterprises (SMEs) need to adapt their own solutions to the new standards (e.g., FIWARE), rather than building from the scratch their systems as the current FI-PPP Use Case projects do (projects FI-STAR, FI-CONTENT, FISPACE, FINENCE, FITMAN referred to above). The challenges that they have to face in order to port or adapt their applications and build an IoT service infrastructure compliant environment are huge. This includes, characterization of the granularity of their systems to define system key operations and mapping to available services offered by the various virtualized service providers as cloud offerings. An almost orthogonal issue refers to porting applications across cloud providers thus dealing with the "vendor lock-in" problem. The adoption of IoT-A promotes the interoperation of IoT solutions by enabling interoperability at the communication level (how IoT and cloud system interoperate) and at the service level (how services are integrated) as discussed in [3]. Next list presents a discussion of related initiatives and EU projects.

- FP7 IoT-A[8] project created the architectural foundations of the Future Internet of Things, allowing seamless integration of heterogeneous IoT technologies into a coherent architecture. IoT-A ARM enabled the interoperability of IoT systems, outlining guidelines for technical design of its protocols, interfaces, and algorithms, as well as the corresponding mechanism for its efficient integration into the service layer of the Future Internet.

[2]https://www.fi-ppp.eu.

[3]http://www.fi-star.eu.

[4]http://www.fi-content.eu.

[5]http://www.fispace.eu.

[6]http://www.finence.eu.

[7]http://www.fitman.eu.

[8]http://www.iot-a.eu.

- FP7 RERUM[9] is a new project that aims to incorporate the concepts of security, privacy, and reliability by design in the IoT architecture, with a specific focus on smart city applications.
- FP7 IoT@Work[10] focused on harnessing IoT technologies in industrial and automation environments to realize the Plug-and-Work of production units. The IoT@Work project adopted a Capability-Based Access Control mechanism for managing access control to its Event Notification Service (ENS) middleware.
- FP7 OpenIoT[11] promoted semantic approaches to naming, addressing, and discovery. Its solution is based on the creation of a distributed directory service which will include semantically annotated resources.
- FP7 iCore[12] addressed the issues of abstracting the technological heterogeneity that derives from the vast amounts of heterogeneous objects, while enhancing reliability. iCore developed a cognitive framework comprising virtual objects, composite virtual objects and functional blocks for user/stakeholder perspectives.
- IERC[13] is an initiative of the EU to bring together projects that are working in the IoT for defining a common vision of the technology in view of global development.
- FI-PPP[14] is a EU initiative that aims to (i) increase the effectiveness of business processes and infrastructures supporting applications in areas such as transport, health, and energy and (ii) to derive innovative business models that strengthen the competitive position of European industry in sectors such as telecommunication, mobile devices, software and services, and content provision and media.
- FIWARE[15] offers specifications and tools to build FI applications from a variety of GEs with essential functionalities, interfaces, and APIs. The project will utilize such tools to implement its service and platform components, developing of "cloud of public services" that increase interoperability and enable incorporation of resources and services independent of their location.
- FI-PPP FITMAN[16] aims to provide 10 industry-led use case trials in the domain of Smart, Digital and Virtual Factories of the Future. FITMAN Trials will test and assess the suitability, openness, and flexibility of FIWARE GEs in several manufacturing sectors such as automotive, aeronautics, white goods, furniture, textile/clothing, LED lighting, plastic, construction, and manufacturing assets management.

[9]http://www.ict-rerum.eu.

[10]http://www.iot-at-work.eu.

[11]http://openiot.eu.

[12]http://www.iot-icore.eu.

[13]http://www.internet-of-things-research.eu/.

[14]http://fi-ppp.eu.

[15]http://www.fiware.org.

[16]http://www.fitman-fi.eu.

- ETP EPoSS[17] is an industry-driven policy initiative, defining R&D and innovation needs as well as policy requirements related to Smart Systems Integration and integrated Micro- and Nanosystems.
- 5G-PPP[18] is joint initiative of the EU ICT industry and the EC to develop the infrastructure for the next generation of communication networks, which includes IoT as one of its main components, from which the 5G extracts the dense machine to machine communication scenario requirements.

10.3 Correlation Between FIWARE and IoT-A

We intent to build a Cloud-based IoT architecture that it is based on IoT-A and FIWARE conceptual models. The IoT-A project introduces an IoT Architecture Reference Model (ARM) for generating reference architectures based on domain specific requirements as in [3]. The reference model consists of a set of concepts, axioms, and relationships that formulate an abstract framework for understanding the relationships among the entities of the IoT. This allows generation of more than one reference architectures that can be used as blueprints for designing concrete architectures.

The relationship between the IoT-A reference architecture, the Cloud-based IoT architecture and the Cloud-based IoT actual system implementation is illustrated in Fig. 10.1, as an adaptation of a figure presented in [8]. The Cloud-based IoT architecture based on the utilization of (a) the IoT-A as the reference architecture to conceptualize essential features and (b) the FIWARE as the implementation and the environment for developer's communities to test and validate in large-scale experiments.

The Cloud-based IoT reference architecture will provide a set of key building blocks that will be designed as application independent components to propose a solution. Moreover, it will be implemented as platform independent allowing various implementations across different platforms and will rely on the FIWARE platform aiming to large-scale demonstrations and validation driven to innovative use scenarios. Figure 10.2 illustrates the Cloud-based IoT functional model.

Initially, architecture development starts with definition of model referred to as Cloud-based IoT ARM consists of several sub-models that are as follows:

- **Cloud-based IoT Domain Model**: This component accomplishes the composition of the main concepts of the IoT-like Devices, IoT Services and relations between them that are technology and use case independent. It defines user types along with their roles and interactions with other Domain concepts. In Cloud-based IoT the domain model of IoT-A is realized by means of UML [9] diagrams (as it is typical in IoT-A) or XML [10] XML Schemas or even better

[17] www.smart-systems-integration.org.

[18] http://5g-ppp.eu.

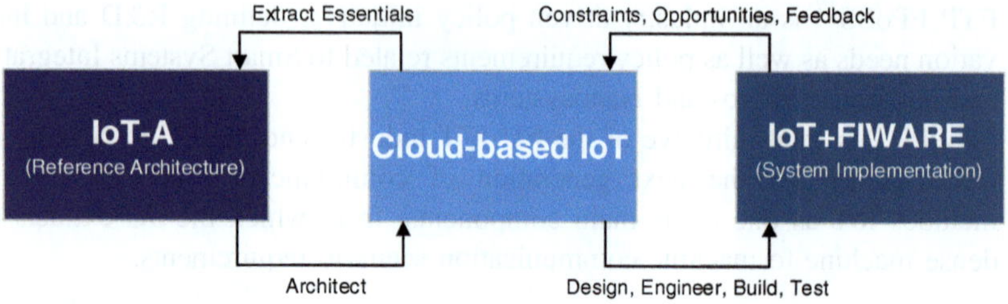

Fig. 10.1 Relationship between reference architecture, architecture, and implementation

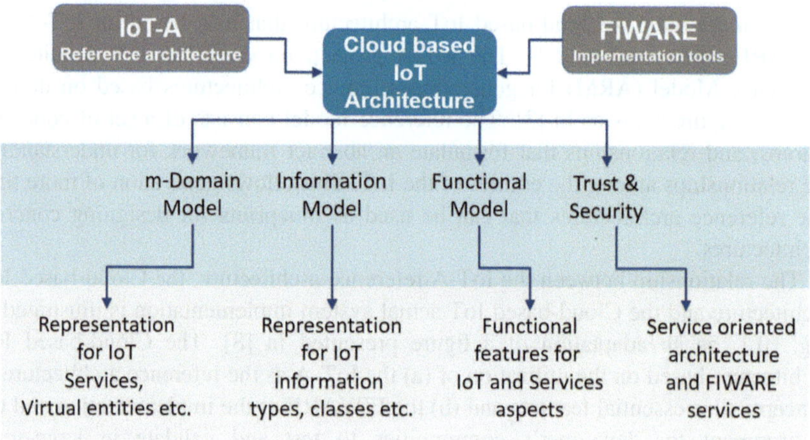

Fig. 10.2 The Cloud-based IoT functional model

for expressing their semantic meaning, in terms of classes, class hierarchies, and of their interrelationships (in terms of data or object properties and constraints applying on data and object properties) using ontologies and RDF [11] or OWL [12] representations of ontologies. Entity classes (e.g., mobile phones, body sensors, humidity, temperature sensors) or their instantiations (e.g., specific sensor products) are instantiated to the Domain Model (e.g., the ontology) and are linked to the services associated with them (which in turn can be ontologies). The domain model is therefore, a blue-print of all entities pertaining to the IoT platform at any time and can be queried (e.g., using SPARQL [13] for an ontology representation) for obtaining information about supported devices and services provided by Cloud-based IoT platform. The domain model will bring important benefits to the Cloud-based IoT as follows:

– Represents the virtual entities (users, devices, systems etc.), their relationships and identifies their attributes. The domain model provides a structural view of the domain that can be complemented by other dynamic views, such as the Cloud-based IoT use case models.

  - In Cloud-based IoT the domain model will give an important advantage as it describes and constrains the scope of the use case domain as can be effectively used to verify and validate the understanding of the problem domain among various stakeholders.

- **Cloud-based IoT Information Model**: This model is complementary to the domain model. It defines a minimal abstract framework (i.e., a meta-model) for explaining common information about elements or concepts defined in the IoT Domain (or of concepts associated with concepts in the Domain Model but not represented there, e.g., applicability of concepts). Similarly to the Domain Model, in Cloud-based IoT, the information model is realized in UML or as an ontology. In Cloud-based IoT we may opt of a unified and uniform representation of both models, e.g., by a single ontology.
- **Cloud-based IoT Functional Model**: This will identify functionality groups (FGs) which are grounded in key concepts of the IoT Domain Model. The Functionality Groups provide the functionalities for interacting with the instances of concepts or managing the information related to concepts. In Cloud-based IoT the IoT Communication model introduces concepts for handling the complexity of communication in heterogeneous IoT environments. Communication also constitutes one FG in the IoT Functional Model. In Cloud-based IoT, the functional model will provide significant advantages as follows:
  - **Process management**: Provides functional concepts necessary to conceptually integrate the IoT world into traditional (business) processes.
  - **Service organization**: Provides the communication hub between several other FGs. The Service Organisation FG is responsible for resolving and orchestrating IoT Services and also deal with the composition and choreography of Services.
  - **Services**: Resources associated to the devices are exposed as IoT Services on the IoT Service level. This has two functional components, (a) the IoT Service that is responsible for handling information from resources and (b) the IoT Service Resolution, which provides all the functionalities needed by the user in order to find and be able to contact IoT Services.
  - **Virtual entities**: It models higher level aspects of the physical world, and these aspects can be used for discovering Services.
  - **Management**: It contains all functionalities related to controlling and administrating an IoT system; will provide benefits such as cost reduction, attending unexpected usage issues, fault handling, and flexibility test.
  - **Communication**: It is the low-level communication aspects and the various protocols.
- **Cloud-based IoT Trust, Security and Privacy (TSP) Model**: Introduces the relevant functionalities and interdependencies of Trust, Security and Privacy. As in the case of communication, TSP constitutes one FG in the Functional Model. In Cloud-based IoT the functional model and all its Functional Groups will be implemented as services formulating the generic SOA [14] architecture of the

platform. More specifically, as the Cloud-based IoT platform will be deployed on FIWARE, each service will be a FIWARE enabler implementing a REST API [15].

Regarding functionality features, the architecture is divided into the **front-end** and **back-end system**. The front-end is the user and IoT devices layer that includes the following:

- **The Cloud-based IoT Core**: is the abstractions hub that performs abstractions of the IoT resources at two levels: (i) at the physical devices level, translating them to abstract virtual entities and (ii) at the services level. The abstraction of IoT resources provides the basis for simplified management, discovery of heterogeneous embedded devices, and assurance that interactions between them and the orchestration of these interactions take place in a uniform manner.
- **The Cloud-based IoT Modular OS (Cloud-based IoTOS)**: framework includes a modular and composable operating system specifically designed for IoT with main characteristics its real-time support and portability. The Cloud-based IoTOS will include schedulers, tasks, routines, and communication modules, fully embracing the Cloud-based IoT-core abstraction models, providing abstractions for sensors and controllers. The Cloud-based IoTOS include also a software layer with compilers and linkers that automate these phases for the hardware platforms of the Cloud-based IoT architecture. The automated procedures will also include the output of executables that will be loaded automatically on the hardware devices.

The back-end is the user and IoT devices layer that includes the following:

- **The Cloud-based IoT Middleware (Platform)**: It supports the deployment of the applications, playing the role of the intermediate layer between the applications and the operating system of the embedded devices. The middleware will be secure and privacy-preserving by design aligned with the overall approach of the Cloud-based IoT architectural framework. The Middleware will be designed as an Event-Driven Architecture and it will incorporate mechanisms for access control, authentication, encryption, and identity management. It will also implement intelligent filtering and classification of events, as well as messaging and notification for human users (context awareness mechanisms). The platform will include process and device management, service organization and management, virtual entities management, security and trust and communication protocols. The platform will host services and will be based on FIWARE to offer an elastic environment that can scale on demand providing computational power, storage, and network according to the IoT platform's needs and demands.
- **The Cloud-based IoT Interfaces**: These are graphical user interfaces that aim to facilitate the management of IoT resources and allow the interaction of the developers/designers with the overall system of IoT. The interfaces will be the front-end for different kinds of users (designers, developers, engineers, programmers) to design IoT applications and to browse, request, and allocate resources to services.

10.4 Implementation: Technological Roadmap

Implementation of the Cloud-based IoT model utilizes the main Functional Groups (FGs) of the IoT-A Functional model illustrated in Fig. 10.3 [3]. It is comprised by several modules that interact with each other.

Each functional group can be realized in terms of FIWARE GEs that correspond to its key functionalities. The role and adequacy in of these GEs in implementing Cloud-based IoT indented functionality will be analyzed in the course of the Cloud-based IoT project. One possible result of this analysis is that certain GEs are not appropriate while other are missing from FIWARE and these will have to be redesigned, implemented, and suggested for inclusion to FIWARE catalogues. Figure 10.4 shows the implementation plan and the association between Cloud-based IoT modules (corresponding to IoT-A FGs) and FIWARE GEs.

In Fig. 10.4, each FG is associated with one GE (the Services FG is associated with three GEs). These are as follows:

- **The IoT Process Management**: This module provides the functional concepts necessary to conceptually integrate the IoT world into traditional (business) processes. In Cloud-based IoT a possible direction is the already implemented solution of FIWARE namely as the IoT broker GE. This GE retrieves and aggregates information from IoT devices acting as a middleware component that separates IoT applications from devices. The GE is based on NGSI[19] and closes the gap between information-centric applications and device-centric IoT installations by communicating simultaneously with large quantities IoT gateways and devices in order to obtain exactly the information that is required by the running IoT applications.Using the broker all IoT devices can be abstracted as NGSI entities on a higher level hiding the complexity of the Internet of Things from developers.
- **The Service Organisation Module**: This is a major module that acts as a communication hub between several other modules. The Service Organisation is responsible for resolving and orchestrating IoT Services and also deal with the composition and choreography of Services. It contains three functional Components as follows:
  - Service Orchestration that is used for orchestrating services. Further it resolves the appropriate services that are capable of handling the IoT User's request.
  - Service Composition that resolves services that are composed of IoT services and other services in order to create services with extended functionality (e.g., the combination of a humidity sensing Service and a temperature Service could serve as input for an air-conditioning)
  - Service Choreography, which supports service brokerage.

[19]http://technical.openmobilealliance.org/Technical/technical-information/release-program/current releases/ngsi-v1-0.

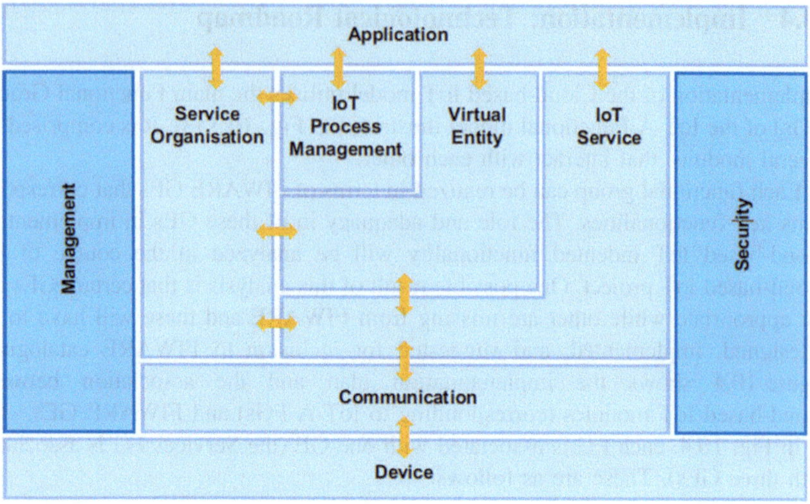

Fig. 10.3 Cloud-based IoT functional model [3]

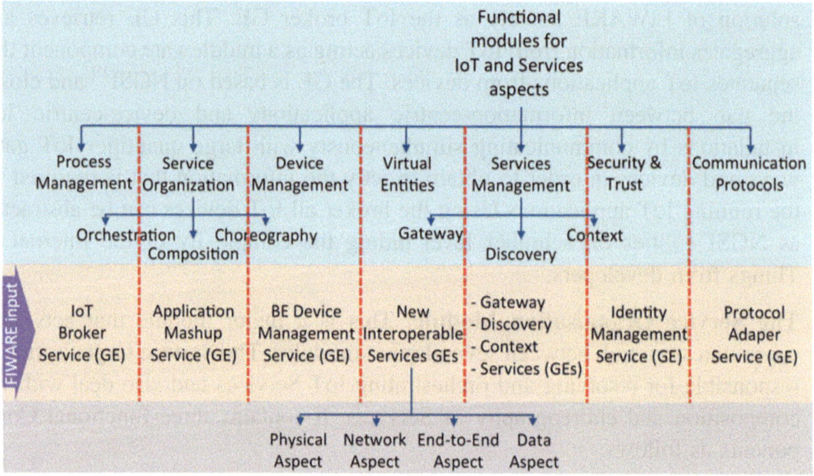

Fig. 10.4 Cloud-based IoT implementation plan (association among Cloud-based IoT modules and FIWARE GEs)

In Cloud-based IoT the Service Organisation module could be mapped to the *FIWARE Application Mashups GE*. This GE is responsible for developing Web application mashups combining information and data from various services. Also, it covers the required functionality for the Service Composition and Service Orchestration functional components. In Cloud-based IoT we will further implement a service Choreography component that will handle the service brokerage concept allowing services to subscribe to other services.

- **The Service Module**: The Resources associated to the devices are exposed as IoT Services on the IoT Service level. This has two functional components, (a) the IoT Service FC that is responsible for handling information from resources and (b) the IoT Service Resolution FC which provides all the functionalities needed by the user in order to find and be able to contact IoT Services. In Cloud-based IoT this module will be realized in FIWARE by various services as follows:
  - The Gateway Data Handling GE is designed to provide a common access in real time to all data, for any kind of sensors and "Things." Using a simple local XML storage, this enabler can save and locally store relevant processed data, as close as possible to the processed entities. It offers filtering, aggregating, and merging real-time data from different sources. Transforms data into events in a way that applications need only to subscribe to events (data), which is relevant to them.
  - The IoT Discovery GE allows context producers to register their IoT Objects in linked-data format, and in turn allows context consumers to discover them using a set of search techniques. Focuses on semantically annotated IoT descriptions and supports querying via SPARQL. The IoT Discovery GE offers a device discovery mechanism but it does not offer a service resolution mechanism. In Cloud-based IoT we will provide a service resolution component based on an ontology of services that will enable the look-up and discovery of services.
  - The Context Broker GE that acts as a mediator between consumer producers (e.g., sensors) and context consumer applications (e.g., a Smartphone using data from sensors). The context broker allows user to register context producer applications, update context information, receive notification on changes on context information or with a given and query context information.
- **Virtual Entity Module**: It models higher level aspects of the physical world, and these aspects can be used for discovering Services. Contains functions for interacting with the IoT System on the basis of VEs, as well as functionalities for discovering and looking up Services that can provide information about VEs, or which allow the interaction with VEs. Currently, there is no GE in FIWARE to handle the association of virtual entities of the Virtual Entity module to the services in the Service module. In Cloud-based IoT we will use ontologies to model virtual entities and associate them to services. Service discovery on this level (for services that provide information on VEs) is another thing that is not implemented by current FIWARE GEs. In Cloud-based IoT we will implement a service discovery component for looking up Services that can provide information about VEs
- **Management Module**: It contains all functionalities related to controlling and administrating an IoT system. The module includes: (a) Cost reduction: control the cost of a system based on user/use cases etc., (b) attending unexpected usage issues: provides strategies and actions for handling unforeseen situations (e.g., link

down, device malfunctioning, etc.), (c) fault handling: addresses the unpredictability of the future behavior of the whole system, (d) flexibility test: Handle changes in the user requirements without rebuilding the system. In Cloud-based IoT, the Backend Device Management GE is the FIWARE implementation that provides an Admin REST API for M2 M application developers and a device communication API for device (sensor/actuators/gateways) communication which currently implements the SensorML and Lightweight SensorML following protocols [16] The GE Collects data from devices (status etc) and translates them into NGSI events available at a context broker. Application developers can use data and send commands through the broker. An open source Reference Gateway called "FIGWAY"[20] is also offered for Raspberry PI and Z-wave devices. The Backend Device Management offers administrator capabilities for devices through the FIGWAY gateway but cannot address any of the system goals other than cost reduction. Moreover, The GE currently works only with FIGWAY meaning that it can collect data and manage only specific devices controlled by the Raspberry PI and Z-wave gateways. It should be extended to support any kind of gateway and support all protocols other than CoaP (e.g., XMPP, MQTT, REST). In Cloud-based IoT we will design and implement components to handle unexpected usage issues, system faults and changes in user requirements

- **Security Module**: It contains all functionalities and interdependencies of Trust, Security and Privacy. The Identity Management GE provides secure and private authentication from users to devices, networks and services, authorization and trust management, user profile management, and privacy-preserving disposition of personal data. Cloud-based IoT will explore further this module.
- **Communication Module**: It contains the low-level communication aspects, and the FIWARE implementation of the Protocol Adapter GE allows pluging of devices using on CoaP over 6LowPan protocol into the FIWARE ecosystem and have access to them. It currently supports only the IBM mote hardware running the moterunner operating system.[21] The protocol Adapter is currently working with specific devices over a specific platform. Cloud-based IoT will explore technological frameworks with available devices over a common platform.

10.5 Deployment: FIWARE Utilization

The Cloud-based IoT Platform Deployment and Instantiation to Use Cases (UC) will be deployed over FIWARE cloud. FIWARE nodes (such as the one deployed at the Technical University of Crete—TUC) can support the deployment of the Cloud-based IoT platform and services solution to the cloud. TUC hosts the

[20]https://github.com/telefonicaid/fiware-figway/.

[21]http://www.zurich.ibm.com/moterunner/.

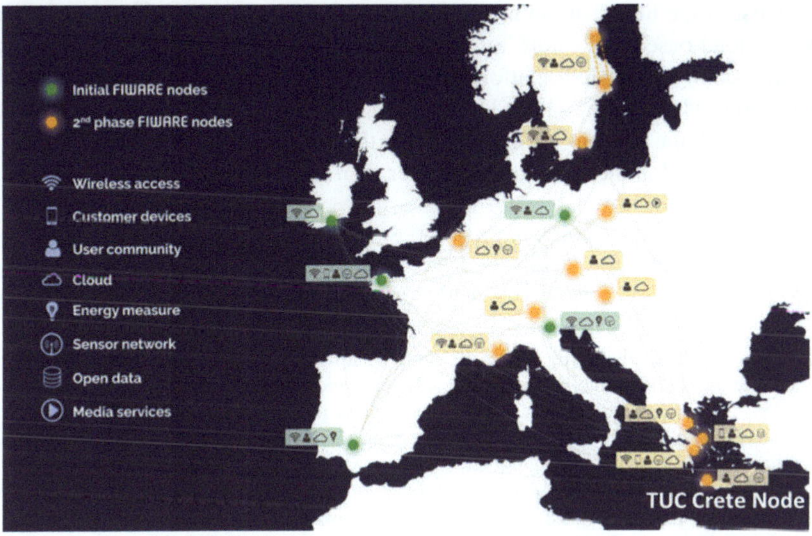

Fig. 10.5 FIWARE nodes around Europe

"Crete" (openstack cloud infrastructure) in the FIWARE Lab[22] Infrastructure Federation, a sustainable pan-European open federation of test infrastructures. Figure 10.5 demonstrates the FIWARE nodes around Europe.[23]

The FIWARE Lab infrastructure federation comprises 19 nodes and as such it can cope with large trial deployments and can serve the various needs of a broad set of Future Internet users and experimenters. Recently, TUC "Crete Node" has received the Silver label complying with the Silver quality validation criteria of its infrastructure. The Crete node hosts all available FIWARE GEs that will be utilized in the development of the Cloud-based IoT platform and for the instantiation of the UCs. To ensure data privacy and security, we opt to provide secure VPN connection over Cloud-based IoT services.

10.6 Concluding Remarks

Existing IoT architectures have been conceptualized and implemented to address certain challenges based on single domain and single use case-oriented requirements, thus not considering issues of openness, scalability, interoperability, and use case independence. As a result they, are less principled, lacking standards and are vendor or domain specific. Most importantly, they are hardly replicable in the sense

[22]http://www.fiware.org/lab.

[23]https://www.fi-xifi.eu/home.html.

that, most of the times, the same architecture cannot be used in more than one use cases. Cloud-based IoT aims to go beyond that, by considering an open and dynamic configurable IoT platform that can be perfectly suitable for smart city applications where there is a need to seamlessly connect the different IoT silos. This goes beyond the current solutions' aims that are mainly vertically closed, so forming many "Intranets of Things" rather than an "Internet of Things" .

References

1. Sotiriadis S, Petrakis GME, Covaci S, Zampognaro P, Georga E, Thuemmler C (2013) An architecture for designing future internet (FI) applications in sensitive domains: expressing the software to data paradigm by utilizing hybrid cloud technology. In: 13th IEEE international conference on bioinformatics and bioengineering (BIBE 2013), Chania, Greece, 10–13 Nov 2013
2. Zorzi, M et al (2010) From today's intranet of things to a future internet of things: a wireless-and mobility-related view. IEEE Wirel Commun 17(6):44–51
3. Lange S, Nettsträter A, Haller S, Carrez F, Bassi A, Bauer M, Bui N, Carrez F, Giacomin P, Haller S, Ho E, Jardak C, De Loof J, Magerkurth C, Nettsträter A, Serbanati A, Thoma M, Walewski JW, Meissner S (2013) Final architectural reference model for the IoT v3.0. http://www.iot-a.eu/public/public-documents
4. Zahariadis T, Papadakis A, Alvarez F, Gonzalez J, Lopez F, Facca F, Al-Hazmi Y (2014) FIWARE Lab: managing resources and services in a cloud federation supporting future internet applications. In: Proceedings of the 2014 IEEE/ACM 7th international conference on utility and cloud computing (UCC'14). IEEE Computer Society, Washington, DC, USA, pp 792–799
5. Stravoskouyfos K, Preventis A, Sotiriadis S, Petrakis E (2014) A survey on approaches for interoperability and portability of cloud computing services. In: The 4th international conference on cloud computing and services science (CLOSER 2014), Barcelona, Spain, 3–5 April
6. Stravoskoufos K, Sotiriadis S, Petrakis E (2016) IoT-A and FIWARE: bridging the barriers between the cloud and IoT systems design and implementation. In: 6th international conference on cloud computing and services science (CLOSER 2016). Rome, Italy, 23–25 Apr 2016
7. Galis A, Gavras A (2013) The future internet: future internet assembly 2013 validated results and new horizons. Springer
8. Muller G (2008) A reference architecture primer. http://www.gaudisite.nl/info/ReferenceArchitecturePrimer.info.html
9. Rumbaugh J, Jacobson I, Booch G (2004) Unified modeling language reference manual. The Pearson Higher Education
10. Bray, T et al (1998) Extensible markup language (XML). World Wide Web Consortium Recommendation REC-xml-19980210. http://www.w3.org/TR/1998/REC-xml-19980210 16
11. Klyne G, Carroll JJ (2006) Resource description framework (RDF): concepts and abstract syntax
12. Bechhofer S (2009) OWL: Web ontology language. In: Encyclopedia of database systems. Springer, US 2008–2009
13. Harris S, Seaborne A, Prud'hommeaux E (2013) SPARQL 1.1 query language. W3C Recommendation 21

14. Newcomer E, Lomow G (2004) Understanding SOA with web services (independent technology guides). Addison-Wesley Professional
15. Masse, M (2011) REST API design rulebook. O'Reilly Media, Inc.
16. Botts M, Robin A (2007) OpenGIS sensor model language (SensorML) implementation specification. OpenGIS implementation specification OGC 7.000

Part III
Use Cases

Chapter 11
Traffic Management for Smart Cities

Andreas Allström, Jaume Barceló, Joakim Ekström, Ellen Grumert, David Gundlegård and Clas Rydergren

11.1 Introduction

One of the key components in smart cities of the future is the use of Advanced Traffic Management Systems (ATMS) and Advanced Traveler Information Systems (ATIS) for efficient management and control of traffic flows. The purpose of the ATMS/ATIS is to improve the overall traffic system performance, e.g. reducing emissions, noise, and travel times.

In order to manage and control traffic flows, the conditions of the road traffic have to be captured. The road traffic state can be described using speed, flow, and density on a specific segment of the road. The length of the segment might vary depending on the geometry of the road. When estimating the traffic state, different types of traffic models are commonly used. However, the models can not include all aspects of the real system, and in order to have a good representation of reality the models have to be combined with measured data of the traffic state, e.g., traffic counts and speed/travel time measurements.

Today, most existing ATMS/ATIS rely on fixed point (Eulerian) measurements from loop and radar detectors. Eulerian sensors can collect observations in terms of flow, speed, and occupancy,[1] but are unable to provide any trajectory-based measurements (Lagrangian measurements), such as direct trip observations or travel times on routes, which can contribute even further to the understanding of the behavior of the traffic flow. Already, cities generate large amounts of space-time location data

[1]Road density is commonly approximated from measured occupancy.

A. Allström · J. Ekström (✉) · E. Grumert · D. Gundlegård · C. Rydergren
Communications and Transport Systems, Department of Science and Technology, Linköping University, Campus Norrköping, 601 74 Norrköping, Sweden
e-mail: joakim.ekstrom@liu.se

J. Barceló
Department of Statistics and Operations Research, Universitat Politècnica de Catalunya, Jordi Girona 1-3, 08034 Barcelona, Spain

V. Angelakis et al. (eds.), *Designing, Developing, and Facilitating Smart Cities*,
DOI 10.1007/978-3-319-44924-1_11

from different systems, such as cellular networks, social networks, and participatory sensing. When Eulerian sensors are combined with Lagrangian sensors available in connected vehicles and user devices, the possibility of observing large-scale mobility patterns will dramatically change. Massive amounts of Lagrangian sensors enable a new era of road traffic sensing, making it possible to directly observe trips for a much larger penetration than before. These observations enable a new dynamic understanding of experienced travel times, as well as departure time, mode and route choices, which relate to the travel demand. If detailed data is available, activity patterns on individual level can also be captured.

In an ATMS/ATIS framework we are interested in making use of the new data sources for improving the overall system performance, and additionally give added value to users of traffic information services, in terms of mode and route options. However, the collected data will not automatically improve overall traffic state estimations and support traffic management decisions. Therefore, traffic models are needed, which given a current traffic state can predict near future traffic conditions. In the ATMS/ATIS framework we are, however, not only interested in predicting the future, but also to use the available data and models to shape it. This can be done by evaluating control strategies, with respect to some system performance measurement, using a traffic model.

The amount of sensor data available for traffic estimation and prediction has increased dramatically the last years, but also the number of different sensor types has increased. Large amount of sensor data from heterogeneous sensors makes it important to use efficient methods for fusion. Better estimates of dynamic travel demand together with improved sensor data also makes it possible to calibrate and estimate boundary conditions to more advanced traffic models in real time. More advanced and better calibrated traffic models enable better possibilities for traffic control, but also makes methods for combining models with sensor data, i.e., data assimilation, more important.

A key component of the proposed estimation and prediction framework includes a coherent mapping between travel demand models, traditionally used for long term planning, and dynamic traffic models used for data assimilation, fusion, and short-term prediction. This chapter gives an overview of emerging sensors and related models together with assimilation and fusion approaches for state estimation and prediction. Two case studies for traffic state estimation and prediction are presented, and finally a framework for an ATMS/ATIS is outlined.

11.2 Sensors

Traditionally, Eulerian sensors have been used for road traffic observations. Sensors like loop detectors and radars are widespread in many major roads in cities throughout the world. These sensors observe or estimate one or several of flow, speed, and occupancy for a specific location with a very high penetration rate. The last decades significant efforts have been made in the area of langrangian sensors and in many cities travel times or point speed observations from Lagrangian devices has been used for real-time traffic state estimation, see e.g. [56, 69]. However, the sensors that enable a bridging between traffic demand models and real-time traffic prediction models are the sensors that maintain a user or vehicle identity over large periods of time and distance. These sensors are for example cellular network data, Bluetooth/Wi-Fi sensors, automatic license plate recognition systems and to some extent GPS-equipped vehicles. These are all Lagrangian sensors and independent of the traffic data type (e.g. travel times or travel demand) that is to be estimated, the performance depends on four main properties of the probe system; penetration level, sampling strategy, measurement type, and measurement accuracy.

The penetration level is the number of equipped vehicles compared to the total number of vehicles in the area of interest. The sampling strategy refers to the frequency with which the probe measurements are recorded and how often these are sent to the traffic information server. The measurement type is mainly related to the type of sensors that the probe is equipped with, e.g. GPS or accelerometer, but also what kind of data that is supported by the transmission protocol. The measurement accuracy refers to which accuracy that can be achieved in each measurement, e.g. positioning accuracy. The importance of the different properties is depending on what type of traffic data that is estimated, but also what kind of traffic application that will make use of the data.

Lagrangean sensors can be used to estimate most traffic data types, like speed, flow, density, travel time, and incidents. Some types of traffic data can be measured directly, and some needs to be estimated based on models that relate the different kinds of traffic data. Common for most probe systems available today is that the probes are measuring only the state of the probe itself. Few attempts are yet to try to measure the state of the road section that it traverses, but a good exception can be found in [94]. This means that the traffic state of the road section needs to be estimated based on the measurements made by the probes. Few probe systems are developed explicitly for traffic estimation applications, and thus the sampling strategy, measurement type, and measurement accuracy are not decided with traffic estimation applications in mind but for, e.g. a fleet management application. Common for most probe systems is that speed and position are available. Travel times can then easily be inferred from consecutive position measurements given that an identity of the probe is included, and traffic flow and density can be estimated from point speed or travel times using the fundamental diagram [104].

An important aspect of all vehicle probe systems is that we typically cannot control when and where we get access to state measurements. For critical traffic applications this can be problematic. Fixed sensors are placed where traffic planners believe it will be most important and hence we get constant high quality measurements in these places. The main advantage of probe data is the fact that we utilize already existing data, which makes it cost efficient compared to fixed installations.

11.2.1 Cellular Network Data

Since the first trial to use cellular network data for road traffic estimation, the CAPITAL project [98], a lot of progress has been made. The CAPITAL project failed due to poor cellular location accuracy. However, since then the available data in the network, as well as the methods to process data, has changed dramatically. The data available in cellular networks related to road traffic estimation is described in detail in [96]. As for today, numerous projects have shown positive results and indicate a large potential for the data source [7, 100].

Early projects using cellular network data aimed at travel time estimation, mainly based on handover events [7, 48]. Lately, there has been an increasing interest in estimating OD matrices based on this data [21, 103]. A limiting factor of this data type has historically been the difficulty to get access to the data from the cellular operators. Recently Orange released a cellular network data set for research purposes and the interest was very high from researchers and practitioners all over the world [15]. This might increase the knowledge about the potential of the data source, but possibly also make it easier for other cellular operators to share data.

The penetration level for cellular network data depends on how the cellular network data is collected. The main aspect is related to whether cell phones that are (1) making phone calls, (2) have a data connection or (3) are idle, provide data. For (3) the penetration level is equal to the operators' market share, for (1) it is equal to a few percent of the vehicles and for (2) the penetration rate is somewhere in between, depending on the users' data usage. The sampling strategy strongly depends on which type of cellular network and which interface that is monitored, see e.g. [47] for details.

Most of the research so far in the area has been using call detail record data, that is used for billing purposes in the operators' network. However, this data is only a fraction of the data available in the cellular network. If the personal integrity aspect of using cellular network data can be handled efficiently, it is likely that cellular network data will become an important data source in the near future, both within travel time, traffic flow, and OD estimation.

11.2.2 GPS-Equipped Devices

Alternative traffic data is available from different probe client types, e.g. navigation systems, fleet management clients and insurance black-box systems. The client type will affect the characteristics of the data that is collected and might introduce bias in the traffic estimations. For example, professional drivers with good knowledge of the traffic system tend to avoid congested parts of the road network and can hence cause underestimation of the traffic congestion. Some probe client types, e.g. some municipal vehicles, are only available at certain time, other types, e.g. taxis, can use dedicated bus/taxi lanes and hence indicate lower travel times than for regular vehicles. Another possible problem is that some client types, e.g. heavy vehicle fleet management clients and buses are more restricted in terms of speed limits. The probe client type needs to be considered when estimating the traffic state.

The device type can also affect the characteristics of the traffic data. We can for example be relatively sure that a navigation device is located in a vehicle, which is not necessarily true for a cell phone device. Some devices have poor GPS-receivers and the location accuracy reported from these clients can potentially affect the results of travel time estimations, especially if the travel time is calculated over a relatively short distance. The main type of measurements from GPS-equipped vehicles are point speed and travel times. Some devices and/or client types does not support point speed measurements, which reduces the amount of traffic state-related information that can be extracted from each probe significantly. If the id of the probe is collected and maintained, it is still possible to collect travel times, otherwise we are left to try to estimate density based on probe locations, which with most realistic penetrations rates is challenging to do with reasonable accuracy.

The sampling strategy of the probes is also very important for the road traffic estimation result. A low sampling frequency causes two main problems. The first is that there is an inherent delay of the measurement equal to the sampling time when travel times are calculated. The second problem is that path inference becomes more difficult the longer the sampling time is. This is especially problematic in dense urban areas with a large number of route choices. Map matching and path inference for low sampling devices are further described in e.g. [62, 91]. The location sampling strategy for GPS-equipped devices is typically time-based sampling with a typical reporting interval between 30 s and 2 min. This type of sampling together with a high measurement accuracy enables a large number of traffic-related applications.

11.2.3 Automatic Vehicle Identification

Automatic Vehicle Identification (AVI) is an aggregate name for data collection techniques where vehicle identities are captured in selected locations. The most common systems are based on license plate recognition (LPR) and reidentification of Bluetooth or Wi-Fi physical addresses. Bluetooth is for example used in hands-free devices and for communication between different devices in vehicles. The measurement type for AVI systems is similar as for cellular network data, i.e., space-time tuples with an identifier of the vehicle or the Bluetooth/Wi-Fi device. In Europe, conducted field trials, using Bluetooth technique, indicate a penetration rate around 30 % [13, 70].

The measurement error relates to the location error of where the device is captured. For Bluetooth, it is possible to adjust the coverage of the sensors collecting Bluetooth data and research has shown that a lower coverage, which also means a lower number of captured vehicles, decrease the error in the travel time estimation [90]. Also, identifying the MAC address takes about 5 s on average but may take up to 10 s in some extreme cases [61], which will affect the travel time estimation. However, a number of field trials have reached the conclusion that travel times estimated from Bluetooth detectors, deployed along a motorway or arterial, are comparable to those estimated from GPS and toll tag readers [51, 67, 70, 90].

11.3 Traffic State Estimation and Prediction for ATMS/ATIS

The ability to predict the short-term evolution of the current traffic state is an important basis for further traffic management and control applications. Here we focus on methods applicable to ATMS/ATIS, in a traffic network equipped with traffic sensors supplying real-time traffic data. After an appropriate management of the data (e.g. filtering outliers), suitable statistical and traffic models are applied to either generate the local information necessary to estimate and predict the short-term evolution of the current traffic state and associated variables, or to generate the input data to more sophisticate traffic models to support wide area management and control policies.

As a consequence of the advent of new sensor technology, which makes massive amount of new traffic data available, there is a need for new techniques and methodologies to combine the vast amount of data from different sources. A plausible technological scenario will combine

- Point detection with discrete time resolution, as for instance conventional inductive loop detectors, and radars, measuring flows (veh/h), occupancies (time %), spot speeds (km/h), traffic mix (% light, heavy vehicles).
- Point detection with continuous time resolution. For examples magnetometers measuring time in/time out on the detector from which flow counts, spot speeds, occupancies, or traffic mix can be estimated.
- Advanced AVI detectors recording time tag, vehicle/device identification and downstream re-identification, providing sample counts of travel time measurements.
- Continuous time-space detection. For example, by tracking GPS devices, or connected cars, supplying time tag, position (X, Y, Z coordinates) local speed and heading direction.
- Additionally every cell phone can, besides sharing the position data from its GPS, supply spatiotemporal observations associated with signaling in the cellular network.

From a methodological point of view, the main change from using new sensor technologies, consist on having to deal with massive amount of heterogeneous data. Thus, in order to efficiently manage and utilize such massive amount of data in an ATMS/ATIS, complex statistical techniques (e.g. Kernel smoothing, Kalman filtering and Bayesian networks) have to be combined with mathematical models of the traffic system. This process is synthesized in the methodological diagram in Fig. 11.1.

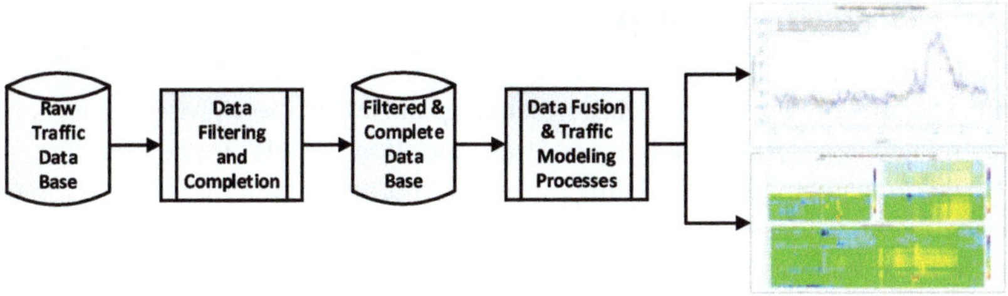

Fig. 11.1 Main methodological components for short-term prediction of the traffic state

Approaches for traffic state estimation and prediction can be divided into non-parametric and parametric prediction models. Nonparametric models require both parameters and model structure to be determined from data. Thus, they rely on a vast amount of historical data. Traffic dynamics can be captured although no knowledge of the traffic processes as such is included. Nonparametric models have the property that only already occurred traffic states can be predicted. Parametric models, on the other hand, include parameters with a predetermined structure. Still, the parameters must be calibrated according to empirical data. Parametric models have the property of only describing traffic phenomena which follows from the predetermined relationship between model parameters. Also, they rely on boundary conditions, e.g. traffic demand, which need to be predicted for the entire prediction horizon.

The remainder of this section gives an overview of models, filtering of traffic data, assimilation of traffic data with traffic models and fusion of heterogeneous types of data.

11.3.1 Nonparametric Models

Nonparametric models are created from large amount of historical data, and make use of big data analysis approaches, such as linear time series, K-nearest neighbors, locally weighted regression, Fuzzy logic, Bayesian networks, and Neural networks. In [58], an overview of commonly used methods is given. It is important to recognize that although nonparametric models can capture the traffic dynamics even though no knowledge of the traffic processes as such is needed, they all inherit the property that only traffic states already occurred can be predicted. Thus, they are appropriate for predicting recurring traffic conditions, but less appropriate for nonrecurring traffic conditions, such as congestion due to road work, events or incidents. Also, since they lack any knowledge of the traffic processes, they are of less interest when one wants to evaluate traffic management strategies.

11.3.2 Parametric Models

Parametric models applied to traffic planning and estimation problems are commonly related to as “traffic models” or “transportation models” in the literature, and are based on mathematical modeling of the transportation system. Different models describe the transportation system with different level of detail, depending on their application areas. Here we will focus on models appropriate for application within an ATMS/ATIS. Thus we will solely describe dynamic modeling approaches, which can describe both the spatial and temporal evolution of congestion. Static modeling approaches can still be relevant for an ATMS/ATIS in terms of providing an initial estimate of the OD matrix.

Dynamic traffic assignment approaches decompose the problem of route choice and network loading into two main components

1. A method for determining the flow rates on the paths in the network.
2. A dynamic network loading method, which determines how these path flows give raise to time-dependent link volumes, link travel times and route travel times.

The difference between pure dynamic traffic assignment (DTA) and dynamic user equilibrium (DUE) depends on how the first component is implemented, but both are based on a route choice model. In DTA, paths and path proportions are selected at each time interval according to a discrete choice model, that does not guarantee equilibrium, while DUE uses an iterative procedure to solve the variational inequality formulation of equilibrium [26].

In the network loading phase, the evolution of congestion can be modeled in detail by microscopic approaches, in which each single driver-vehicle unit is described in detail, with position, speed, acceleration, and driver behavior. When the density of a road segment increase, the interactions among the vehicles will result in lower speeds and either density or mean travel time can be used as a measure of the level of congestion. There are also macroscopic approaches, in which the dynamic traffic conditions are described by aggregated measures (e.g. flow, density, and speed) as a function of both space and time. Macroscopic models are based on mathematical relationships between flow, density, and speed, and these relationships are used to describe how the traffic state will evolve in both space and time. This group of models include the Lighthill-Whitham and Richard (LWR) model [74], and its discretized version, the cell transmission model (CTM) [30]. A third group of dynamic modeling approaches can be placed somewhere between microscopic and macroscopic models, and are referred to as mesoscopic models. They usually model single vehicles or group of vehicles but with simplified description of the interactions between them.

While dynamic modeling approaches provide mechanisms for high-resolution modeling of both the spatial and temporal distribution of congestion, it is important to recognize that they heavily rely on the availability of high-resolution data. Both in terms of boundary conditions and calibration. Boundary conditions are commonly the OD matrix, and calibration of models can be done using both Eulerian or Lagrangian measurements.

11.3.3 Filtering, Fusion, and Assimilation

Filtering, fusion, and assimilation are all referring to the process of improving the accuracy of a traffic state estimate by interpolation and removal of noise and bias in measurements. The main difference between the different concepts relates to the core methodology of improving the estimate. In plain filtering, there is typically only one modality of measurement included. In fusion, the focus is to combine measurements of different modalities and in assimilation the focus is to combine measurements with the output of a mathematical model. However, whereas assimilation can include multiple measurement types, fusion typically includes a model for the system evolution. In general, to turn the raw traffic data into a state estimate, a number of steps needs to be performed

Step 1. Removing outliers and erroneous measurements, which could induce large errors and biases in the estimate.

Step 2. Filling gaps in data, since malfunctioning sensors and the removal of outliers will generate gaps, these gaps must be filled in order to have complete coherent series of observations.

Step 3. Fusing measurements from different sources, to extract the richest homogeneous information from the heterogeneous data available.

Step 4. Combining the measurements with a system or process model.

In some literature, especially in the area of weather predictions and hydrology, Step 4 is referred to as assimilation [35]. For traffic state estimation, typically Steps 1 and 2 are jointly implemented as a preprocessing step and Steps 3 and 4 are implemented as a joint fusion and assimilation step. In the first step, knowledge of the physical process that is observed and the sensor that produces the measurements can be used to remove outliers and obviously faulty measurements. In the second step, the outliers are replaced with reasonable values. Although both fusion and more complex process models can be used in this step, often very basic system models are used separately for sensors with different modality. In Step 3, sensors with different modality are fused together to improve the estimate. The different modalities require models for translation between them, these are often referred to as measurement models, and an example can be the relationship between flow and density, i.e. the fundamental diagram. Different sensors are also separated in space and their observations are separated in time, which requires models for the evolution of the state in both the spatial and temporal domain. The macroscopic models described in Sect. 11.3.2 are good examples of models that describes the traffic state evolution in space and time and can be used for assimilation.

The Kalman filter was first presented in [65, 66] and has successfully been used for all steps described above. The right part of Fig. 11.1 depicts two examples, the output of a Kalman filter to estimate outliers and replace missing data for traffic flows (upper right part of Fig. 11.1), and a space-time reconstruction using an assimilation technique (lower right part of Fig. 11.1). For a more in depth description of Kalman filtering, and its extensions, in traffic management and control, the reader is referred to [5, 75].

In [36] a survey is presented which shows that for traffic state estimation and prediction a large variety of methods and models have been evaluated. Autoregressive models, Bayesian frameworks, Kalman filters, and neural networks are some of the methods that are mentioned. However, their conclusion is that none of the proposed methods produce accurate estimations and predictions except for some special conditions. The explanation for this is that the dynamics in traffic cannot be formalized by a single procedure. In order to achieve accurate results for a variety of network configurations and data sets available, a combination of different approaches is recommended.

A Single-Constraint-At-A-Time (SCAAT) Kalman filter that uses the single most recent speed measurement from any available sensor, in this case loop detectors and GPS probe data, is applied in [20]. The state is updated based on the characteristics of that particular sensor and the accumulated state estimation from the previous step. The results are promising but the limitation is that only speed measurements can be used.

In [6], different methods for fusing speed from loop detector data with travel times collected with Bluetooth are compared. This was done both on collected data and simulated data. Among the evaluated methods were neural network, measurement fusion Kalman filter, SCAAT Kalman filter, fuzzy integral, ordered weighted averaging, and a simple convex combination. The results show that most data fusion techniques improve the accuracy over single-type sensor approaches. Moreover, the measurement fusion Kalman filter, which is a multisensor multi-temporal Kalman filter [81], and the simple convex combination perform well in all scenarios and does often significantly improve the accuracy of the estimations.

The adaptive smoothing method is presented in [99], which is a nonlinear spatiotemporal low-pass filter that uses the information from loop detectors to reconstruct the spatiotemporal traffic dynamics for a certain road network. The proposed method is based on a number of parameters and some of them are traffic flow related such as the propagation velocity in congestion and free-flow conditions. This method was later extended and generalized in [77] to also handle other types of data. The so-called Extended Generalized Treiber-Helbing filter (EGTF) can handle all kinds of data, no matter of their respective spatial and temporal resolution, as long as the data provides a mean to distinguish between free flow and congestion. In [76] the method is compared with piecewise, linear and quadratic speed-based interpolation methods. As earlier concluded in [59], using kinematic wave theory in combination with filtering technique results in more accurate travel time estimations in comparison with interpolation methods.

In [104] the cell transmission model for velocities (CTM-v) is used together with an ensemble Kalman filter (EnKF) to fuse speed measurements from loop detectors and probe data. This method is modified for prediction and is described in more detail in a case study presented in Sect. 11.4.2. Later [80] evaluated almost 1000 scenarios where the CTM-v and EnKF was used to fuse probe and loop detector data and estimate travel times. The results clearly showed that when complementing loop detector data with probe vehicle data, better estimates for travel times were obtained. However, this implementation of the CTM-v and the EnKF could only handle point speed measurements, not travel times. In [46], the use of EnKF was extended to allow for fusion between point speed and travel time measurements and this is further described in the case study in Sect. 11.4.2.

11.3.4 Time-Dependent OD Estimation

The use of traffic models within an ATMS/ATIS raises an additional problem concerning the traffic data processing: the need to estimate time-dependent traffic patterns formulated in terms of OD matrices, which in recent years has become a field of intense research, see for instance [12, 22, 60, 63]. Taking into account the dynamic nature of traffic phenomena, it has been quite natural to develop estimators based on variants of Kalman filter approaches, exploiting real-time traffic measurements. However, although in general all these approaches have proven their robustness in terms of convergence to sound solutions, most of them have so cumbersome computational requirements that they are not applicable to support real-time decisions. Among the various factors determining the computational performance, the quality of the initial OD estimate proved to be critical [11]. A solution to the initialization problem could be to adjust OD matrices exploiting available traffic measurements and using a static off-line approach. Among the static off-line procedures to adjust OD matrices to traffic measurements, the bi-level optimization methods provide the most consistent results [78]. These approaches formulate and solve an upper level optimization problem. At the upper level the objective function is usually a distance measure including both link counts, measured by sensors and estimated from the updated OD matrix, and the OD information itself, measured by the difference between the updated OD matrix and a target matrix (e.g. from a travel survey). At the lower level a traffic assignment is carried out, that estimates the values of the traffic variables as a function of the adjusted OD at the upper level of the current iteration. Figure 11.2 illustrates the logic of the process and the information exchange between both levels. The objective function at the lower level depends on what type of assignment that is carried out (e.g. dynamic or static), and is for both the dynamic and static cases a non-differentiable function.

In [28], a stochastic perturbation stochastic approximation (SPSA) method is proposed for solving the upper level problem with a DUE assignment carried out to account for the dynamics of the congestion propagation at the lower level. This approach has later on been improved in [23]. In the formulation proposed by [19] the bi-level problem to solve is

$$
\begin{aligned}
F(\tilde{g}^k, v^k, \tilde{tt}^k) = & \gamma_1 F_1(g^k, \tilde{g}^k) + \gamma_2 F_2(v^k, \tilde{v}^k) + \gamma_3 F_3(tt^k, \tilde{tt}^k) \\
\text{subject to} \quad & (\tilde{v}^k, \tilde{tt}^k) = \text{assignment}(\tilde{g}^k) \\
& \tilde{g}^k \geq 0.
\end{aligned}
$$

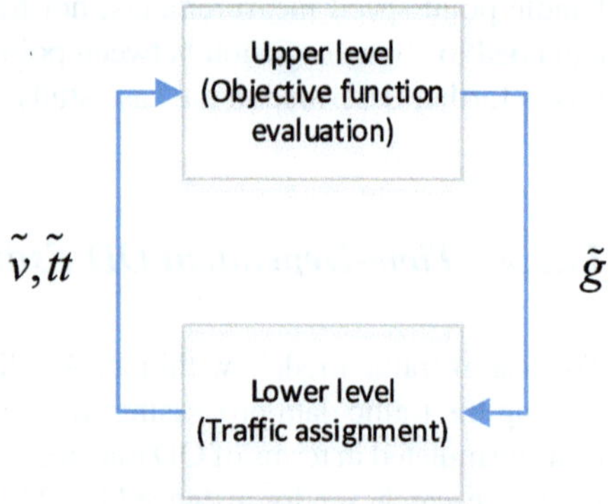

Fig. 11.2 Computational scheme of the bi-level OD estimation approach

where $\tilde{g}^k$, $\tilde{v}^k$, $\tilde{tt}^k$ are, respectively, the estimates of the OD matrix g, the traffic volumes v, and the travel times between pairs of Bluetooth detectors tt, at iteration k. The functions F_1, F_2 and F_3 are distance functions. The output of this off-line procedure generates an off-line data base of target OD matrices that can be used to initialize an online procedure which estimate the real-time OD matrix to be used in dynamic traffic model.

11.4 Case Studies of Traffic State Estimation and Prediction

We will present two case studies where new sources of data are used for providing information of both the current and the future traffic state. The first one is related to online time-dependent OD matrix estimation, here with an example from the city of Vitoria in Spain, combining traditional loop detector data with AVI Bluetooth detectors. The second one is related to fusion of different data sources and assimilation with a traffic model, to provide both real-time traffic state estimation and prediction for a part of the Stockholm motorway system in Sweden.

11.4.1 Online OD Estimation in Vitoria

Application of dynamic traffic assignment models in an ATMS/ATIS rely on known information of demands, in terms of OD matrices. Section 11.3.4 outline a methodology for off-line OD estimation. Here we provide an example of how such an historical (known) OD matrix can be used for updating a current OD matrix, using traffic counts and AVI Bluetooth detectors.

11.4.1.1 A Kalman Filtering Approach

A Kalman filter approach, with the recursive linear Kalman filter state-space formulation [11, 12], is adapted to exploit traffic counts collected by AVI sensors and conventional detection technologies, and travel times observed by the AVI sensors. The formulation uses deviations of OD path flows as state variables, calculated with respect to DUE-based historic OD path flows. A subset of the most likely OD path flows identified from a DUE assignment is used. The number of paths to take into account is a design parameter. A list of paths going through each sensor is automatically built for each AVI sensor from the OD path description, AVI sensor location and the network topology.

The proposed approach initially assumes flow counting detectors and AVI sensors located in a cordon at each possible point for flow entry and AVI sensors located at intersections in urban networks covering links to/from the intersection. Flows and travel times are available from AVI sensors for any time interval greater than 1 s. Trip travel times from origin entry points to sensor locations are measures provided by the detection layout. Therefore, they are no longer state variables but measurements, which simplify the model and make it more reliable.

The time-varying dependencies between measurements (sensor counts of Bluetooth-equipped vehicles) and state variables (deviates of Bluetooth-equipped OD path flows), are used for estimating discrete approximations of travel time distributions. Since the approach uses the AVI travel time measurements from Bluetooth-equipped vehicles, the nonlinear approximations can be replaced by estimates from a sample of vehicles. Then no extra state variables for modeling travel times and traffic dynamics are needed, since sampled travel times are used to estimate discrete travel time distributions, see [11, 12] for details.

11.4.1.2 Results

Computational experiments were conducted with urban networks of various sizes, see [4] for details. For the sake of completeness we report here those for the medium-size network from the city of Vitoria (Spain), depicted in Fig. 11.3, including 57 centroids, 3249 OD pairs, 2800 intersections, and a modeled network of about 600 km. This network resembles a reasonable sized real-life network, with representative congestion levels and route choice dimension as found in many large urban areas. Two different sets of sensors have been identified in the Vitoria network

- 389 standard loop detectors, located as depicted in the left-hand side of Fig. 11.3, providing flows, speeds and occupancies, related to all detected vehicles at the loop.
- 50 AVI sensors, located as shown in the right-hand side of Fig. 11.3. Notably, the AVI sensors are deployed following a layout strategy, whose details can be found in [9], to optimize the capturing of Bluetooth-equipped vehicles and provide effective travel time measurements between AVI sensors.

Fig. 11.3 The Vitoria network, Basque Country, Spain: Loop detector sensors layout (*left*) and AVI Bluetooth detectors layout (*right*)

Table 11.1 Method 5—Average Theil's coefficient, NRMSE and R^2 values for quartile groups of OD pairs according to a priori OD flows

Initial matrix	Theil's coefficient					NRMSE (%)					R^2 (%)				
	Q1	Q2	Q3	Q4	All	Q1	Q2	Q3	Q4	All	Q1	Q2	Q3	Q4	All
D7	0.16	0.15	0.14	0.16	0.15	45.1	36.6	27.4	28.3	35.1	23.4	15.9	68.1	77.4	88.5
D8	0.15	0.13	0.12	0.15	0.14	40.8	26.3	24.3	26.5	31.9	25.3	17.0	74.5	76.5	88.0
D9	0.15	0.13	0.13	0.15	0.14	39.1	36.0	24.4	26.5	31.5	26.8	18.0	75.6	72.8	86.3

Various methods were tested in the referenced paper [4] depending on different values for the design factors, such as initial a priori OD estimates, and penetration rates of Bluetooth-equipped vehicles. Results for the modified Kalman filtering approach (denoted Method 5 in [4]), assuming 100 % Bluetooth penetration rate and prior OD scenarios D7 (low-demand), D8 (medium demand), and D9 (high-demand), are presented in Table 11.1 for selected goodness of fit measures: normalized root mean squared error (NRMSE), Theil's U coefficient and R^2. Good performance of the modified Kalman filtering approach is reflected in high R^2 fit (above 85 %) that is obtained for the overall OD pairs and demand levels, but mostly for the most important OD flows (i.e. those in the 4th and 3rd quantiles). The fit of true versus estimated OD flows for all considered OD pairs (for the aggregated 1-h period) and for a scenario initialized with prior demand levels D7, D8, and D9 show coefficients of determination of the simple regression line of almost 90 %. Figure 11.4 presents the evolution of estimated OD flows (for OD pairs 221 and 343 in Fig. 11.4) per departure time interval for prior high OD demand scenario (D9), demonstrating also how Method 5 is able to recover from an initial point characterized by overestimated prior OD flow with respect to the true OD flow.

Overall, results presented show that the potential of information from advanced traffic measurements to improve OD demand estimation can be fully exploited only using OD estimation methods capable to correct for biases in spatial and temporal OD pattern given by prior OD matrix, especially in congested networks. In this

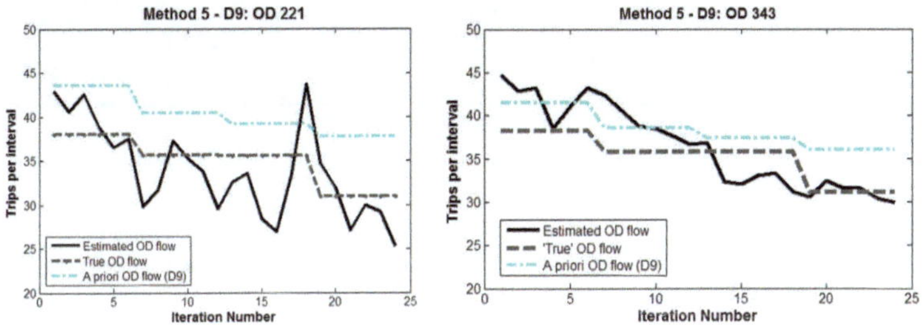

Fig. 11.4 Filtered values throughout 1 h for prior OD scenario D9 for the modified Kalman filtering approach

respect, the computational experiments presented in this work prove the robustness and quality of the OD estimates exploiting AVI measurements. Computational times of less than 2 min on standard laptop and MATLAB implemented algorithms make them applicable for real-time operations.

11.4.2 Traffic State Estimation and Prediction in Stockholm

To provide an accurate estimation of the current traffic condition in Stockholm, a Kalman filtering approach has been adopted for estimation and short-term prediction of the traffic state. The filtering approach can be used both for fusion of different data sources, and for assimilation of traffic sensor observations, or predictions of such observations, with macroscopic traffic flow model outputs. Here we present the outline of the methodology, and an evaluation of traffic state estimation and prediction for a 7 km long section of the Stockholm motorway, just north of central Stockholm, for traffic going towards the central parts. The motorway section is illustrated in Fig. 11.5, in which each subsection is a link with different characteristics and thus different fundamental diagram parameters. For this section three different type of sensors are available

- 20 fixed point radar detectors, measuring speed and flow. In Fig. 11.5 these detectors are marked by "+".
- 1500 equipped taxis (for the whole of Stockholm county), sending their position every 1–2 min.
- AVI/Bluetooth measurements from seven subsections, providing mean travel times. In Fig. 11.5 the approximate locations of the Bluetooth detectors are marked by "BT".

In the evaluation process, the AVI measurements is mainly used for calibration and evaluation purposes. In this section, we give the outline of the core components, and present the main results from the evaluations. For further details on both model components, the calibrations procedure and the results we refer the interested reader to [3, 46].

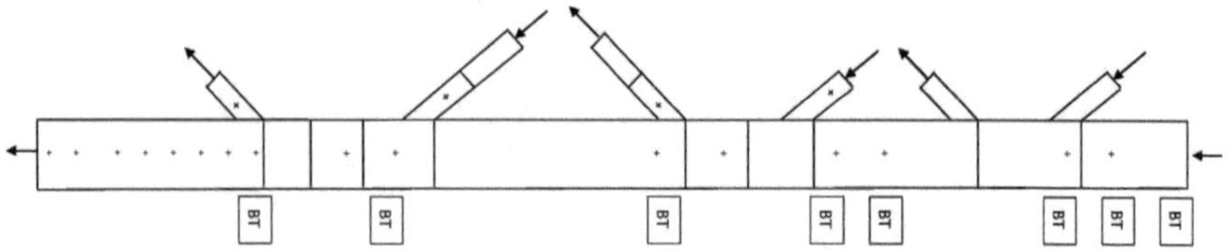

Fig. 11.5 The Stockholm motorway section, with radar detector marked by "+" and Bluetooth detectors marked by "BT"

11.4.2.1 Core Components for Traffic State Estimation and Prediction

An EnKF approach is used as core for data fusion and assimilation with a macroscopic traffic model. This approach was developed for data fusion and traffic state estimation within the Mobile Millennium project [14, 104], and adopts the CTM-v as system model in the EnKF. Data fusion and assimilation based on the CTM-v and the EnKF is appealing for many reasons; it is developed to run in real-time and for large networks, it can fuse different types of Eulerian point speed measurements, and it enables the possibility to include Lagrangian travel time measurements.

The state-space model of the system is

$$\begin{aligned} v^n &= M(v^{n-1}) + \eta^n \\ y_k^n &= h_k(v^n) + \chi_k^n \\ \eta^n &\propto (\mu_{\text{mod}}, Q^n) \\ \chi_k^n &\propto (\mu_{\text{obs}}, R_k^n) \end{aligned}$$

where v^n is the state vector in time step n, including the speed for each part of the road network (cell) according to the spatial resolution of the system. $M(\cdot)$ is the system model, here the CTM-v. y^n is the observation vector in time step n and $h_k(\cdot)$ is the observation model for observation type k. η and χ are the possibly time-varying model uncertainty (with mean μ_{mod} and covariance Q) and observation noise (with μ_{obs} and covariance R), respectively.

The observation model for Eulerian speed observations is simply a mapping of speed from the point speed sensor location to a cell in the road network. For Lagrangian sensors measuring travel time, a corresponding naïve solution would be to allocate the travel time to the cell traversed. This would, however, assume that the travel time is equally distributed over the traversed stretch. A sounder method would be to estimate the most likely speed profile for the travel time observation, given some external input and use that speed profile as input to the traffic estimation. The external input can be information about the road network, traffic models, and other observations (see e.g. [46]).

While the standard Kalman filter is designed for linear problems, the EnKF can handle both nonlinear and non-differentiable state-space equations, which is the case implied by the use of CTM-v. For further reading about the EnKF we refer to [34].

The CTM-v is based on the kinematic wave theory and the discretization [30] of the LWR model [74]. Traditionally, these models use density as a measure of state. However, due to the characteristics of the data available in the Mobile Millennium project, new models were developed that use velocity as the state. The new models are referred to as the LWR-v and the CTM-v. The velocity state estimation problem is solved using the EnKF. For a more in depth description of the models the reader is referred to [104, 105]. The CTM-v require boundary conditions as input, in terms of inflow. This could either be determined using the OD estimation procedure presented in Sects. 11.3.4 and 11.4.1, or using a nonparametric model. Here a simple approach has been applied by taking the mean of historical inflows, clustered by day of week, for 15 min periods.

11.4.2.2 Results

Two studies on traffic state estimation and prediction have been carried out using the proposed framework.

In [46] the framework is used for data fusion of travel time measurements from taxis with fixed point speed measurements from radar sensors. The results presented in Table 11.2 show mean average percentage error (MAPE) values, with AVI travel times used as ground truth, for different combinations of radar and taxi data. MAPE values are computed for four days in March 2013. The inclusion of taxi travel times improves the travel time for all evaluated days.

In [3] the framework is used for short-term travel time prediction. Two approaches are evaluated. One is based on running the CTM-v forward in time, from an estimated traffic state, the second approach is a hybrid between a nonparametric prediction of

Table 11.2 Travel time MAPE for different combinations of taxi and radar measurements

Scenario	March 19	March 20	March 21	March 22
CTM-v model only. Sources and sinks predicted using historic observations	0.701	0.583	0.262	0.667
CTM-v model and taxi observations	0.233	0.228	0.213	0.263
CTM-v model and two radar observations in start and end of test site	0.134	0.154	0.178	0.151
CTM-v model, two radar observations and taxi observations	0.119	0.126	0.154	0.126
CTM-v model and all radar observations	0.049	0.067	0.058	0.053
CTM-v model, all radar observations and taxi observations	0.049	0.052	0.055	0.047

Table 11.3 Travel time MAPE for EnKF estimation, and for CTM-v and hybrid prediction

Prediction/Estimation method	Horizon			
	0	5	15	30
Estimation (%)	4.8	–	–	–
CTM-v prediction (%)	5.1	9.2	10.1	10.9
Hybrid prediction (%)	5.5	6.8	7.7	7.9

radar sensor measurements and the CTM-v model. Here the EnKF framework is not only used for estimating the current traffic state, but also for combining the two type of predictions. Results, in terms of MAPE values comparing predicted travel times with AVI measurements from the March 21, 2013, are provided in Table 11.3. Note that a prediction horizon of 15 min corresponds to a trip taking place 15 min into the future, making use of predicted travel times 15–30 additional minutes into the future, depending on the level of congestion in the network. For 5, 15, and 30-min prediction horizons, the hybrid prediction has shown better results in comparison with only using the CTM-v. For the presentation of the complete results from the study we refer to [3].

11.5 Active Traffic Management

If the main objective is to support traffic management and information decisions, the traffic state estimation and prediction techniques discussed in Sect. 11.4 are not sufficient. Early research on traffic management and control topics, see for example [16, 17, 29, 68], propose decision support systems whose architectures have two main components: a rule-based knowledge system and a dynamic traffic simulation model, usually a mesoscopic or macroscopic model [8]. The integrated corridor management program, prompted by the U.S. Federal highway administration, has contributed to the consolidation of these architectures, whose guidelines can be found in [2].

11.5.1 Proposal of an Logical Scheme for an ATMS/ATIS

The ATMS/ATIS logical scheme is depicted in Fig. 11.6, and is split into four components:

A. The off-line generation of candidate target matrices for the initialization of the online procedure. It combines an initial OD matrix or similar practical applications with the historical traffic data clustered in profiles for different times of the day (e.g. raining Tuesday from 6:00 am until 8:00 am, and so on). This is likely to come from a static model OD demand calibration procedure. For most metropolitan areas, such models should already be available as they have been part of the strategic planning tools for several decades. A heuristic procedure, as the one described in [10], split the static demand in time slices in order to create an initial OD demand in terms of the traffic profiles. These time slices, along with a network model, are the input to the described OD matrix adjustment process in Sect. 11.3.4. The repetitions of the process for the different time intervals generate a database of candidate target matrices off-line.

B. An online selection of the most likely OD matrix for the initialization, given the current traffic conditions. The real-time traffic data from the traffic monitoring system provides input used for identification of the profile that best fits the current situation resulting in the most likely OD matrix for that time interval is selected.
C. The real-time estimation and prediction of the OD matrix. The selected OD matrix is the initial matrix used by a Kalman filter model, as is illustrated in Sect. 11.4.1, to estimate and predict the expected OD matrix for the next time period. This OD matrix will be the input for the traffic state estimation process, to estimate the network state and to predict its short-term evolution.
D. Online event detection and selection of management strategies. These strategies are evaluated in a traffic model, based on performance indicators such as travel time and emissions. The management strategies are based on control and/or harmonization of the traffic volumes. A more detailed description of commonly used strategies are given in Sect. 11.5.2

11.5.2 Traffic Control Strategies

Control of traffic volumes (demand-based strategies) and homogenization of traffic flows (supply-based strategies) are two key concepts for improving the traffic situation in a congested traffic network. Concepts based on these control strategies could be part of the management strategies database included in component D in Fig. 11.6. The aim (measured in terms of performance indicators) could for example be to reduce emissions, to increase safety or to maximize throughput. The two concepts address different problems.

Control of traffic volumes can be done by:

1. Pricing and information strategies aiming to shift the demand to less congested time periods, or to other modes of transportation. Examples of such strategies are congestion pricing [32, 33] and real-time traffic information systems [73].
2. Rerouting the traffic to improve system performance, either by providing real-time information direct to the car drivers [84], or using variable message signs.
3. Gating the inflow to areas sensitive to congestion using perimeter control, for example using signal timings (see e.g. [31]).

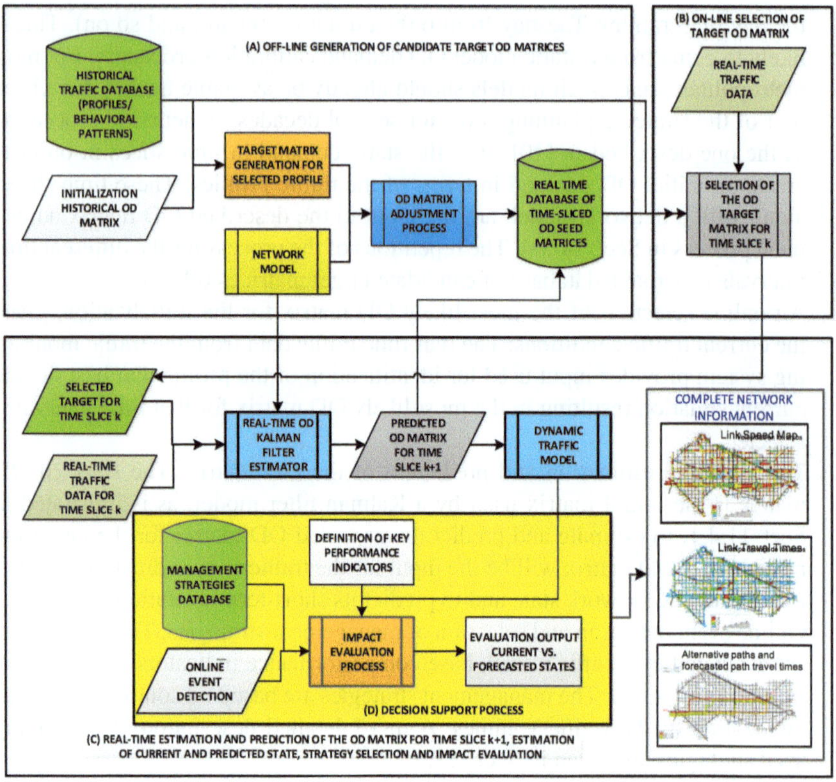

Fig. 11.6 Logical scheme of an ATMS/ATIS

Example of a combination of demand strategy 2 and 3 above is illustrated in Fig. 11.7, which include both rerouting and gating.

When the traffic flow on the road is approaching capacity levels, and above capacity levels, the conditions on the road usually become unstable resulting in incidents, big variations in speed levels, and resulting in stop-and-go conditions. Further, it is a well-known phenomenon that the capacity in unstable conditions tend to drop, resulting in lower throughput levels of the traffic flow than what can actually be achieved. Several empirical studies has identified the drop, see e.g. [27, 95, 110]. Flow harmonization can be used to keep the traffic flow stable and thereby prevent the capacity drop to occur. Also, when an incident has occurred flow harmonization can be used to prevent further breakdown and to keep the traffic flow stable under the incident situation. Flow harmonization can be achieved by

1. Ramp metering, which distribute the inflow to a motorway in such a way that the total flow on the motorway does not exceed its capacity. It can also be viewed as a perimeter control for motorways, allowing the traffic at the motorway to flow, and storing the vehicles in the secondary network. However, ramp metering systems commonly discontinue the ramp metering whenever the queue from the ramp

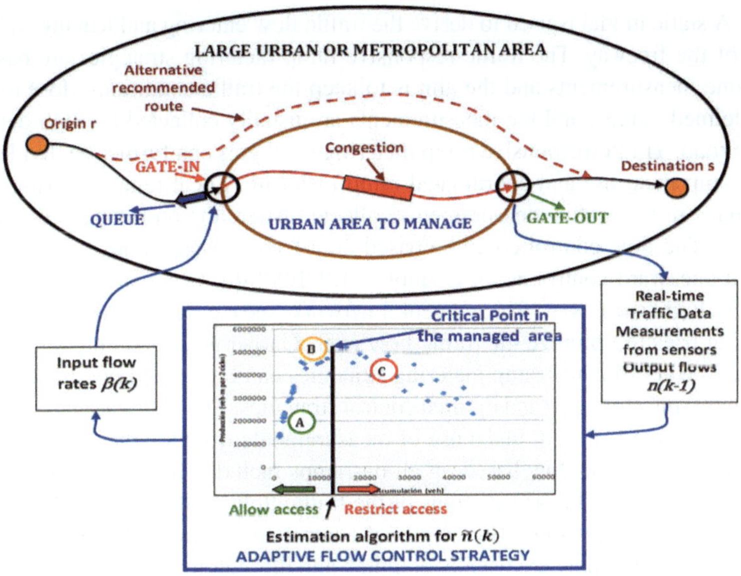

Fig. 11.7 Flow control in an urban area through rerouting and perimeter control

spills over to the surrounding traffic network, which is not the case for perimeter control.

2. Variable speed limits, which by altering the speed limits on roads close to congested areas or to an incident, tries to keep the flow levels as high as possible by preventing a breakdown before it has happened.

In the remainder of this section we provide an overview of traffic control strategies based on gating and variable speed limits.

11.5.2.1 Traffic Control on Motorways

Ramp Metering

Ramp metering tries to limit the on-ramp flow entering a freeway by the use of green-red signal timing. The aim is to keep the traffic flow as close to capacity levels as possible by allowing as much vehicles as possible without causing a drop in capacity. This is similar to perimeter control strategies, and thereby ramp metering can be considered as a local perimeter control strategy.

Numerous ramp metering strategies have been presented in the literature. See [87] for an overview of the different strategies. The ramp metering strategies can be either fixed-time or traffic-responsive. The fixed-time ramp metering strategies derive signal patterns off-line based on historical data which is dependent on the time of the day, and the freeway is divided into segments consisting of at a maximum one on-ramp. A static model is used to derive the traffic flow entering and leaving each segment of the freeway. The traffic-responsive ramp metering strategies are based on real-time measurements and the aim is to keep the traffic conditions close to some user-defined values, and the measurements are usually collected through detectors on the road. Traffic-responsive ramp metering strategies can further be divided into local ramp metering and coordinated ramp metering. Local ramp metering strategies make use of traffic measurements collected where the on-ramp is entering the freeway. The flow control could be based on a feed-forward control, see e.g. [79], or feed-backward control as for example ALINEA [86]. Other strategies make use of fuzzy logic such as in [97] and neural network, see e.g. [108]. Coordinated ramp metering tries to optimize the traffic flow over a larger area including multiple on-ramps. The coordinated ramp metering strategies can be divided into multivariable ramp metering strategies and optimal control strategies. The multivariable strategies, see for example [42, 85], make use of measurements from a larger area and try to coordinate the traffic flow based on all on-ramps included in the area. These algorithms are based on linear equations of the traffic flow models and might therefore be inefficient during congestion. Optimal control strategies are better in modeling the traffic flow but they are usually complex and difficult to implement in field. See [43, 52, 88, 109] for example of optimal control strategies. Due to the complexity of optimal control strategies a third category of coordinated ramp metering has evolved, called heuristic rule-based strategies. An overview of such strategies can be found in [18]. More recent examples are presented in [50, 89].

Variable Speed Limit Systems

Many of the existing variable speed limit (VSL) systems [101, 102] are mainly focusing on safety, i.e. incident detection systems. Detector stations are used to measure flow and/or speed. The VSL is communicated to the vehicles via a VSL sign. When an incident occur, measured as very low speed and/or flow levels, the system is triggered by applying lower speed limits. Thereby, the risk of further breakdown is limited and the system tries to resolve the existing congestion. In incident detection systems the speed limit is usually lowered substantially. The speeds in the VSL system can be recommended, as in the Swedish motorway control system [102], or mandatory as in the UK system [57].

For flow harmonization systems the main purpose is to keep the flow levels as high as possible by preventing a breakdown before it has happened. In this case the speed limit is not necessarily activated in low speed and/or flow situations. Activation can occur when there is a high diversity in speed levels among the vehicles, or at high speed levels with high traffic flow. The new speed limit is based on an evaluation of the risk of a breakdown. This can be done by prediction of the future traffic states using current and/or historic information. The speed limit is lowered in small steps compared to the more abrupt change in the incident detection systems.

Combined systems making use of both flow harmonization and incident detection exists in implemented systems, such as in the system used in the United Kingdom [57]. Flow harmonization systems only exist the literature, see for example [24].

VSL systems presented in the literature can be categorized as predictive models and reactive models. A common approach for the predictive models is to make use of model predictive control [37, 82]. In predictive models the traffic evolution is described with a dynamic traffic model, a history of past control actions and an optimization cost function, in order to predict the future traffic states. The cost function is constructed based on the aim and the purpose of the system. The aim could be to minimize the total travel time spent as in [53], or to minimize the emission levels, or a combination of the two as in [106]. In the reactive models only current information about the traffic state given from road detectors, road design, and current operational conditions are needed in order to determine a new speed limit. Examples of reactive models can be found in [24, 54, 72].

A simulation-based study [45] of four reactive VSL systems show that the accuracy of the information from detector stations, the prediction of the traffic states, the estimation of when an incident occur and over which stretch the incident reaches are essential for the performance of the VSL system. Further, reactions to changed conditions on the road which are not necessarily related to an incident, such as high variations of speed levels, high flow levels, are also important for the efficiency of the system.

Data from connected vehicles can for VSL applications provide a new source of data. For example, speed, distance to an updated VSL and position on the road, could be used in order to improve the VSL system performance. Also, by having a traffic system with a high rate of connected vehicles the VSL systems does not have to be dependent on detector stations and VSL signs, which are expensive to maintain. Instead, the systems could use communication between vehicles and between vehicles and the infrastructure (and some control center) to give individualized speed limits to the drivers [44]. This could presumably maximize the throughput on the road even further.

Combinations of Ramp Metering and Variable Speed Limits

Ramp metering is an efficient tool to use to increase efficiency during congestion since it limits the flow to the capacity levels of the freeway. The main drawback is the spill-back to other arterial roads due to queue propagation from the on-ramp. This is especially observable during heavy congestion. On the other hand, VSLs cannot reduce the inflow. The goal is to make the best of the traffic situation by trying to maximize efficiency and/or reduce the effect of an incident without changing the conditions of the inflow levels. During moderate congestion the VSLs can usually increase the efficiency, especially if it is set to harmonize the traffic flow rather than to increase safety. Although, for heavy congestion, and for long time periods with congestion, the traffic system tends to become unstable even though VSLs are applied.

Therefore, a combination of ramp metering and VSLs could potentially increase the efficiency on the freeway even further, see e.g. [25, 55, 93, 107]. In order to maximize the throughput even more than when only one measure is applied, the measures can be applied simultaneously, or after each other.

11.5.2.2 Perimeter Control in Urban Networks

Perimeter control has been suggested for reducing congestion in urban areas and to improve travel times. The nonlinear nature of congestion, where a few additional vehicles on a road can create a traffic breakdown, is difficult to deal with using local control algorithms, since they only consider local measurements but not the overall network (or subnetwork) performance. Instead, using signal timings in order to coordinate and control the traffic flow on the border of a controlled region, or in the case of multiple controlled regions, in-between regions, overcrowding of a specific region can be avoided. Thus, the traffic flow can be maintained at capacity. The concept of perimeter control for urban regions was first presented for a single region in [31], and later extended to two regions in [40, 49] and multiple regions in [1, 92].

In perimeter control algorithms, macroscopic fundamental diagrams (MFD), sometimes referred to as network fundamental diagrams, plays an important role. They can be viewed as an area-wide extension of fundamental diagrams, and in [39] it was shown that such a MFD relationship can be found for neighborhood-sized sections of a city. Traditionally, fundamental diagrams describe the relationship between flow, density, and speed at a specific road segment. The MFD instead describe the relationship between network (or subnetwork) wide densities and mean flows (or out flows from the subnetwork). Used in an automatic control loop, the MFD provide a bridge between the control, usually signal timing, and the traffic state. The MFD can provide this, without requiring microsimulation of the traffic, which makes it appealing for large-scale online applications.

A prerequisite to applying perimeter control is a suitable way of defining subnetworks, for which a well-behaved MFD can be found [41] (i.e. low scatter MFD). A determining factor for finding an MFD has shown to be areas of a city with homogenous densities on the roads. One methodology for finding such areas, based on network partitioning, is suggested in [64]. To determine the current traffic state in the subnetworks, both fixed location sensors (commonly loop detectors), as well as vehicle probes [38, 83], can be used. This can include both Eulerian and Lagrangian observations, and estimation of MFDs based on the two types of observations are compared in [71]. Their results highlight that a better estimation of the MFD can be achieved by combining the two types of data.

The key concept of perimeter control includes sensors and a control algorithm. The sensors provide real-time measurements of the density and the control algorithm makes use of these measurement, through the MFD, to determine the most suitable signal timings. Two main control algorithms have been proposed; multivariable feedback regulators, which is an optimal open-loop control, and model predictive control algorithms, which is an optimal closed-loop control. A linear-quadratic multivariable feedback regulator is presented in [1], and model predictive control algorithms are proposed in [40, 92]. While an open-loop control cannot take the error between model and reality, i.e. the MFD scatter, into account, a closed-loop control approach, like the model predictive control approach, can take this error into account. The model predictive control approach also has the benefit of handling noise in both travel demands and the MFDs. For a model predictive control approach a prediction of future traffic demand is required, which could for instance be provided by the Kalman filtering approach presented in Sect. 11.4.1.

11.6 Conclusions

A large number of heterogeneous sensors for road traffic observation are already in place in many cities around the world. The large amount of data from these sensors does not by itself improve information to the road users (e.g. travel time information) or provide means for traffic control to the road authorities. It is through the use of filtering techniques and models that this data enables new possibilities for online estimation and prediction of the traffic state, and for wide-area control in urban areas.

The emerging road traffic sensors described in this chapter enable a better understanding of long-term travel demand and travel pattern dynamics, but also makes it possible to estimate and predict travel demand and wide-area traffic densities more accurately in real time. This will enable the introduction of wide-area traffic control, which is a research area that has gained a lot of interest in recent years. Also, by making use of new data sources shared by communication between vehicles, and between vehicles and the infrastructure, so called cooperative systems, there is a great potential to improve the performance of existing traffic control strategies as well.

References

1. Aboudolas K, Geroliminis N (2013) Perimeter and boundary flow control in multi-reservoir heterogeneous networks. Transp Res Part B 55:265–281. doi:10.1016/j.trb.2013.07.003
2. Alexiadis V, Sallman D, Armstrong A (2012) Traffic analysis toolbox volume XIII: integrated corridor management analysis, modeling, and simulation guide. FHWA-JPO-12-074. Federal Highway Administration
3. Allström A, Ekströ J, Gundlegård D, Ringdahl R, Rydergren C, Bayen AM, Patire AD (2016) A hybrid approach for short-term traffic state and travel time prediction on highways. Transp Res Rec: J Transp Res Board 2554. doi:10.3141/2554-07
4. Antoniou C, Barceló J, Breen M, Bullejos M, Casas J, Cipriani E, Ciuffo B, Djukic T, Hoogendoorn S, Marzano V, Montero L, Nigro M, Perarnau J, Punzo V, Toledo T, van Lint H (2016) Towards a generic benchmarking platform for origin-destination flows estimation/updating algorithms: design, demonstration and validation. Transp Res Part C 66:79–98. doi:10.1016/j.trc.2015.08.009
5. Antoniou C, Ben-Akiva M, Koutsopoulos HN (2010) Kalman filter applications for traffic management. In: Krodić V (ed) Kalman Filter. INTECH
6. Bachmann C, Abdulhai B, Roorda MJ, Moshiri B (2013) A comparative assessment of multi-sensor data fusion techniques for freeway traffic speed estimation using microsimulation modeling. Transp Res Part C 26:33–48. doi:10.1016/j.trc.2012.07.003
7. Bar-Gera H (2007) Evaluation of a cellular phone-based system for measurements of traffic speeds and travel times: a case study from Israel. Transp Res Part C 15(6):380–391. doi:10.1016/j.trc.2007.06.003

8. Barceló J (2010) Models, traffic models, simulation, and traffic simulation. In: Barceló J (ed) Fundamentals of traffic simulation. Springer, New York, NY, pp 1–62
9. Barceló J, Gilliéron F, Linares M, Serch O, Montero L (2012) Exploring link covering and node covering formulations of detection layout problem. Transp Res Rec: J Transp Res Board 2308:17–26. doi:10.3141/2308-03
10. Barceló J, Montero L, Bullejos M, Linares M (2014) A practical proposal for using origin-destination matrices in the analysis, modeling and simulation for traffic management. In: 93rd TRB annual meeting compendium of papers, 14-3793
11. Barceló J, Montero L, Bullejos M, Linares M, Serch O (2013) Robustness and computational efficiency of Kalman filter estimator of time-dependent origin-destination matrices. Transp Res Rec: J Transp Res Board 2344:31–39. doi:10.3141/2344-04
12. Barceló J, Montero L, Bullejos M, Serch O, Carmona C (2013) A Kalman filter approach for exploiting bluetooth traffic data when estimating time-dependent OD matrices. J Intell Transp Syst 17(2):123–141. doi:10.1080/15472450.2013.764793
13. Barceló J, Montero L, Marqués L, Carmona C (2010) A Kalman-filter approach for dynamic OD estimation in corridors based on bluetooth and wi-fi data collection. In: Proceedings of the 12th WCTR. Lisabon, Portugal
14. Bayen A, Butler J, Patire A (2011) Mobile Millennium final report. Institute of Transportation Studies, University of California, Berkeley, California Center for Innovative Transportation
15. Blondel VD, de Cordes N, Decuyper A, Deville P, Raguenez J, Smoreda Z (2013) Mobile phone data for development-analysis of mobile phone datasets for the development of Ivory Coast
16. Boero M (1999) Case studies of systems: the KITS model. Workshop on intelligent traffic models. Delft University, ERUDIT
17. Boero M, Kirschfink H (1999) Case studies of systems: the ENTERPRICE model. Workshop on intelligent traffic models. Delft University, ERUDIT
18. Bogenberger K, May AD (1999) Advanced coordinated traffic responsive ramp metering strategies. PATH working paper, vol iii, no 2
19. Bullejos M, Barceló J, Montero L (2014) A DUE based bilevel optimization approach for the estimation of time sliced OD matrices. In: Proceeding of international symposium of transport simulation 2014
20. Byon YJ, Shalaby A, Abdulhai B, El-Tantawy S (2010) Traffic data fusion using SCAAT Kalman filters. In: Proceedings of the transportation research board 89th annual meeting
21. Caceres N, Wideberg JP, Benitez FG (2007) Deriving origin destination data from a mobile phone network. IET Intell Transp Syst 1(1):15–26. doi:10.1049/iet-its:20060020
22. Calabrese F, Diao M, Di Lorenzo G, Ferreira J Jr, Ratti C (2013) Understanding individual mobility patterns from urban sensing data: a mobile phone trace example. Transp Res Part C 26:301–313. doi:10.1016/j.trc.2012.09.009
23. Cantelmo G, Cipriani E, Gemma A, Nigro M (2014) An adaptive bi-level gradient procedure for the estimation of dynamic traffic demand. IEEE Trans Intell Transp Syst 15(3):1348–1361. doi:10.1109/TITS.2014.2299734
24. Carlson R, Papamichail I, Papageorgiou M (2011) Local feedback-based mainstream traffic flow control on motorways using variable speed limits. IEEE Trans Intell Transp Syst 12(4):1261–1276
25. Carlson RC, Papamichail I, Papageorgiou M (2014) Integrated feedback ramp metering and mainstream traffic flow control on motorways using variable speed limits. Transp Res Part C 46:209–221. doi:10.1016/j.trc.2014.05.017
26. Chiu YC, Bottom J, Mahut M, Paz A, Balakrishna R, Waller T, Hicks J (2011) Dynamic traffic assignment: a primer. Transp Res E-Circ E-C153
27. Chung K, Rudjanakanoknad J, Cassidy MJ (2007) Relation between traffic density and capacity drop at three freeway bottlenecks. Transp Res Part B 41:82–95
28. Cipriani E, Florian M, Mahut M, Nigro M (2011) A gradient approximation approach for adjusting temporal origin-destination matrices. Transp Res Part C 19(2):270–282. doi:10.1016/j.trc.2010.05.013. Emerging theories in traffic and transportation and methods for transportation planning and operations

29. Cuena J, Hernández J, Molina M (1995) Knowledge-based models for adaptive traffic management systems. Transp Res Part C 3(5):311–337. doi:10.1016/0968-090X(95)00013-9
30. Daganzo CF (1994) The cell transmission model: a dynamic representation of highway traffic consistent with the hydrodynamic theory. Transp Res Part B 28(4):269–287. doi:10.1016/0191-2615(94)90002-7
31. Daganzo CF (2007) Urban gridlock: macroscopic modeling and mitigation approaches. Transp Res Part B 41(1):49–62. doi:10.1016/j.trb.2006.03.001
32. Ekström J, Engelson L, Rydergren C (2014) Optimal toll locations and toll levels in congestion pricing schemes: a case study of Stockholm. Transp Plann Technol 37(4):333–353. doi:10.1080/03081060.2014.897129
33. Ekström J, Kristoffersson I, Quttineh NH (2016) Surrogate-based optimization of cordon toll levels in congested traffic networks. J Adv Transp doi:10.1002/atr.1386 (Accepted)
34. Evensen G (2003) The ensemble Kalman filter: theoretical formulation and practical implementation. Ocean Dyn 53(4):343–367. doi:10.1007/s10236-003-0036-9
35. Evensen G (2009) Data assimilation: the ensemble Kalman filter. Springer, Science & Business Media
36. Faouzi NEE, Leung H, Kurian A (2011) Data fusion in intelligent transportation systems: progress and challenges—a survey. Inf Fusion 12(1):4–10. doi:10.1016/j.inffus.2010.06.001. Special Issue on Intelligent Transportation Systems
37. García CE, Prett DM, Morari M (1989) Model predictive control: theory and practice—a survey. Automatica 25(3):335–348
38. Gayah V, Dixit V (2013) Using mobile probe data and the macroscopic fundamental diagram to estimate network densities. Transp Res Rec: J Transp Res Board 2390:76–86. doi:10.3141/2390-09
39. Geroliminis N, Daganzo CF (2008) Existence of urban-scale macroscopic fundamental diagrams: some experimental findings. Transp Res Part B 42(9):759–770. doi:10.1016/j.trb.2008.02.002
40. Geroliminis N, Haddad J, Ramezani M (2013) Optimal perimeter control for two urban regions with macroscopic fundamental diagrams: a model predictive approach. IEEE Trans Intell Transp Syst 14(1):348–359. doi:10.1109/TITS.2012.2216877
41. Geroliminis N, Sun J (2011) Properties of a well-defined macroscopic fundamental diagram for urban traffic. Transp Res Part B 45(3):605–617. doi:10.1016/j.trb.2010.11.004
42. Goldstein N, Kumar K (1982) A decentralized control strategy for freeway regulation. Transp Res Part B 16(4):279–290. doi:10.1016/0191-2615(82)90012-1
43. Gomes G, Horowitz R (2006) Optimal freeway ramp metering using the asymmetric cell transmission model. Transp Res Part C 14(4):244–262. doi:10.1016/j.trc.2006.08.001
44. Grumert E, Ma X, Tapani A (2015) Analysis of a cooperative variable speed limit system using microscopic traffic simulation. Transp Res Part C 52:173–186
45. Grumert E, Tapani A, Ma X (2016) Evaluation of four control algorithms used in variable speed limit systems. In: TRB 95th annual meeting compendium of papers, 16-2880. Washington, U.S.A
46. Gundlegård D, Allström A, Bergfeldt E, Bayen AM, Ringdahl R (2015) Travel time and point speed fusion based on a macroscopic traffic model and non-linear filtering. In: 2015 IEEE 18th international conference on intelligent transportation systems, pp 2121–2128. doi:10.1109/ITSC.2015.343
47. Gundlegård D, Karlsson JM (2006) Generating road traffic information from cellular networks—new possibilities in UMTS. In: 2006 6th International conference on ITS telecommunications, pp 1128–1133. doi:10.1109/ITST.2006.288805
48. Gundlegård D, Karlsson JM (2009) Handover location accuracy for travel time estimation in GSM and UMTS. IET Intell Transp Syst 3(1):87–94. doi:10.1049/iet-its:20070067
49. Haddad J, Geroliminis N (2012) On the stability of traffic perimeter control in two-region urban cities. Transp Res Part B 46(9):1159–1176. doi:10.1016/j.trb.2012.04.004
50. Hadi MA (2005) Coordinated traffic responsive ramp metering strategies—an assessment based on previous studies. In: 12th World congress on intelligent transport systems

51. Haghani A, Hamedi M, Sadabadi K, Young S, Tarnoff PJ (2010) Freeway travel time ground truth data collection using Bluetooth sensors. In: Proceedings of the transportation research board 89th annual meeting. Transportation Research Board, Washington DC
52. Hegyi A, De Schutter B, Heelendoorn J (2003) MPC-based optimal coordination of variable speed limits to suppress shock waves in freeway traffic. In: American control conference, 2003. Proceedings of the 2003, vol 5, pp 4083–4088. doi:10.1109/ACC.2003.1240475
53. Hegyi A, De Schutter B, Hellendoorn J (2005) Optimal coordination of variable speed limits to suppress shock waves. IEEE Trans Intell Transp Syst 6(1):102–112
54. Hegyi A, Hoogendoorn S, Schreuder M, Stoelhorst H, Viti F (2008) Specialist: a dynamic speed limit control algorithm based on shock wave theory. In: 11th International IEEE conference on intelligent transportation systems. Beijing, China, pp 827–832
55. Hegyi A, Schutter BD, Hellendoorn H (2005) Model predictive control for optimal coordination of ramp metering and variable speed limits. Transp Res Part C 13(3):185–209. doi:10.1016/j.trc.2004.08.001
56. Herrera JC, Work DB, Herring R, Ban XJ, Jacobson Q, Bayen AM (2010) Evaluation of traffic data obtained via GPS-enabled mobile phones: the mobile century field experiment. Transp Res Part C 18(4):568–583. doi:10.1016/j.trc.2009.10.006
57. Highways Agency (2007) M25, Control motorway, summary report. Technical report. Department for Transportation, London, UK
58. van Hinsbergen C, van Lint J, Sanders F (2007) Short term traffic prediction models, deliverable DIIF-1a in project II-F: travel time prediction on urban networks. TU Delft and Vialis Traffic
59. van Hinsbergen CPI, Zuurbier FS, van Lint JWC, van Zuylen HJ (2008) Using an LWR model with a cell based extended Kalman filter to estimate travel times. In: The 3rd International symposium of transport simulation. Surfer's Paradise, QLD, Australia
60. Hofleitner A (2012) Leveraging geolocalization technologies to model and estimate urban traffic. Ph.D. thesis. Université Paris-Est
61. Huang A, Rudolph L (2007) Bluetooth essentials for programmers. Cambridge University Press
62. Hunter T, Abbeel P, Bayen A (2014) The path inference filter: model-based low-latency map matching of probe vehicle data. IEEE Trans Intell Transp Syst 15(2):507–529. doi:10.1109/TITS.2013.2282352
63. Iqbal MS, Choudhury CF, Wang P, González MC (2014) Development of origin-destination matrices using mobile phone call data. Transp Res Part C 40:63–74. doi:10.1016/j.trc.2014.01.002
64. Ji Y, Geroliminis N (2012) On the spatial partitioning of urban transportation networks. Transp Res Part B 46(10):1639–1656. doi:10.1016/j.trb.2012.08.005
65. Kalman R (1960) A new approach to linear filtering and prediction problems. J Basic Eng 82(1):35–45. doi:10.1115/1.3662552
66. Kalman R, Bucy R (1961) New results in linear filtering and prediction theory. J Basic Eng 83(1):95–108. doi:10.1115/1.3658902
67. Kim K, Chien S, Spasovic L (2011) Evaluation of technologies for freeway travel time estimation: a case study of i-287 in New Jersey. In: Proceedings of the transportation research board 90th annual meeting
68. Kirschfink H, Riegelhuth G, Barceló J (2003) Scenario analysis to support strategic traffic management in the region Frankfurt Rhein-Main. In: 10th World conference on intelligent transport systems. Madrid
69. Kong QJ, Zhao Q, Wei C, Liu Y (2013) Efficient traffic state estimation for large-scale urban road networks. IEEE Trans Intell Transp Syst 14(1):398–407. doi:10.1109/TITS.2012.2218237
70. Lahrmann H, Pedersen SK, Christensen LT (2010) Bluetooth detektorer som ny cost-effektiv sensor i vejtrafikken. Trafikdage på Aalborg Universitet
71. Leclercq L, Chiabaut N, Trinquier B (2014) Macroscopic fundamental diagrams: a cross-comparison of estimation methods. Transp Res Part B 62:1–12. doi:10.1016/j.trb.2014.01.007

72. Lee C, Hellinga B, Saccomanno F (2006) Evaluation of variable speed limits to improve traffic safety. Transp Res Part C 14(3):213–228
73. Li JQ, Zhou K, Zhang L, Zhang WB (2012) A multimodal trip planning system with real-time traffic and transit information. J Intell Transp Syst 16(2):60–69. doi:10.1080/15472450.2012.671708
74. Lighthill M, Whitham G (1955) On kinematic waves. II. A theory of traffic flow on long crowded roads. Proc R Soc Lond Ser A: Math Phys Sci 229(1178):317–345
75. van Lint H, Djukic T (2012) Applications of Kalman filtering in traffic management and control. In: Mirchandani PB, Smith JC, Greenberg HJ (eds) 2012 Tutorials in operations research: new directions in informatics, optimization, logistics, and production. INFORMS, pp 59–91. doi:10.1287/educ.1120.0099
76. van Lint J (2010) Empirical evaluation of new robust travel time estimation algorithms. Transp Res Record: J Transp Res Board 2160:50–59. doi:10.3141/2160-06
77. van Lint J, Hoogendoorn SP (2010) A robust and efficient method for fusing heterogeneous data from traffic sensors on freeways. Comput-Aid Civil Infrastruct Eng 25(8):596–612. doi:10.1111/j.1467-8667.2009.00617.x
78. Lundgren JT, Peterson A (2008) A heuristic for the bilevel origin-destination-matrix estimation problem. Transp Res Part B 42(4):339–354. doi:10.1016/j.trb.2007.09.005
79. Masher D, Ross D, Wong P, Tuan P, Zeidler P, Peracek S (1975) Guidelines for design and operating of ramp control systems. Technical Report NCHRP 3-22, SRI Project 3340, Standford Research Institute, SRI, Menid Park, CA
80. Mazare PE, Tossavainen OP, Bayen A, Work D (2012) Tradeoffs between inductive loops and GPS probe vehicles for travel time estimation: a Mobile Century case study. In: Proceedings of the transportation research board 91st annual meeting. Washington D.C
81. Mitchell H (2007) Multi-sensor data fusion: an introduction. Springer, New York
82. Morari M, Lee JH (1999) Model predictive control: past, present and future. Comput Chem Eng 23(4–5):667–682
83. Nagle A, Gayah V (2014) Accuracy of networkwide traffic states estimated from mobile probe data. Transp Res Rec: J Transp Res Board 2421:1–11. doi:10.3141/2421-01
84. Pan J, Popa IS, Zeitouni K, Borcea C (2013) Proactive vehicular traffic rerouting for lower travel time. IEEE Trans Veh Technol 62(8):3551–3568. doi:10.1109/TVT.2013.2260422
85. Papageorgiou M, Blosseville JM, Hadj-Salem H (1990) Modelling and real-time control of traffic flow on the southern part of boulevard peripherique in paris: Part I: Modelling. Transp Res Part A 24(5):345–359. doi:10.1016/0191-2607(90)90047-A
86. Papageorgiou M, Hadj-Salem H, Blosseville JM (1991) ALINEA: a local feedback control law for on-ramp metering. Transp Res Rec: J Transp Res Board 1320:58–64
87. Papageorgiou M, Kotsialos A (2000) Freeway ramp metering: an overview. In: Intelligent transportation systems, 2000. Proceedings. 2000 IEEE, pp 228–239. doi:10.1109/ITSC.2000.881058
88. Papamichail I, Kotsialos A, Margonis I, Papageorgiou M (2010) Coordinated ramp metering for freeway networks: a model-predictive hierarchical control approach. Transportation Research Part C 18(3):311–331. doi:10.1016/j.trc.2008.11.002. 11th IFAC symposium: the role of control
89. Papamichail I, Papageorgiou M (2008) Traffic-responsive linked ramp-metering control. IEEE Trans Intell Transp Syst 9(1):111–121. doi:10.1109/TITS.2007.908724
90. Porter JD, Kim D, Magaña M (2011) Wireless data collection system for real-time arterial travel time estimates. Report No. OR-RD-11-10 OTREC 10-16. Oregon State University
91. Rahmani M, Koutsopoulos HN (2013) Path inference from sparse floating car data for urban networks. Transp Res Part C 30:41–54. doi:10.1016/j.trc.2013.02.002
92. Ramezani M, Haddad J, Geroliminis N (2015) Dynamics of heterogeneity in urban networks: aggregated traffic modeling and hierarchical control. Transp Res Part B 74:1–19. doi:10.1016/j.trb.2014.12.010
93. Schelling I, Hegyi A, Hoogendoorn S (2011) SPECIALIST-RM: integrated variable speed limit control and ramp metering based on shock wave theory. In: 2011 14th International

IEEE conference on intelligent transportation systems (ITSC), pp 2154–2159. doi:10.1109/ITSC.2011.6083116

94. Seo T, Kusakabe T, Asakura Y (2015) Estimation of flow and density using probe vehicles with spacing measurement equipment. Transp Res Part C 53:134–150. doi:10.1016/j.trc.2015.01.033
95. Srivastava A, Geroliminis N (2013) Empirical observations of capacity drop in freeway merges with ramp control and integration in a first-order model. Transp Res Part C 30:161–177
96. Steenbruggen J, Borzacchiello MT, Nijkamp P, Scholten H (2011) Mobile phone data from GSM networks for traffic parameter and urban spatial pattern assessment: a review of applications and opportunities. GeoJournal 78(2):223–243. doi:10.1007/s10708-011-9413-y
97. Taylor C, Meldrum D, Jacobson L (1998) Fuzzy ramp metering: design overview and simulation results. Transp Res Rec: J Transp Res Board 1634:10–18
98. Transportation Studies Center (1997) Final evaluation report for the CAPITAL-ITS operational test and demonstration program
99. Treiber M, Helbing D (2002) Reconstructing the spatio-temporal traffic dynamics from stationary detector data. Cooper Transp Dyn 1:3.1–3.24
100. Valerio D, D'Alconzo A, Ricciato F, Wiedermann W (2009) Exploiting cellular networks for road traffic estimation: a survey and a research roadmap. In: IEEE 69th Vehicular technology conference, 2009. VTC Spring, pp 1–5 (2009). doi:10.1109/VETECS.2009.5073548
101. van den Hoogen E, Smulders S (1994) Control by variable speed signs. Results of the Dutch experiment. In: Seventh international conference on road traffic monitoring and control, vol 391. London, UK, pp 145–149
102. van Toorenburg JAC, de Kok ML (1999) Automatic incident detection in the motorway control system MTM. Technical report, Bureau Transpute, Gouda, Holland
103. Wang P, González M, Hunter T, Bayen A, Schechtner K (2012) Understanding road usage patterns in urban areas. Sci Rep
104. Work DB, Blandin S, Tossavainen OP, Piccoli B, Bayen AM (2010) A traffic model for velocity data assimilation. Appl Math Res eXpress 2010(1):1–35. doi:10.1093/amrx/abq002
105. Work DB, Tossavainen OP, Blandin S, Bayen AM, Iwuchukwu T, Tracton K (2008) An ensemble Kalman filtering approach to highway traffic estimation using GPS enabled mobile devices. In: 47th IEEE conference on decision and control, 2008. CDC 2008, pp 5062–5068. doi:10.1109/CDC.2008.4739016
106. Zegeye S, De Schutter B, Hellendoorn J, Breunesse E (2011) Reduction of area-wide emissions using an efficient model-based traffic control strategy. In: 2011 IEEE forum on integrated and sustainable transportation systems, FISTS 2011. Austria, Vienna, pp 239–244
107. Zegeye S, De Schutter B, Hellendoorn J, Breunesse E, Hegyi A (2012) A predictive traffic controller for sustainable mobility using parameterized control policies. IEEE Trans Intell Transp Syst 13(3):1420–1429. doi:10.1109/TITS.2012.2197202
108. Zhang HM, Ritchie SG (1997) Freeway ramp metering using artificial neural networks. Transp Res Part C 5(5):273–286. doi:10.1016/S0968-090X(97)00019-3
109. Zhang L, Levinson D (2004) Optimal freeway ramp control without origin-destination information. Transp Res Part B 38:869–887
110. Zhang L, Levinson D (2004) Some properties of flows at freeway bottlenecks. Transp Res Rec 1883:122–131

Chapter 12
Smart Grid for the Smart City

Riccardo Bonetto and Michele Rossi

12.1 Introduction

Modern cities are becoming more and more dependent on the reliability and efficiency of the electrical distribution infrastructure. In the past few years, a great effort has been devoted to the creation of an integrated infrastructure that combines a resilient power distribution system, distributed generation devices based on renewables (as, for example, photovoltaic panels and wind turbines), a reliable and secure communication system, and real-time energy pricing policies. The resulting infrastructure is called *smart grid* and constitutes the backbone of the *smart city*. As noted in [22], "The *smart city* is all about how the city "organism" works together as an integrated whole and survives when put under extreme conditions" and, moreover: "the energy infrastructure is arguably the single most important feature in any city. If unavailable for a significant enough period of time, all other functions will eventually cease." Hence, developing and implementing a fully functional *smart grid* infrastructure is a priority for future *smart cities*.

In order to implement a *smart grid*, at least the following three technical domains must cooperate, namely: (i) power electronics, (ii) information and communication technology (ICT), and (iii) economics.

From a power electronics standpoint, the diffusion of distributed energy resources (DERs) poses a number of interesting challenges that the future power distribution grid must face to evolve into a *smart* power distribution grid. On the one hand, the presence of grid connected DERs (i.e., DC power generators connected to the main power distribution system through grid-tie inverters or capacitor banks) must be carefully handled to maintain acceptable electrical power quality. In particular, voltage sags (overvoltages) due to the switch off (on) of groups of DERs must be avoided. Moreover, harmonic distortion must be minimized, since it is one of the major factors degrading the overall power quality [18]. On the other hand, if the presence

R. Bonetto · M. Rossi (✉)
University of Padova, Padua, Italy
e-mail: rossi@dei.unipd.it

V. Angelakis et al. (eds.), *Designing, Developing, and Facilitating Smart Cities*,
DOI 10.1007/978-3-319-44924-1_12

of grid tie DERs is fully accepted and embraced, these devices can be exploited to ameliorate the overall power grid performance and to provide ancillary services.

The coordination of DERs relies on three basic tiles: (i) measurement devices (i.e., synchrophasors measuring real-time voltages and currents) distributed across the power distribution system, (ii) a communication infrastructure that allows sharing the collected data among the active agents in the grid (i.e., the DERs and the utility), and (iii) suitable algorithms that use electrical measurements to automatically coordinate the DERs' actions in order to enhance the grid performance. The synergy between power electronics and IT solutions supported by a communication infrastructure constitutes the basement of a *smart grid*.

Once the *smart grid* basement has been set up, the catalyst that will boost the cooperation between the grid agents lies in economics. Guaranteeing economic advantages to those agents that contribute to the electrical grid efficiency is, indeed, a key factor to turn the *smart grid* concept into reality, i.e., to make it profitable for customers and utilities.

The aim of this chapter is to present the main electrical, IT, and economic open challenges to the *smart grid* and to discuss some of the solutions that have been proposed in the scientific literature. Finally, some real-world examples of *smart grids* will be considered and the adopted solutions will be analyzed.

The rest of this chapter is organized as follows: in Sect. 12.2, we introduce the electrical background, the communication scenario, and we formally describe the considered power grids. In Sect. 12.3, we discuss open issues related to the electrical optimization of smart grids, identifying some promising solutions from the literature. In Sect. 12.4, we highlight some key points to spur the adoption of smart grid technologies and their widespread diffusion. In Sect. 12.5 we elaborate on state-of-the-art gaming approaches to promote electrically efficient behaviors by power grid customers. In Sect. 12.6, we focus on data mining techniques, whereas in Sect. 12.7 we discuss relevant real-world deployments, and the benefits that are obtained by them. In Sect. 12.8 we present our concluding remarks.

12.2 Notation

In this section, the electrical scenario and the notation that will be used throughout the chapter are introduced.

12.2.1 Notions of Electrical Distribution

In order to define the electrical scenario that is considered in this chapter, a brief description of the traditional power production and transmission system is provided. Moreover, the main differences between the North American and European systems are introduced. The reader already familiar with these subjects may skip to the next section.

The traditional power distribution system has a well established hierarchical structure. Electrical power is produced in dedicated power plants that can be based on several energy sources (i.e., thermal power plants, nuclear reactors, biomass, gas turbines, and water turbines are the most common sources of energy). The power generator uses the mechanical energy obtained by these sources to operate a rotor that produces three sinusoidal currents oscillating at the same frequency (called *utility frequency*, 50 Hz in Europe and 60 Hz in the U.S.) but with a 120-degrees phase shift between each other. These currents are called 3-phase current. The 3-phase voltage at the power generator is then raised to a transmission level which depends on the length of the transmission lines and is aimed at reducing the distribution power losses. The produced power flows through high voltage transmission lines until it reaches transmission or local distribution substations where step down transformers reduce the voltage. Distribution substations connect the high voltage transmission lines to the medium voltage distribution systems. From there, distribution lines carry medium voltage power to medium to low voltage transformers where the voltage is reduced to 110 V_{rms} in the U.S. or 230 V_{rms} in Europe[1] and is used to feed houses, small businesses, and so on

In the past few years, many end users started installing small power generation devices based on renewables in their properties. At first, the locally generated power was used directly by the generator's owner to feed relatively small home appliances off-the-grid. As this technology started spreading, main power suppliers commenced buying the generated power which was converted into AC current by inverters. At this point, the power flow was not unidirectional anymore.

Bidirectional flows open new new opportunities as energy is generated in a distributed fashion by the end users and can be utilized to compensate for high peaks in the demand, etc., but it also leads to important challenges as this bidirectional flow model, if not properly handled, can lead to grid instability.

12.2.2 *Electrical Details*

In this chapter, single phase low voltage (i.e., 110 V_{rms}/230 V_{rms}) power grids are considered. These networks result from the extraction of one of the phase wires and the neutral from the three-phase low voltage distribution grid, and provide electrical power to small neighborhoods. The considered grids are connected to the main distribution grid through a special node called the point of common coupling (PCC) which, if needed, can be equipped with intelligent control algorithms and can act as a coordinator for the grid end users. The reason for this choice lies in the fact that, along main power transmission paths, solutions for monitoring and guaranteeing voltage—and, in turn, power quality—are already in place by means of capacitor banks, transformers, circuit breakers and so on [18]. In the considered grid scenarios, i.e., in residential neighborhoods, the coordinated cooperation among end users

[1] rms stands for root mean square and equals the peak value divided by $\sqrt{2}$.

equipped with small power generation devices can greatly ameliorate the overall power quality of the area and result in economic advantage for the community (by reducing the power distribution losses) as well [10]. It is the authors' opinion that these are the grids in which the deployment of the *smart grid* technology will have a greatest impact.

End users of the considered power grids can be divided into two main categories: (i) *consumers*, and (ii) *prosumers* (end users that are at the same time energy producers and consumers).

Consumers do not own any kind of power generation device, hence they completely depend on the power grid for feeding their electrical appliances. These users will be called *loads* in the rest of the chapter. Loads are usually classified based on their behavior. In particular, there are (1) constant impedance loads, (2) constant power loads, and (3) constant current loads. Constant impedance loads exhibit a constant impedance with respect to voltage variations, hence when variations occur to the supplied voltage, the absorbed current will vary as well resulting in a variable power demand. Examples of constant impedance loads are incandescent light bulbs and water heaters. Constant power loads absorb a constant power despite voltage variations, i.e., the current they need varies inversely with the applied voltage. Examples of constant power loads are motor drives for which the torque changes inversely with the rotation speed. Finally, constant current loads absorb a constant current despite voltage variations, hence their impedance must vary. Constant current loads are quite rare and can be found in a few applications as, for example, magnetic ballasts (which, by the way, are being replaced by the most efficient electronic ballasts whose characteristics resemble more those of a constant power load) and airport runways lighting. It is worth noting that loads rarely exactly fall into one of the above categories. The most common scenario is that where loads are a mixture of the three categories. The characteristics of this mixture strongly depend on the type of activity of the end user associated with the load (i.e., household, small business, industrial, ...) and on the season. For example residential loads during summer usually are 70 % constant power and 30 % constant impedance [54].

Prosumers are end users equipped with power generators. The power generation can come from renewables (i.e., mostly wind and solar energy) or, for example, from bidirectional electric vehicle (EV) chargers. In the latter case, the energy source is the EV battery. All these sources generate DC current that is fed to an inverter which, in turn, delivers AC power to the user. Initially, generators based on renewables were only used to fullfil the owners' power demand. The diffusion of grid-tie inverters,[2] however, allowed the owners to start selling the produced energy to the utility. As powerline communication has been developed, signaling from the utility to the end users began to turn the grid into a *smart* entity. By means of grid-tie inverters, *prosumers* can inject some of their excess energy (if any) into the grid. This energy can either be bought by the utility or, where a control infrastructure is in place, be used to provide ancillary services improving the overall performance of the grid. In this

[2]Grid-tie inverters are devices that allow to interface with the main distribution line and inject the generated power directly into the grid.

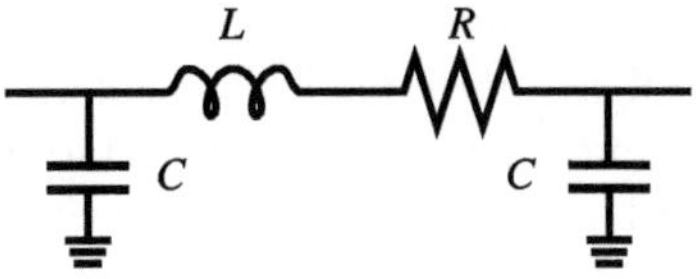

Fig. 12.1 *Π* line section example

chapter, *prosumers* are modeled as AC current generators with adjustable current phase, and connected in parallel to a load which will be called the "associated load."

The coordination of *prosumers* and (if possible) loads by means of a communication infrastructure and intelligent control strategies is a much advocated scenario, since it can relieve the main utility from some of its workload and at the same time greatly ameliorate the quality of the delivered power, especially during peak hours. This condition appears even more important if we consider the fact that the biggest source of greenhouse gas emissions are the electricity production plants [1].

The distribution lines are the last component of the considered grids. Usually, transmission and distribution lines are theoretically modeled as π-sections [23], whose characteristic impedances vary with respect to their position in the grid, i.e., their distance from the PCC. A π-section example is shown in Fig. 12.1. The *RL* series represents the actual path of the current, while the capacitors represent the intrinsic shunt capacitance of the two conductors. For short transmission lines (i.e., shorter than 80 km) the shunt capacitance becomes so small that can be neglected [5], and, hence, these lines can be represented as *RL* series. Due to the inductive and capacitive components, the impedance of the distribution lines vary according to the utility frequency. As the distance from the PCC grows, so does the impedance of the distribution lines. This is due to the fact that thinner (thus cheaper) cables are used as the delivered current decreases, hence the lines connecting the end users to the grid are the ones with the greatest resistive characteristic, while the line connecting the PCC to the grid is the one with the smallest one.

12.2.3 Communication Infrastructure

The cooperation among the *prosumers*, the loads, and the utility is a key feature for the *smart grid*. Such a cooperation is however possible only if a communication infrastructure supports the operations of the grid agents. The requirements for the communication infrastructure greatly vary depending on the control algorithms that are to be supported. For this reason, there is no single "one-fit-all" technology which meets all the *smart grid* requirements. For example, some algorithms require that local data (i.e., real time currents and voltages) be frequently sent to a central control unit, which then dispatches the control actions to the end users [9]. Other algorithms are fully distributed and require that only few agents, within the same area, communicate among themselves sharing local measurements and estimating the impedances of the lines connecting them [7, 50]. In the first case, a wireless-based solution (for example, exploiting cellular networks) might be the right choice since it allows the central control unit to act as a base station, broadcasting the control actions to the local users. In large and distributed networks, instead, powerline communication (PLC)-based solutions might be the ones to go for. Note that the PLC technology besides enabling communication over distribution lines, can also be utilized for the estimation of their impedance. Moreover, the PLC network topology is the same as that of the electrical grid which utilizes it. Hence, as a side product, PLC can also be exploited to get topology estimates, which may then be a valuable information to certain smart grid algorithms [20, 55].

12.2.4 Formal Grid Representation

In this section, the mathematical notation used in the rest of this chapter is introduced. Let the *smart grid* agents (i.e., the end users and the PCC) be represented by the set $\mathcal{N}$ in (12.1),

$$\mathcal{N} = \{n_0, \dots, n_N\} \subseteq \mathbb{N}\,. \tag{12.1}$$

End users will be called nodes from now on. Moreover, let n_0 be the PCC.

The distribution grid is represented as a graph $(\mathcal{N}, \mathcal{E})$, where the node set $\mathcal{N}$ contains the vertexes and $\mathcal{E}$ is the set of arcs connecting pairs of vertexes, representing the distribution lines. Each arc (i,j) is weighted by the quantity $Z_{(i,j)} \in \mathcal{Z} \subseteq \mathbb{C}$ representing the impedance associated with the distribution line (i,j) at the utility frequency, having real and imaginary part $R_{(i,j)}$ and $X_{(i,j)}$, respectively. Equations (12.2) and (12.3) represent the *load* and *prosumer* sets, $\mathcal{L}$ and $\mathcal{G}$ respectively.

$$\mathcal{L} \subseteq \mathcal{N}\,, \tag{12.2}$$

$$\mathcal{G} \subseteq \mathcal{N}\,. \tag{12.3}$$

Loads and *prosumers* (which from now on will be called distributed generators, DGs) are subsets of the node set. Each load is identified by L_i, $i \in \{1, \dots, N\}$ where i identifies the index of the node n_i to which L_i refers to. Similarly, DGs are identified by G_j, $j \in \{1, \dots, N\}$. The PCC is considered as a special DG, hence it is always true that $n_0 \in \mathcal{G}$. $\mathcal{L}$ and $\mathcal{G}$ are disjoint set. Moreover, let $L, G \leq N$ be the cardinalities of $\mathcal{L}$ and $\mathcal{G}$, respectively. In this representation, nodes can be considered as buses from a power systems point of view, hence the two terms will be used interchangeably in this chapter. As said before, loads can be classified into three main categories (i.e., constant power, constant impedance, and constant current). If needed, the load type will be specified by $L_i^{(k)}$, $k \in \{1, 2, 3\}$, where $k = 1$ identifies a constant power load, $k = 2$ identifies a constant impedance load and $k = 3$ identifies a constant current load.

In order to provide a full electrical description of each network element, four quantities are associated with each node, load, and DG. These quantities are the current ($I \in \mathbb{C}$), the voltage ($V \in \mathbb{C}$), the active ($P \in \mathbb{R}$), and the reactive ($Q \in \mathbb{R}$) power, as shown in (12.4).

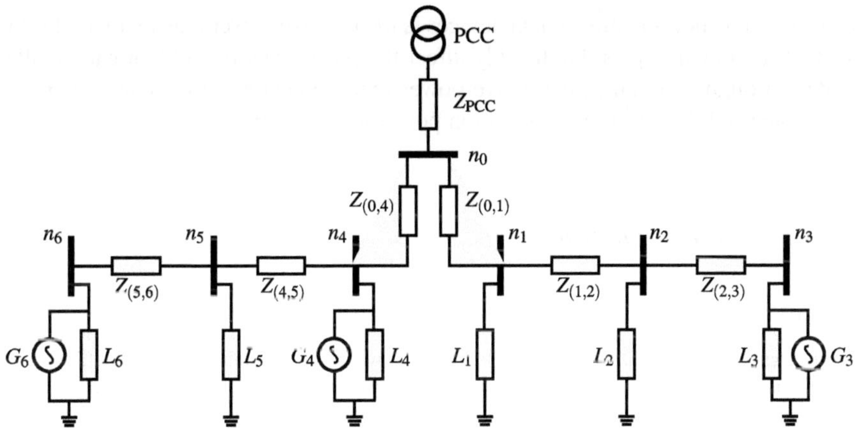

Fig. 12.2 Grid example

$$\begin{cases} V_{n_i}, I_{n_i}, P_{n_i}, Q_{n_i} & \forall\, i \in \mathcal{N} \\ V_{L_i}, I_{L_i}, P_{L_i}, Q_{L_i} & \forall\, L_i \in \mathcal{L} \\ V_{G_i}, I_{G_i}, P_{G_i}, Q_{G_i} & \forall\, G_i \in \mathcal{G} \\ V_{Z_{(i,j)}}, I_{Z_{(i,j)}}, P_{Z_{(i,j)}}, Q_{Z_{(i,j)}} & \forall\, (i,j) \in \mathcal{E} \end{cases} \tag{12.4}$$

Figure 12.2 shows a power grid example. End users are connected to the distribution lines (identified by $Z_{(i,j)}$) through the nodes (also called buses) $n_1, \ldots, n_6$. Consumers are represented through electrical loads, namely L_1, L_2, and L_5. Prosumers are represented through the parallel connection of an electrical load and a generator, namely (L_3, G_3), (L_4, G_4), and (L_6, G_6). Node n_0 represents the connection point between the PCC and the distribution grid.

12.3 Electrical Grid: Open Problems and Solutions

As stated in [18], the quality of the power delivered by the electric utilities has become one of the main concerns for the end users and the energy provider. This is due to a number of reasons, the main ones being that (i) loads have become more sensitive to power quality variations, (ii) there is a worldwide quest for increasing the efficiency of the distribution system, and (iii) end users are more aware of what happens in the distribution grid, and hence they pay more attention to the quality of the power being delivered. The *smart grid*, thanks to the interconnection of measurement units (i.e., synchrophasors), DGs, and control algorithms, may have a great impact on the improvement of the quality of the power delivered to the end users. An important goal, which can be reached through the interconnection and the subsequent control of the aforementioned DG devices is the reactive power management. In this section, two main topics that heavily affect the power quality and that can be alleviated through an intelligent reactive power management are investigated, namely, (i) voltage stability, and (ii) distribution power loss reduction.

12.3.1 Voltage Stability

According to the IEEE/CIGRE definition:

> the term voltage stability refers to the ability of a power system to maintain steady voltages at all buses in the system after being subjected to a disturbance from a given initial operating condition. It depends on the ability to maintain/restore equilibrium between load demand and load supply from the power system. Instability that may result occurs in the form of a progressive fall or rise of voltages of some buses. A possible outcome of voltage instability is loss of load in an area, or tripping of transmission lines and other elements by their protections leading to cascading outages that in turn may lead to loss of synchronism of some generators.

With regards to voltage stability, DGs can have a positive impact by stabilizing the voltages in the portions of the grid where they operate. The ability of a power system to maintain the grid voltage levels stable and within the operating limits under different load conditions is measured through the so-called voltage stability margins. One of the most common voltage stability margin is the Saddle Node Bifurcation (SNB). This voltage margin is based on the difference between the active power actually delivered measured at any bus of the grid and the active power corresponding to the SNB point of this bus Power–Voltage curve. Figure 12.3 shows a voltage

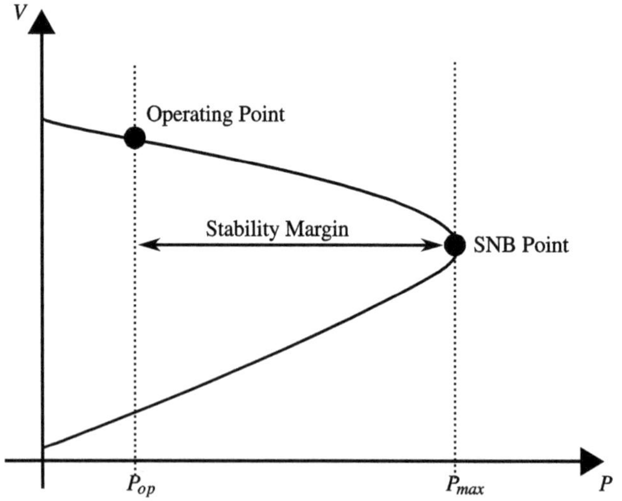

Fig. 12.3 Saddle Node Bifurcation voltage stability margin example

stability margin example. It can be noticed that as the active power delivered by the considered bus grows (which is identified by P_{op} and corresponds to an increase of the load in the considered portion of the grid), the voltage at the same point gradually decreases. As the maximum deliverable active power (P_{max}) is met, the voltage suddenly drops and the delivered active power starts decreasing as well. This condition may lead to a *voltage collapse* and, consequently, to blackouts. When adequately coordinated/placed/sized, by injecting power into the grid, DGs can reduce the power P delivered at the bus, hence a voltage rise happens and the stability margin increases. In [3], for example, the impact of DGs on the voltage stability in power grids is analyzed. In that paper, the authors show that when the power demand in the grid increases DGs do not cause voltage stability issues when operating within acceptable voltage conditions. Moreover, in [33] it is shown that, when the amount of power injected by the DGs is controlled, their distributed generation can enhance the overall system performance in terms of steady state voltage profile and voltage stability. As expected, this enhancement becomes more pronounced as the number of DGs in the grid grows. Given these encouraging results, great effort has been devoted to studying the placement of DGs, their dimensioning, and control policies to provide voltage stability in distribution grids.

In [28], the impact of reactive power management on the voltage stability of the grid is evaluated. Moreover, a heuristic management algorithm is proposed. In this work, the DGs are used to inject a controlled amount of reactive power into the grid in order to minimize the voltage fluctuations measured at the buses. To determine the optimal amount of reactive power that each DG has to inject, an optimization problem is formulated. In (12.5) the objective function of the proposed optimization problem is shown. This function measures the weighted sum (by means of the parameters ω_1, ω_2) of the total voltage deviation with respect to the specified voltage V_i^{spec} that each node should achieve and the total reactive power injected by the DGs. The voltage deviation has to be minimized ($\omega_1 > 0$), while the reactive power has to be maximized ($\omega_2 < 0$) to relieve the main supplier from some of the reactive power demand. The constraints for this optimization problem depend on the physical limitations of the devices installed in the grid. The combined utility function is:

$$U = \omega_1 \sum_{i=1}^{N} \left(|V_{n_i}^{spec} - V_{n_i}| + \sqrt[3]{|V_{n_i}^{spec} - V_{n_i}|} \right) + \omega_2 \sum_{i \in \mathcal{G}} Q_{n_i} \,. \tag{12.5}$$

To minimize Eq. (12.5), the authors use a centralized approach requiring little communication. The voltage at each bus, together with the injected reactive power must be sent to a central controller which, in turn, uses a genetic optimization approach. The resulting optimal reactive power that each DG has to inject is then dispatched. Quantitative results demonstrate the effectiveness of controlled power injection (i.e., injection of reactive power by the DGs) in ameliorating the voltage stability margin of all the buses of the considered power grid [28].

Another important aspect is the placement and sizing of DGs. Some of them are installed directly by the end users and, in turn, their size and placement cannot be controlled by the utility operator (actually, these facts can be influenced by incentives that are beyond the scope of this chapter). In future *smart grids*, it is expected that the utility operators will install smart power production units based on renewables acting as DGs. Hence, the problem of finding optimal positions and sizing for these DGs is well founded. Several approaches to accomplish this task have been proposed so far. In this chapter, two of them are briefly discussed. It is worth recalling that placement and sizing of DGs are not dynamic procedures. They require knowledge of the grid topology and its load distribution. Moreover, the computation of the optimal parameters is done once for all (offline) when DGs are installed. In [41], the authors used the voltage stability margin proposed in [49], which differs from the SNB presented above. This margin, referred to here as *SI*, turns out to be useful when placing and sizing DGs, as it requires to solve the power flow equations of the grid only once, considering pairs of buses connected by one distribution line. In (12.6), the computation of *SI* for a generic bus j connected to a bus i is shown.

$$SI(n_j) = 2V_{n_i}^2 V_{n_j}^2 - V_{n_j}^4 - 2V_{n_j}^2 (P_{n_j} R_{(i,j)} + Q_{n_j} X_{(i,j)}) - |Z_{(i,j)}|^2 (P_{n_j}^2 + Q_{n_j}^2). \quad (12.6)$$

Solving the power flow equations allows to know the voltages of all the nodes and the currents of all the branches. Hence, the quantities P_{n_j} and Q_{n_j} can be easily calculated $\forall n_j \in \mathcal{N}$. The bus with the lowest *SI* is the most prone to voltage collapse, and hence is the best candidate for the placement of a DG. The sizing of DGs highly depends on the grid load conditions and in [41] is determined heuristically. This method results in a considerably increased stability margin for all the considered scenarios, although, if no attention is paid to the reduction of power losses, these may actually increase as a result of placing the DGs.

In [2], the stochastic nature of the power production from DGs based on renewables and of the power demand from the loads are considered and a mixed integer nonlinear maximization problem is formulated. First, a statistical analysis of the power production (i.e., solar irradiance and wind speed) and load power demand is performed based on a data set spanning over three years. The result of this analysis is a 24 h average day for each one of the four seasons (hence, 96 possible combinations) for power production and demand. DGs and loads are then represented according to this model. Let $Pr[\xi]$ be the probability of the output power and load demand $\xi \in \{1, \ldots, \Xi\}$, where Ξ is the number of possible combinations. Moreover, let V_P^{DG} and $V_P^{\text{No DG}}$ be the voltage profiles of the grid with and without DGs, respectively. A utility function is defined as in (12.7),

$$U = \frac{\sum_{\xi=1}^{\Xi} \frac{V_P^{\text{DG}}}{V_P^{\text{No DG}}} Pr[\xi]}{96}, \quad (12.7)$$

which is subject to several constraints representing the physical limitations of the power grid (i.e., voltage limits, device capacities, etc.) and the maximum number of available DGs. Simulation results show that solutions exist (i.e., placement and sizing of the DGs) for which the stability margin is greatly improved. These results are also confirmed by [41].

An interesting point of [2] is the probabilistic approach to the modeling of the power production of DGs and the power demand of the loads. This approach allows for the simulation of many dynamic power systems using random data generators, whose probability distributions are based on real-world data, see, e.g., [9, 35]. It is the authors' opinion that this approach is of high value for the development of future *smart grids*. Using algorithmic techniques to synthetically generate a large number of distribution grids [40] allows studying the impact of new solutions on a large number of scenarios without the burden of collecting new data for each, see [11]. Moreover, being able to generate the electrical grid parameters, according to randomly generated data (whose probability distributions are based on real-world data sets) may have a great impact on the quantitative assessment of new control schemes.

12.3.2 Power Distribution Loss Reduction

Power distribution losses are the result of current flowing through the distribution lines. Considering a distribution line with equivalent impedance Z and a current I flowing through it, the dissipated power is $P = I^2R$ due to Joule's effect. When delivering power to the end users, distribution losses have many downsides for the customers and the energy producer. From the customers point of view, the main downside is that the dissipated power must be paid even though it is not consumed. From the producer's point of view, power dissipation shortens the life of distribution lines because of the heat generated by the Joule's effect. Moreover, the losses force the generators to produce more power, hence having higher working regimes. In distribution systems, line impedances can be changed only by actually replacing the distribution lines, and this solution is economically impractical. Hence, the only viable way to reduce power distribution losses is to reduce the amount of current flowing through the distribution lines. Since the dissipated power is proportional to the square of the current flowing through the lines, even a small reduction of the current can result in a substantial improvement on the efficiency of the power delivery process. To this end, three approaches are possible: (i) move the power supply closer to the power's destination, hence reducing the path length the current travels to supply the loads, (ii) reduce the amount of current flowing through the line, and (iii) a mixture of (i) and (ii). These three methods can be implemented exploiting the DGs operating in a grid-connected way. The first mentioned way requires that the DGs inject a certain amount of current based on the total amount of power needed in the area where it is economically convenient for them to operate. Operating in this way, the path the current travels before reaching its destination is shortened and hence also the power distribution losses are reduced. The second way is based on the fact that the cosine of the phase shift between voltage and current (called power factor) equals the ratio between the active and the apparent powers (i.e., the power that is actually used by the load and the delivered power, respectively). The power factor is formally expressed through (12.8):

$$\cos\phi = \frac{P}{S}\,. \tag{12.8}$$

A first result coming from (12.8) is that achieving a unitary power factor (i.e., $\phi = 0$) assures that all the delivered power can actually be converted into work. Moreover, from (12.8), and recalling that $S = VI$, (12.9) holds,

$$P = S\cos\phi = VI\cos\phi\,. \tag{12.9}$$

Equation (12.9) tells that, in order to reduce the current flowing through the distribution lines while keeping the voltage fixed, the power factor must be raised as close as possible to 1. Recalling that the apparent power S results from $S^2 = P^2 + Q^2$, it holds that a unitary power factor means that no reactive power flows through the lines. This condition is not acceptable, since many loads (for example electric motors and, in general, every load that has inductive or capacitive components) need a certain amount of reactive power. Once again, DGs can be coordinated to supply small areas with the correct amount of reactive power, hence moving the apparent power delivered by the main supplier as close as possible to the active power that is actually needed. The third way is the most effective, yet far looking, approach. We can implement this by having the DGs (1) to locally supply a fraction of the active power required by the end users and, at the same time, (2) to allow their cooperation to keep the power factor (as seen by the main power supplier) as close as possible to 1, by selectively injecting reactive power into the grid. Through this approach, power distribution losses can be drastically reduced. In the following, three methods exploiting the cooperation between DGs to reduce the power distribution losses are presented and discussed. In [52] and [51] the authors propose a lightweight control scheme for reactive power control solely based on local measurements. In [51], a local control scheme for reactive power injection is introduced. Considering a DG G_i with an associated load L_i, the proposed scheme compensates for the needed reactive power Q_{L_i} by injecting Q_{G_i}, as defined in (12.10),

$$Q_{G_i} = \min(Q_{G_i}^{max}, Q_{L_i})\,, \tag{12.10}$$

where $Q_{G_i}^{max}$ is the maximum reactive power that the DG's inverter can generate. In [52], voltage control is taken into account as well. Considering a node n_i connected to a node n_j through line $Z_{(i,j)}$, in order to reduce the voltage variation at node n_i, the combined power flow $R_{(i,j)}P_{(i,j)} + X_{(i,j)}Q_{(i,j)}$ must be minimized. To do so, (12.11) must hold,

$$Q_{G_i} = \min\left(Q_{G_i}^{max}, Q_{L_i} + (P_{L_i} - P_{G_i})\frac{R_{(i,j)}}{X_{(i,j)}}\right)\,. \tag{12.11}$$

Fig. 12.4 Grid example

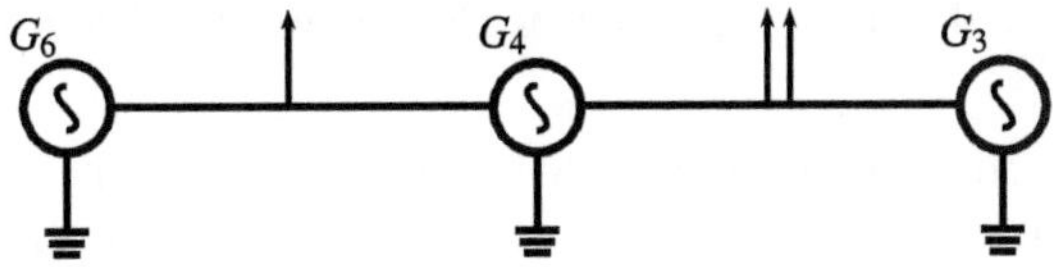

To reduce the power distribution losses, while minimizing voltage variations, the authors combine (12.10) and (12.11) through a weighted sum. Their results show that even these simple schemes can have a great impact on the reduction of power distribution losses. More refined schemes fully exploiting a communication infrastructure to achieve cooperation between end users have been proposed. For example, in [50] a fully distributed power loss minimization algorithm where DGs are coordinated through a token ring control approach is proposed. Considering Fig. 12.4, the voltage $V^*_{G_i}$ that G_i should reach in order to operate in the best condition with regards to the power loss minimization task is given by (12.12),

$$V^*_{G_i} = \frac{\sum_{k \in \mathscr{C}_i} \frac{R_{(i,k)}}{Z^2_{(i,k)}} V_{G_k}}{\sum_{k \in \mathscr{C}_i} \frac{R_{(i,k)}}{Z^2_{(i,k)}}}. \tag{12.12}$$

where $\mathscr{C}_i$ is the set of DGs forming a cluster with DG G_i, as explained in [10]. By sharing their actual voltage with their neighbors and estimating the impedance of distribution lines, DGs are in the position of evaluating the optimal voltages they should achieve. These voltages are then reached through iterative current injections, as dictated by (12.13) and (12.14),

$$Re(\Delta I_{G_i}) = \frac{R^{eq}_i (Re(V^*_{G_i}) - V_{G_i}) + X^{eq}_i Im(V^*_{G_i})}{Z^{eq2}_i}, \tag{12.13}$$

$$Im(\Delta I_{G_i}) = \frac{-X^{eq}_i (Re(V^*_{G_i}) - V_{G_i}) + R^{eq}_i Im(V^*_{G_i})}{Z^{eq2}_i}, \tag{12.14}$$

where Z^{eq}_i is the equivalent Thevenin impedance seen from node n_i, and V_{G_i} is assumed to be real. The results proposed in this work show that by coordinating the end-users equipped with DGs, the power distribution losses can be further reduced with respect to the previous two cases (i.e., where the DGs are operated independently, with no coordination).

Many other distributed algorithms for the reduction of power distribution losses through the coordinated operation of DGs have been proposed. It is beyond the scope of this chapter to analyze and discuss all of them in details. If interested, the reader can further check [6, 8], as these represent two technically sound approaches.

12.4 Pricing: Open Problems and Solutions

Economics is expected to play a major role in the *smart grid* infrastructure. A noticeable effort has been devoted to developing efficient demand response algorithms that the energy provider can utilize to shape the end-users energy demand. Moreover, new electricity market models and energy pricing strategies tailored to enforce and promote the cooperation between the *prosumers* and the utility to boost the grid stability and efficiency are emerging. Agents implementing demand response algorithms usually exploit a communication infrastructure to dispatch real time energy price information. The price is used by end users to determine whether some appliances should be utilized or not. Some form of reward (as for example economic benefits) might also be implemented in the demand response algorithms to incentivize the end users towards adopting a good behavior. Market models and pricing strategies that enforce the cooperation among *prosumers* are paramount elements for the techniques introduced in the previous section to work. As a matter of fact, end users that decide to install energy production plants based on renewables have to face non negligible initial investments and maintenance costs. In order for these users to cooperate to the whole grid's benefit, some form of economic incentive must be put in place to reward cooperative behaviors. Although very appealing, these aims are still under debate and a widely accepted solution is yet to come, as stated in [17].

12.4.1 Demand Response

As noticed in [44], many benefits may come from embracing the demand response paradigm. These benefits range from savings on the electricity bills for the end users involved (and also by end users that are not involved, as a consequence of reduced wholesale market prices), increased reliability and stability of the grid (and, hence, increased customer satisfaction), enhanced market performance, an increased number of choices for electricity cost management, and increased system security. The implementation of demand response algorithms requires the presence of control and measurement devices as, for example, communication devices and synchrophasors. These devices are also much needed for implementing the electrical optimization techniques discussed in the previous section, hence it becomes clear that the role of ICT, in terms of hardware and control software, and the utilization of the data they allow to collect and share are key enablers for the *smart grid* infrastructure.

Examples of demand response schemes can be found in [34, 37, 39, 43]. All these schemes set up mathematical optimization problems to determine the best demand response policy according to the system condition and the to desired outcome. Possible desirable outcomes are the economic benefit of the end users with respect to the case where no demand response is in place, diminished production cost for the energy producer, diminished peak workload for the energy producer, and combinations of the previous ones. The results presented in the mentioned papers lead to the conclusion that demand response is effective in achieving the desired goals.

12.4.2 Electricity Markets

The development of new electricity market models that account for the presence of DGs and that incentivize the end users, not only to install DGs (hence to become *prosumers*) but also to cooperate to the efficiency of the power grid as a whole is currently an open issue. Work has been done in defining new mathematical frameworks that model a multi-utility electricity market (i.e., a market in which multiple energy providers compete in selling their energy to a specific group of end users). For example, in [48] a model allowing for the forecast of electricity prices in a multi utility scenario has been developed. In this work, the authors notice that electricity markets differ from other markets (as, for example, the oil market) and hence they develop a price forecasting model different from the standard day ahead prediction. The model proposed in [48], however, does not account for the presence of small generation units (i.e., DGs) willing to sell small, yet potentially crucial for the grid's efficiency, amounts of energy to the utility or directly to other end users. In [12, 25, 30] market models integrating the presence of distributed energy resources are introduced. However, no strategy for the energy trading between end-users is devised, which is nowadays a much advocated scenario, see, for example, [27] and also the European project "Peer to Peer Smart Energy Distribution Networks (P2P-SmartTest)".

12.5 Serious Games

Several researches have been conducted on the use of gamification as a tool to promote a wise use of energy in households or work places.

In [45], aspects of gamification were applied to motivate users to adopt and develop proactive behaviors in an intelligent environment scenario. The authors use indicators (extracted from sensor readings) to assess the energy efficiency of different rooms in a building. This information is then processed through a reasoning (context-based) engine that builds recommendations, in order to promote energy virtuous behavioral changes. To promote competition among users, with the general objective of improving energy sustainability indicators, the authors exploit gamification elements such as: game points, levels, achievements, and leaderboards. Their results demonstrate that gamification helps stimulate the competitiveness among users, resulting in a desire to achieve the global objective with more determination and proactively. Other works like [31, 47] exploit a cooperative game among coworkers in order to achieve energy savings in their workplace. [47] focuses on the design requirements of a pervasive game and indicates three main points: unobtrusiveness, cooperation, and privacy. The game the authors came up with just sends feedback to the users about their performance, without requiring an active participation. The game sets goals and quests in order to get points. Results shows that the cooperative aspect of the game was the main driving force for the success of the experiment. In [31], a rather large contractor enterprise (over 300 employees in five locations) was involved in a collaborative game. Teams were organized by work units so that employees were on a team with the coworkers they worked with most regularly. A website was set up, where users could claim points, submitting actions, sharing photos and stories. Over the course of a six-month game there were small monthly cash prizes for individuals in the lead to keep the users involvement high. In the context of households' energy consumption, [13] identifies raising collective awareness as the key aspect to enable behavioral changes and motivate people in making sustainable decisions about energy consumption. The authors state that feedback systems and social connectivity constitute the essential elements to motivate the participation of players and their engagement, the same conclusions were drawn in [16]. They design a mobile service based on social mediated interaction through a game design, investigating competitive and a cooperative approaches, concluding that if no other involvement strategies (i.e., monetary rewards) are given, players tend to prefer the competitive aspect. The authors of [24] developed a game with a stimulating user interface for the definition and management of flexibilities in the use of home appliances, embedding a scoring system and social competition aspects to promote participation. However, the study just evaluates the learnability and the ease of use of the developed framework, without investigating the long-term user engagement.

A recent paper [26] provides a comprehensive discussion of gamification approaches, analyzing their effectiveness in terms of (O1) motivational affordances, (O2) psycological outcomes, and (O3) behavioral outcomes. The final conclusions are that gamification does seem to work but with some caveats. Methodological limitations were in fact identified in previous studies such as: small sample sizes (e.g., twenty users), the lack of validated psychometric measures, some experiments lacked control groups, many experiments only present descriptive statistics, the timeframes for the trials were in most cases very short, no single study used multilevel measures jointly considering O1, O2 and O3. Hence, although gamification has received a great deal of attention lately, and is often perceived as a modern and effective way of engaging users, most has still to be understood. At the same time, we lack well established theoretical and validation frameworks, it is not clear whether successes were in some cases due to the fact that the experiments/projects have addressed a population segment with certain qualities and so on.

12.6 Big Data Analytics

New technologies are starting to permeate through the power grid, encompassing energy generation, transmission, and distribution. Renewable energy sources such as wind, sun, and geothermal are being widely adopted, not only by power utilities but also by end users (e.g., through solar panels or small wind turbines). Phasor Measurement Units (PMUs) are being used over long-distance transmission networks for detection and prevention of failures and, at the same time, smart metering technology is being massively installed at end user's premises to monitor, in real time, the energy consumption of households. A great deal of communication technologies is also required to permit the communication between end users (e.g., local energy systems in households) and the electrical utility.

All of this entails a massive amount of information that is being gathered and of control signals that are being distributed to the customers according to the *demand response* paradigm. Control actions are devoted to adjusting the energy consumption from the customers so as to match the energy supply with the demand. Note that this is especially important in the presence of renewables, as these often have an erratic behavior which does not necessarily follow the actual load (the energy request). It is thus clear that modern smart grids are progressively becoming *Cyber-Physical Systems (CPS)*, where communication technologies, machine learning, and adaptation are key elements.

Various experimental initiatives involving data mining are flourishing worldwide. Among many, we cite the Los Angeles Smart Grid Project [46], which is sponsored by the US Department of Energy. This initiative was launched in 2010 and is set to transform the Los Angeles municipal utility into a Smart Grid. Its main goals are the following: (i) install smart meters to thousands of customer premises, (ii) implement and experiment with demand response mechanisms, (iii) develop scalable machine-learning algorithms trained over large amounts of data to *forecast the demand* at intervals of 15 min (for single buildings and aggregate structures) within a few hours or the next day. The *demand response* control loop roughly entails the following steps:

1. Monitor and send data from households to aggregators and from here to the utility servers (utilizing a cloud-based infrastructure).
2. Ingest, store, share and visualize the data.
3. Forecast power demand and renewable energy income from the distributed sources.
4. Decide the best *demand response* strategy based on: (1) current demand, (2) predicted demand, (3) current generation capacity, (4) predicted generation capacity from renewables

Reliable forecast algorithms are key to a successful application of this control strategy and we believe that the problem of forecasting demand and energy resources is still open. Various algorithms are emerging such as [4, 21, 29]. In [29], the authors propose a new evolutionary kernel regression technique assessing its performance with real power consumption data. This is a non-parametric approach based on a Kernel-based estimator from Nadaraya and Watson [36, 53]. The authors of [4] propose a data mining approach to predict the peak load of a consumer. To that end, they use support vector regression with online learning. Gajowniczeka and Ząbkowskia [21] proposes and compares other approaches exploiting Artificial Neural Networks (ANN) and Support Vector Machines (SVMs). Their solutions can forecast electricity demand for individual households with good accuracy. Besides the prediction of demand, as pointed out in [21], the identification of sources of energy consumption within buildings is another key issue for the automated planning of energy schedules. This leads to the appliance recognition problem [32, 42], which is another form of data mining, whose value is more localized and inherent to the energy management within the building.

Besides forecasting, other relevant application of data mining techniques lie in the detection of events in large Smart Grids. In [38], the authors use data from Phasor Measurement Units (PMUs) to detect line events in a wide-area power grid. In their paper, they show that machine learning (decision trees with various feature types) can be effectively used to perform line-event detection with performance very close to that attained by a domain expert's hand-built classification rule (when applied to a signal located near a fault). Along similar lines, [14] focuses on Dynamic Vulnerability Assessment (DVA) for self-healing and adaptive reconfiguration of the Smart Grid based on time series analysis. Specifically, the authors apply some data mining techniques to time series (Multichannel Singular Spectrum Analysis, MSSA, and Principal Component Analysis, PCA) as well as SVMs to find hidden patterns in electric signals, which then allow for an effective classification of the system vulnerability status.

As a last example of data mining, we cite [15], where the authors deal with the detection of attacks in network and system management tools for power systems. They analyze a flood attack and a buffer overflow attack causing Denial of Service (DoS) using a testbed-driven, experimental approach. Using a real setup, they perform a careful selection of *attack attributes*, testing how these can be used in combination with a large number of classifiers (e.g., Bayes, neural networks, SVMs, rule-based classifiers, decision trees, etc.) for a total of 64 data mining algorithms. Results are very good, indicating that quite a few algorithms, if properly trained, are able to deliver detection rates higher than 99 %.

Overall, we believe that data mining will be one of the most important processing blocks of future Smart Grids. Processes such as the real time detection of activities, forecasting energy demand and generation, but also tracking the presence of users within buildings (e.g., for automated heating, ventilation, and air conditioning) are unavoidable technologies for an effective implementation of demand response algorithms, the adaptation of the distribution network, the prediction of major faults and their prevention.

12.7 Application Examples

Currently, smart grids are being developed all over the world, including Australia, Canada, China, Europe, India, Japan, and the U.S. For a full list of ongoing projects, please refer to [19]. Here, for the sake of brevity, three projects are discussed. The projects considered in this section are, namely, the "Pecan Street Project Inc.," the "Duke Energy West Carolinas Modernization Project," and the "Model City Mannheim Project."

The Pecan Street Project Inc. started in 2009 and was concluded in February 2015. Its aims were to develop and implement an Internet of Energy infrastructure in Austin, Texas (U.S.). Home energy monitoring systems, a smart meter research network, energy management gateways, distributed generation, electric vehicles with Level 2 charge systems and smart thermostats are the elements that constituted its smart grid, which counts 1000 residential users (i.e., homes) and 25 commercial users (i.e., small businesses). A consistent part of the end users is equipped with distributed generation devices (i.e., rooftop PV panels) and electric vehicles. The integration of the aforementioned technologies into the power grid allows the users to monitor their energy consumption in real time at the device level, to control the electricity usage of the appliances and to sell the excess energy back to the grid. The use of level-2 chargers allows utilizing the electric vehicles as additional distributed generators, hence enhancing the energy storage capability and self-sufficiency of the end users. Advanced data acquisition and management structures that transform big energy data into useful information are being developed in the scope of this project. The results published in May 2014, stress the fact that the participation of the end users in the project has been enthusiastic. The possibility of interacting with the grid by acquiring data on consumes and smartly tuning the use of electricity—from the end users perspective—resulted in an increased awareness, satisfaction, and economic benefits as well. The final project report states: "[...] By moving towards decentralized energy production and management, utilities can build greater resilience into the grid, reduce the need for costly upgrades to centralized grid infrastructure and offer more services to residents that increase the value they receive from their utility while creating opportunities for new product markets that will generate economic development and local innovation."

The Duke Energy West Carolinas Modernization Project started in May, 2015. It is aimed at providing the end users with reliable and affordable green energy. The current expected investment amounts to 1.1 billion dollars and the project completion is expected in 2019. Expected achievements include: (i) the modernization and expansion of transmission lines and substations, (ii) the dismission of the Asheville 376 MW coal power plant and its replacement with a new combined natural gas and solar energy power plant, (iii) the removal of the coal ash and the ash basin closure. The achievement of these goals is expected to meet the increased power demand (which is expected to grow by more than 15 % in the next decade), while preserving the environment from the production of greenhouse gases and reducing the water usage by as much as 97 % by 2020. The new production system, moreover, is expected to reduce the production of nitrous oxide, sulfure dioxide, and carbon dioxide.

The Model City Mannheim Project is aimed at integrating the different areas of which a smart grid is made of, namely, electronics, ICT, and economics, in order to create an Internet of energy. This project is based on an intelligent power network working as a market place where power supply and demand interact. An intelligent controller, called "energy butler" and installed at the end users, optimizes the power usage in terms of energy and economic efficiency. This control is a key element in the Mannheim project and allows a full interaction between the end users and the power grid. By embracing this system, the end users have a chance of reducing their environmental impact and of experiencing a different kind of power grid usage model. This system is based on the efficient and fast interaction of all the components in the distribution network. The Mannheim project is based on IP-based communication on a broadband powerline communication infrastructure that exploits the existing electricity grid. The project results have shown that the total power consumption dropped by a considerable amount. Moreover, the end users behavior changed thanks to the real-time energy monitoring, made possible by the energy butler, leading to considerable economic benefits. This project, from the initial 200 end users involved, nowadays involves 1500 end users and is claimed to be scalable and applicable all around the world.

12.8 Conclusions

In this chapter, we have elaborated on the usefulness of smart grid technology for smart cities. Our discussion started from electrical optimization schemes aimed at increasing the quality of power, reducing power losses, and preventing failures or blackouts. We thus delved into the description of market policies, gamification strategies, and data mining, by summarizing the scope and the results of some successful deployments. As expected, smart grids are proven to be of great value for future cities, are expected to provide economical benefits for all the actors involved, while also benefiting the environment through a reduction of CO_2 emissions (as testified by nearly all technical studies and experimental trials).

Techniques such as power injection, load balancing, and demand response seem to be a well studied ground, featuring a variety of centralized and distributed solutions. What instead deserves further investigation is the use of gamification approaches, which looks at an embryonic stage. Although it has potential, its actual effectiveness is still unknown for real installations and a sound and methodical evaluation practice is still to be found.

Data mining is as another very much needed and lively field of research, which is becoming increasingly important and is found less explored than electrical optimization algorithms. Especially, its integration within forecasting techniques, demand response, failure detection/prevention, and communication security are still very much open to future developments.

References

1. United States Climate Action Report 2014 (2014) First Biennial Report of the United States of America. U.S. Department of State
2. Abri Al RS, El-Saadany EF, Atwa YM (2013) Optimal placement and sizing method to improve the voltage stability margin in a distribution system using distributed generation. IEEE Trans Power Syst 28(1):326–334. doi:10.1109/TPWRS.2012.2200049
3. Araujo FB, Prada RB (2013) Distributed generation: voltage stability analysis. In: Proceedings of IEEE conference on PowerTech (POWERTECH). Grenoble, FR
4. Aung Z, Toukhy M, Williams JR, Sanchez A, Herrero S (2012) Towards accurate electricity load forecasting in smart grids. In: Proceedings of international conference on advances in databases, knowledge and data applications. Saint Gilles, Reunion Island
5. Bakshi, MBUA (2007) Electrical power transmission and distribution. Technical Publications

6. Bolognani S, Carli R, Cavraro G, Zampieri S (2015) Distributed reactive power feedback control for voltage regulation and loss minimization. IEEE Trans Autom Control 60(4):966–981. doi:10.1109/TAC.2014.2363931
7. Bolognani S, Zampieri S (2011) Distributed control for optimal reactive power compensation in smart microgrids. In: Proceedings of 50th IEEE conference on decision and control and European control conference (CDC-ECC). Orlando, FL, U.S
8. Bolognani S, Zampieri S (2013) A distributed control strategy for reactive power compensation in smart microgrids. IEEE Trans Autom Control 58(11):2818–2833. doi:10.1109/TAC.2013.2270317
9. Bonetto R, Caldognetto T, Buso S, Rossi M, Tomasin S, Tenti P (2015) Lightweight energy management of islanded operated microgrids for prosumer communities. In: Proceedings of IEEE international conference on industrial technology (ICIT). Seville, ES
10. Bonetto R, Rossi M, Tomasin S, Zorzi M (2016) On the interplay of distributed power loss reduction and communication in low voltage microgrids. IEEE Trans Ind Inf 12(1):322–337. doi:10.1109/TII.2015.2509251
11. Bonetto R, Tomasin S, Rossi M (2015) When order matters: communication scheduling for current injection control in micro grids. In: Innovative smart grid technologies conference (ISGT), 2015 IEEE Power Energy Society. Washington D.C., U.S
12. Cardell JB, Tee CY (2010) Distributed energy resources in electricity markets: the price droop mechanism. In: Proceedings of IEEE conference on communication, control, and computing. Allerton, IL, U.S., pp 58–65
13. Castri R, De Luca V, Lobsiger-Kgi E, Moser C, Carabias V (2014) Favouring behavioural change of households energy consumption through social media and cooperative play. In: Proceedings of behave energy conference. Oxford, UK
14. Cepeda JC, Colomé DG, Castrillón NJ (2011) Dynamic vulnerability assessment due to transient instability based on data mining analysis for smart grid applications. In: IEEE PES conference on innovative smart grid technologies (ISGT Latin America). Medellín, Colombia
15. Choi K, Chen X, Li S, Kim M, Chae K, Na J (2012) Intrusion detection of NSM based DoS attacks using data mining in smart grid. MDPI Energies 5(10):4091–4109
16. Darby S (2006) The effectiveness of feedback on energy. Environmental Change Institute, Oxford
17. De Martini P, Wierman A, Meyn S, Bitar E (2012) Integrated distributed energy resource pricing and control. In: CIGRE grid of the future symposium. Kansas City, MO, U.S
18. Dugan RC, McGranaghan MF, Santoso S, Beaty HW (2012) Electrical power systems quality, 3rd edn. McGraw-Hill Education
19. EIA (2011) Smart grid legislative and regulatory policies and case studies
20. Erseghe T, Tomasin S, Vigato A (2013) Topology estimation for smart micro grids via powerline communications. IEEE Trans Signal Process 61(13):3368–3377. doi:10.1109/TSP.2013.2259826
21. Gajowniczeka K, Ząbkowskia T (2014) Short term electricity forecasting using individual smart meter data. In: Proceedings of international conference on knowledge-based and intelligent information and engineering systems. Gdynia, Poland
22. Geisler K (2013) The relationship between smart grids and smart cities. IEEE Smart Grid Newslett Compendium
23. Glover J, Sarma M, Overbye T (2012) Power system analysis and design, 5th edn. Cengage Learning
24. Gnauk B, Dannecker L, Hahmann M (2012) Favouring behavioural change of households energy consumption through social media and cooperative play. In: Proceedings of joint EDBT/ICDT workshops. Berlin, DE, pp 103–110
25. Guobin X, Moulema P, Wei Y (2013) Integrating distributed energy resources in smart grid: modeling and analysis. In: Proceedings of IEEE Energytech. Cleveland, OH, U.S., pp 1–5
26. Hamari J, Koivisto J, Sarsa H (2014) Does gamification work?—A literature review of empirical studies on gamification. In: Proceedings of IEEE Hawaii international conference on system sciences (HICSS). Waikoloa, HI, US

27. Heinberg R (2004) Powerdown: options and actions for a post-carbon world. New Society Publishers, Nanaimo, BC, CA
28. Kazari H, Abbaspour-Tehrani Fard A, Dobakhshari AS, Ranjbar AM (2012) Voltage stability improvement through centralized reactive power management on the smart grid. In: Proceedings of IEEE conference on innovative smart grid technologies (ISGT). Washington D.C., U.S
29. Kramer O, Satzger B, Lässig J (2010) Hybrid artificial intelligence systems. In: Power prediction in smart grids with evolutionary local kernel regression. Lecture notes in artificial intelligence. Springer
30. Kumar J, Jayantilal A (2011) Models of distributed energy resources markets in distribution grid operations. In: Proceedings of IEEE conference and exhibition on innovative smart grid technologies (ISGT Europe). Manchester, UK, pp 1–6
31. Kuntz K, Shukla R, Bensch I (2012) How many points for that? a game-based approach to environmental sustainability. In: Proceedings of ACEEE summer study on energy efficiency in buildings. Pacific Grove, CA, U.S
32. Laia YX, Laib CF, Huanga YM, Chaob HC (2013) Multi-appliance recognition system with hybrid SVM/GMM classifier in ubiquitous smart home. Elsevier Inf Sci 230(1):39–55
33. Londero RR, Affonso CM, Nunes MVA (2009) Impact of distributed generation in steady state, voltage and transient stability; real case. In: Proceedings of IEEE conference on PowerTech (POWERTECH). Bucharest, RO
34. Ma K, Hu G, Spanos CJ (2015) A cooperative demand response scheme using punishment mechanism and application to industrial refrigerated warehouses. IEEE Trans. Ind Inf PP(99). doi:10.1109/TII.2015.2431219
35. Miozzo M, Zordan D, Paolo Dini P, Rossi M (2014) SolarStat: modeling photovoltaic sources through stochastic Markov processes. In: Proceedings of IEEE ENERGYCON. Dubrovnik, Croatia
36. Nadaraya EA (1964) On estimating regression. Theory Probab Appl 9(1):141–142
37. Negnevitsky M, de Groot M (2013) Market-based demand response scheduling in a deregulated environment. IEEE Trans Smart Grid 4(4):1948–1956
38. Nguyen D, Barella R, Wallace SA, Zhao X, Liangt X (2015) Smart grid line event classification using supervised learning over PMU data streams. In: Proceedings of IEEE green computing conference and sustainable computing conference (IGSC). Las Vegas, NV, US
39. Nguyen DT, de Groot M, Negnevitsky M (2011) Pool-based demand response exchange: concept and modeling. In: Proceedings of IEEE power and energy society general meeting. San Diego, CA, U.S
40. Pagani GA, Aiello M (2012) Power grid network evolutions for local energy trading. arXiv:1201.0962 [physics.soc-ph]
41. Parizad A, Khazali A, Kalantar M (2010) Optimal placement of distributed generation with sensitivity factors considering voltage stability and losses indices. In: Proceedings of IEEE Iranian conference on electrical engineering (ICEE). Isfahan, IR
42. Ruzzelli A, Nicolas C, Schoofs A, O?Hare G (2007) Real-time recognition and profiling of appliances through a single electricity sensor. In: Proceedings of EEE communications society conference on sensor, mesh and ad hoc communications and networks (SECON). Boston, MA, US
43. Safdarian A, Fotuhi-Firuzabad M, Lehtonen M (2014) A distributed algorithm for managing residential demand response in smart grids. IEEE Trans Ind Inf 10(4):2385–2393
44. Siano P (2014) Demand response and smart gridsa survey. Renew Sustain Energy Rev 30(C):461–478
45. Silva F, Analide C, Rosa L, Felgueiras G, Pimenta C (2013) Gamification, social networks and sustainable environments. Int J Interact Multimedia Artif Intell 2(4):52–59
46. Simmhan Y, Aman S, Kumbhare A, Liu R, Zhou SSQ, Prasanna V (2013) Cloud-based software platform for big data analytics in smart grids. Comput Sci Eng 15(4):38–47
47. Simon J, Jahn M, Al-Akkad A (2012) Saving energy at work: the design of a pervasive game for office spaces. In: Proceedings of ACM conference on mobile and ubiquitous multimedia (MUM'12). Ulm, DE

48. Skantze P, Ilic M, Chapman J (2000) Stochastic modeling of electric power prices in a multi-market environment. In: Proceedings of IEEE power engineering society winter meeting, vol 2. Singapore, SG, pp 1109–1114
49. Stott B, Alsac O (1974) Fast decoupled load flow. IEEE Trans Power Apparatus Syst PAS-93(3):859–869. doi:10.1109/TPAS.1974.293985
50. Tenti P, Costabeber A, Mattavelli P, Trombetti D (2012) Distribution loss minimization by token ring control of power electronic interfaces in residential microgrids. IEEE Trans Ind Electron 59(10):3817–3826. doi:10.1109/TIE.2011.2161653
51. Turitsyn K, Sŭlc P, Backhaus S, Chertkov M (2010) Distributed control of reactive power flow in a radial distribution circuit with high photovoltaic penetration. In: Proceedings of IEEE power and energy society general meeting. Minneapolis, MN, U.S
52. Turitsyn K, Sŭlc P, Backhaus S, Chertkov M (2010) Local control of reactive power by distributed photovoltaic generators. In: Proceedings of IEEE international conference on smart grid communications (SmartGridComm). Gaithersburg, MD, U.S
53. Watson GS (1964) Smooth regression analysis. Sankhyā: Indian J Stat Ser A 26(4):359–372
54. Willis HL (2004) Power distribution planning reference book, 2nd edn. Power Engineering (Willis). CRC Press
55. Zhang C, Zhu X, Huang Y, Liu G (2016) High-resolution and low-complexity dynamic topology estimation for PLC networks assisted by impulsive noise source detection. IET Commun 10(4):443–451. doi:10.1049/iet-com.2015.0454

Chapter 13
The Significance of User Involvement in Smart Buildings Within Smart Cities

Mervi Himanen

13.1 Intelligent Concepts, Content and Context

No universally accepted definition of smart building exists. Smart city concept or phenomenon can be approached from various definitions as well. However, a common nominator of all definitions of smart environment is to ensure benefits to the users. Built environment is a venue for all activities and technology is an enabler for provision of smart environment that adds on knowledge to the benefit of users' intelligence.

Next, the definitions of intelligent buildings and smart cities pave our way to descriptions of use cases in current smart built environment. In the following Chapters, the user involvement is seen as an important unused potential for an ever more connected society on smart innovations.[1] There the user does not play the role of feedback or preference provider alone or not even the role of an innovator but acts as an individual who understands and uses one's legal rights to own data—or is served as such a client. The future of smart environment that has been started to happen already will end this article.

The very first intelligent buildings can be named in the 1980s but since then the cases of such buildings are numerous. The latest development in European innovation aims to permanent changes in employing new technology in society and increases the number of uses of advanced technologies all over it. Recent technological advancements and related challenges do not present themselves under laboratory conditions. Still testing and validation phases are needed before digital

[1]A process in which new ideas (technologies, designs, procedures, etc.), and combinations of them, bring about changes in (sub) systems like supply chains, markets, urban regions, etc. This process can be incremental, radical or even disruptive.

M. Himanen (✉)
Relate Partnership, Digital Living International Ltd, Espoo, Finland
e-mail: mervi.himanen@relate.fi; mervi.himanen@digitalliving.fi

V. Angelakis et al. (eds.), *Designing, Developing, and Facilitating Smart Cities*,
DOI 10.1007/978-3-319-44924-1_13

eco systems are ready for commercialisation and global deployment. It is necessary to conduct both laboratory and further extensive large-scale field testing in true environment in cooperation of industry or public sector and academia. The latter ones are often called as living labs.

There are already different lists of smart or smarter cities. At time being, one cannot help considering them representing a very welcomed hype of the concept and full scale implementations limit into a special suburban or a special city service such as intelligent transportation for example.

13.1.1 First Milestones

Intelligent building concept was born in 1980s and development of the more effective ways of using information started from 1970s and the World Wide Web was ready to go in 1992. In information age, communication and information technology dominates chasing after speed in computing the ever-increasing amount of data. This time period, as well as knowledge age, can be considered a period when the human mind is employed as a production factor in ever more extend or alone. Very recent technological advancements have led to the creation of distributed wireless sensor (and actuator) networks that are candidate technologies for improved and networked monitoring and controlling of critical infrastructures and operating embedded technology.

Cloud-based information fusion and knowledge management systems[2] provide access to wider content of open data[3] bases and social media. Or data on cloud can be linked to company intranets or interoffice registers, either role-based or content-based depending on the access rights of the user. As well, data management can be personalised or it concerns daily digital living of a family. Intelligent building concept has spread out to the city and society. IT[4] giants such as MicroSoft, Google or Amazon look for new niche in digitalized built environment industry as well as software developers who sit in startups. They intend to enter the market that has been traditionally occupied by real estate developers, designers, constructors or asset, property and facilities management companies etc. who have not expanded their business to wider or open concepts.

Academia, industry and education work for research and innovation in smart technology. For example, the Intelligent Building Master course at the Reading University has a holistic transdisciplinary approach while for example the former Intelligent Building studies at Temasek Polytechnic in Singapore took the

[2]Duhon [28]: 'Knowledge management is a discipline that promotes an integrated approach to identifying, capturing, evaluating, retrieving, and sharing all of an enterprise's information assets. These assets may include databases, documents, policies, procedures, and previously un-captured expertise and experience in individual workers'.

[3]Data that can be freely used, reused and distributed by anyone.

[4]Information technology

technology push approach. The focus in education has turned towards smart buildings management for example at Carnegie Mellon University in Ohio USA and also at Temasek Polytechnic in Singapore.

The CABA Continental Automated Buildings Association (established in 1988; www.caba.org/) have had since 2010 the CABA Research Program. The scope of it includes market research for both large building technologies and home systems (e.g. 11 publications in 2015). The latest tendency has been that the earlier intelligent building professional and networking organisations have become less active and new ones are evolving such as Memoori Business Intelligence Ltd or ABI Research for example and other information services on cloud.

European and global cities and city regions have created collaborative RTDI-networks (Research Technical Development and Innovation) such as Euro-Cities, Smart Cities and Innovative Regions. The European Innovation Partnership on Smart Cities and Communities (EIP-SCC 2016), European Network of Living Labs (ENoLL) are examples of European-wide collaboration platforms with global reach connecting up innovative ecosystems in industry, and open city or regional context to collaborate in RTDI.

Roughly said, the European definitions of intelligent building highlight the user involvement (cf. the PAS Standards in Section of Standardization) in comparison with the USA and Asian approaches (Himanen [45], pp. 57–70, Onaygil and Guler [70]). In the very beginning of the introduction of the intelligent building concept, Japanese put emphasis on human quality in building and expanded the concept of smart building into the city by TRON project already in 1984.

Despite the importance of integration factor in the first definitions of intelligent building, the conferences organised by the North American based Intelligent Building Institute in 1980s and 1990s discussed user involvement as an important issue—and buildings in a wide context of the city. For example, Mr. Johnson See from Kone Elevators listed the technological, operational, communication and environment aspects in the concept of intelligent building at the 'Seminar on Intelligent Buildings for the Next Millennium' in 1997 in Singapore at Temasek Polytechnic.

At the same conference, Dr. W Green from Consutel Australia named the first generation of intelligent buildings as Technology-Centred Approach and originated from the early 1980s, and the next generation as Organisation-Centred from the late 1980s. He saw that the People-Centred Approach had started already in the early 1990s. He listed five views of stakeholder groups to intelligent building

- *'Public: an exciting and familiar icon representing the company,*
- *Employees: an inspiring, efficient, comfortable and secure workplace,*
- *Owners: a statement of company stature, and a sound investment,*
- *Tenants: an attractive, prestigious, cost-effective business location,*
- *Building managers: an energy efficient and easily managed building'*.

The majority of definitions of smart solutions do not touch the meaning of the human intelligence in this context, except the definition of the author [45] defining the forms of building intelligence based on human intelligence by Gardner (1983 [42]).

Onaygil and Guler [70] conclude that 'One view is that intelligence is considered to be an innate, general cognitive ability underlying all processes of conventional reasoning'.

The use of building intelligence is expected to enable for example

- Energy efficiency without losing indoor environmental comfort,
- Meet space optimisation and expectations of and flexible space design and interplay between private and common space,
- Increased performance of building technology due to integration and low running costs [75],
- Increased working performance of the occupant activity and decreased need to travel,
- Improved comfort level due to amenities, active structures and home automation and
- Well-being thanks to ambient assisting or independent living technologies.

The benefits of smart city are expected to realise via sophisticated use of data which will make the field-specific silos collapse and the society open and transparent, and further on increase possibilities to individual citizens to be active in society. Influences are unpredictable in economy and politics, which we share and on which we have agreed common rules.

To gain better energy efficiency has been in focus in particular, in the European energy policy and consequently in European innovation policy. Integrated intelligent technologies in built environment have foreseen an important tool while buildings and building sector as well as transport use most of the energy and Europe is highly depend on imported energy.

The importance of keeping up well-being of the growing elderly population has pushed forward the development of eHealth (defined by World Health Organization WHO as use of ICT for Health or mHeath as the practice of medicine and public health supported by mobile devices by Wikipedia). In this context, smart housing works as an aided tool and as a smart environment for embedded technology or as a platform for the digital ecosystem. Also, the national policies on social affairs and health consider the use of technology important, for example, in such countries as the United Kingdom, Finland or the Netherlands [63, 79]. The Digital Agenda[5] targets all citizens in this respect of digitalization and considers it as a tool for good quality of living (Digital Agenda [24]).

In summary, the focus in the most advanced European buildings and cities has been on digital solutions of passive energy management and of assisted technology beneficial for all. Recently, an integrated communications infrastructure has started to have impact on intelligent building and smart city development within the innovations of Internet of Things[6] (IoT) technologies. The active end-user involvement has

[5]Set within the seven pillars of the Europe 2020 Strategy which sets objectives for the growth of the European Union (EU) by 2020.

[6]The Internet of Things (IoT) refers to the ever-growing network of physical objects that feature an IP address for internet connectivity, and the communication that occurs between these objects and other Internet-enabled devices and systems.

not yet turned the representatives from building and real estate industry or the city councils on to smart environmental solutions although it has become increasingly important in the competition in attractive and successful cities and regions and when pushed forward from such concepts as servitisation[7], collaborative[8] or digital economy—the latter one still in search of its final form or course.

13.1.2 Intelligent Building Integrating Technologies

The definition of the intelligent building as a building of integrated technologies was introduced by the Intelligent Building Institute Foundation (I.B.I. [50]):

> *'An intelligent building is one which provides a productive and cost effective environment through the optimisation of its four basic elements ñ systems, structure, services, management and the inter-relationship between them. Intelligent buildings help building owners, property managers, and occupants realize their goals in the areas of cost, comfort, convenience, safety, long term flexibility, and marketability. There is no intelligence threshold past, which a building 'passes' or 'fails'. Optimal building intelligence is the matching of solutions to occupant needs. The only characteristic that all intelligent buildings have in common is a structured design to accommodate change in a convenient, cost-effective manner.'*

The idea of integrating the technologies in a wider context popped up after the Internet became a main stream technology in daily work and living. Bluetooth was introduced and the blue ideas of using Internet for integrating everything by such technologies as RFID (Radio Frequency Identification) tags in all products, use of IP addresses and signaling for surveillance although the reach of it in the first place was short.

13.1.3 Technology Push Approach in Smart Buildings

The recently established Smart Building Institute (SBI; www.smartbuildingsinstitute.org/) represent the USA type definition of smart building in their recently published book (Sinopoli [81]):

---

[7]The delivery of a service component as an added value, when providing products. Servitisation supplements the traditional product offerings. The service is usually delivered via mobile devices or internet.

[8]Also known as shareconomy, collaborative consumption or peer economy, a common academic definition of the term refers to a hybrid market model (in between owning and gift giving) of peer-to-peer exchange. The collaborative economy is defined as initiatives based on horizontal networks and participation of a community. It is built on "distributed power and trust within communities as opposed to centralized institutions, blurring the lines between producer and consumer.

"Smart Buildings Systems for Architects, Owners and Builders is a practical guide and resource for architects, builders, engineers, facility managers, developers, contractors, and design consultants. The book covers the costs and benefits of smart buildings, and the basic design foundations, technology systems, and management systems encompassed within a smart building. Unlike other resources, Smart Buildings is organized to provide an overview of each of the technology systems in a building, and to indicate where each of these systems is in their migration to and utilization of the standard underpinnings of a smart building."

They highlight that the technology is the enabler of performance of the building for all stakeholders, but do not especially define factors that create the end-user benefits. Still, one cannot deny that the USA type definition does not apply in Europe as well. Especially the commercial sector, component providers and systems consultants favour it. As well, the Smart Homes Foundation's definition for home sector represented the same approach (van Berlo [15]).

It has been typical that smart building to address smart building concept as integration of the number of systems and provides overall management through one system platform. The idea is that smart buildings are designed for the efficiency of all components of a building, such as lighting, monitoring, safety and security, emergency systems, heating, ventilation and air conditioning systems and car parking management. Smart building technologies lengthen the life span of a building by identifying problems as and when they occur and taking the required corrective measures. It seems that the one system platform strategy—as a killer application by one operator—has not been so far a success in sales.

In 2002 Realcomm published an article on how networks will connect building processes [78].

Despite the high potential of an integrated communications infrastructure, building automation and management systems (BAS/BMS) still consider new entrants leveraging the connectivity of BAS/BMS systems and the SaaS business model [2].

13.1.4 Large Intelligent Buildings

Intelligent Buildings were advocated by UTBS Corporation (United Technology Buildings Systems Corporation) in the USA in 1981, and became a reality in July, 1983 with the inauguration of the City Place Building in Hartford, Connecticut USA (Onaygil and Guler [70]). The UTBS Corporation was responsible for controlling and operating such shared equipment as air-conditioning equipment, elevators and disaster prevention devices. The company further provided each tenant with communication and shred tenant services, such as office automation services, using local area networks (LANs), digital private automatic branch exchanges (PABXs) and computers.

Intelligent Buildings Institute's definition focused on the lack of integration of technologies. For example, architects and engineers at Foster + Partners fulfilled the demand by designing and managing the construction Hongkong and Shanghai Bank Headquarters building in Hong Kong, China in 1979–1986 (Foster [40]). At that time, it was considered one of the most advanced office towers also in the sense of integrated building and office automation, which was installed innovatively into the structures together with the building service facilitation. Foster + Partners introduces themselves today as '... the best design comes from a completely integrated approach from conception to completion and the design teams are supported by numerous in-house disciplines, ensuring the knowledge base to create buildings that are environmentally sustainable and uplifting to use'. One of the Fortner + Partners' designs is a commercial skyscraper, 30 St Mary Axe (widely known informally as The Gherkin) in London in 2013. In addition to its neo-futuristic architecture, the building represents advancement energy efficient and green building carried out by sophisticated building service technology (as, e.g. renewed natural ventilation) and structures.

Many projects have been realised on remarkable office buildings following the similar path than that of the Foster + Partners' example from integrated technology towards sustainability[9] (cf. more Section Energy Efficient Building) and advanced office layout for knowledge work place (cf. more Section User-Oriented Intelligent).

Integrating the design and construction phases close together is a trend to gain more efficient construction projects in time and money (Kruus et al. [53]). It has created new alliances between designers and construction companies.

Another interesting example of the progress from the very first intelligent buildings to today is the Nippon Telegraph and Telephone Corporation head office building buit in the 1980s in Tokyo. It had several floors preserved for servers of telecommunication services provided by the company. Similar situation was for example in Helsinki Finland with the Elisa Oyj's[10] head office, which was designed after the concept of intelligent building in 1997. Soon after, the floors of equipment have become insufficient to satisfy the increased demand for communication capacity in any location. They were replaced by (1) distributed technology in national fibre networks or (2) the huge telecenter buildings facilitating communication in cloud and operated by IT giants such as Google, Cisco. Due to the sensitive apparatus and huge heat load their operation causes demand for the cold or cool climatic condition and earthquake resistance which are favourable qualities for the location of such buildings. As well, the owners of telecenters prefer locations in countries with stabile political situation for the sake of data security.

Despite the huge advantage of the use of cloud-based technologies, the construction sector continues its tradition in practice and stick to the IT islands. Despite the potential of digital economy exists. The author has recently (2015) come across

---

[9]A multifaceted property that describes the extent to which social, economic and environmental objectives are in balance; that economic activity is not declining, that non-renewable resource throughputs are minimised and that society has high capital and is cohesive, equitable and inclusive.

[10]previously Elisa was Helsingin Puhelin Oy originated from a company established in 1882.

with companies in Denmark, Norway and Finland providing smart building management technology in cloud, who report that they can find few customers but the idea does not yet fly—there is largely interest (as towards free uploads) but majority of the market those who are interested do not buy. The market potential in digitalisation of building sector is huge and that promise keeps the providers come and go, or trying again and again with better approach to technology and customer needs.

Academia, industry and education work for research and innovation in smart technology. Further on, the inter-sectorial and multidisciplinarily integrated approach is still to come while for example the smart building market research report together with Global Forecast to 2020 by Markets and Markets [59] outlines in a relatively limited and technology-push-oriented manner that the global smart building market has been segmented into building automation system, networking technologies, applications and regions.

- Building automation systems include physical security, BEMS, communication technology, and parking management system.
- Networking technologies have been segmented into bus technology, power line technology and wireless technology.

The applications that use the smart building services are commercial buildings, industrial, institutional, residential, hospitality, hospitals, airports and others. This market has been further segmented into the regions of North America, Europe, APAC, MEA and Latin America.

The major players of smart building market according to Markets and Markets [59] are ABB, Cisco, Delta Controls, Schneider Electric, Siemens, IBM, General Electric, Johnson Controls, Accenture and Honeywell. However one can ask if these companies are players of the market in smart building or rather the market of such areas as building automation, facilities management or building maintenance? Markets and Markets [59] prognoses that the global smart building market is expected to reach $36398.7 Million in 2020 from $7260 Million in 2015, at a CAGR[11] of 38.0 % during the forecast period.

Many of the above-mentioned companies are originated from building automation business or they are IT giants as Cisco or IBM, of whom the latter one has just recently launched together with Carnegie Mellon University First Cloud-Based Analytics Partnership for Smarter Buildings. It heads also to Smarter Cities within its Smarter Planet initiative (www.ibm.com/smarterplanet/us/en/). Similarly, Ericsson approaches the wider concept of smart environment by the Smart Metering as a Service (SMaaS; www.ericsson.com) which is a complete end-to-end smart metering business process outsourcing solution operated. On the

[11]The Compound Annual Growth Rate (CAGR) is the mean annual growth rate of an investmentover a specified period of time longer than one year.

other hand, for example Honeywell has informed that they stick to the smart building research, development and innovation (in 2014).

The module of building automation which enables the building managers easily adjust the system operation is not a norm of the building automation installation—most probably partly due to the prizing and operational costs of the unit. A user study shows that the building manager, sitting in an office building, and available even in person, was the most favourable form of informing ones wishes on indoor air quality in the end of 1990s (Himanen [45]). Today, the response could be different after we are used to the vivid usage of social media. Nevertheless, the company gave up employing the building managers, after a relatively short period of time (roughly 2 years), in the intelligent building where the study was carried out.

13.1.5 Smart Housing

To integrate the in-house technologies and their operations proves of the technology push approach rather than user demand-driven approach. Also, the development of smart cities has followed this approach of technology push. Actually, the concept of integrated technologies in intelligent building expanded soon to the city context—starting with the TRON project as early as in 1984 in Japan by Professor Ken Sakamura of the University of Tokyo (Anon [7], Lehto[12] [55] and [56]).

The TRON house introduced several advanced technologies for kitchen appliances and their integration into the interior architecture and structural design as well as for storages and active furniture (cf. Lehto 1990a and Lehto 1990b). After the TRON project several demonstration projects have been carried out and show houses built. Among the first ones are such as: INTEGER HOUSE in United Kingdom (1998), ARKKIMEDES HOUSE in Finland in a smart housing fair area (1991), A*STAR house at the National University Singapore campus (2005), and many others, e.g. in the Netherlands, Germany or Canada.

In housing market, there are innumerable commercial applications that represent automation which can be considered as part of the integrated smart home idea such as integrated building services for energy efficiency or home automation, (remote) smart metering, safety and alarm systems (presence and video surveillance), lighting controls, universal remote controls for in-home apparatus, home theatres, smart kitchen appliances, etc. As the modern home integrators run on mobile apps or in Internet they are replacing the old buses and system dependent control units.

Many relative simple features of automation might be missing from housing as for example the home heating and cooling control does not necessarily comprise controls provided by outdoor temperature sensors, not to mention, other outdoor-air qualities, although the vision of smart house already in 1980s comprises the ability

[12]The author (under previous name)

to protect the residents (e.g. by shutting the windows automatically) from impurities from a dangerous release due to failure in the neighboring manufacturing process for example, or increased pollution or pollen level in outdoor air.

Still, only the luxury homes take more sophisticatedly advantage of the smart home potential and the solutions are often tailored for the residents.

According to a new market research report (Markets and markets [60] the smart home market is expected to grow at a CAGR of 17 % between 2015 and 2020, and reach $58.68 billion by 2020. The study characteristics are

- The global smart home market has an exhaustive product portfolio. The smart homes market has been segmented by product into energy management systems, HVAC control, entertainment control, security and access control. The energy management system is further segmented into smart devices and lighting control solutions. Different security and access control product solutions included in the report are intrusion detection system, video surveillance, motion sensors, touch screen and keypads.
- The smart homes market is split into four regions: North America, Europe, APAC, and RoW. North America includes the U.S., Canada, and Mexico, while Europe is divided into the Germany, France, the U.K., and others; APAC smart homes market includes China, Japan, India, and others while RoW is divided into the Latin America, the Middle East and Africa.
- The study comprised the competitive landscape of main players, which covers key growth strategies followed by all prominent players. Some of the big players profiled in the market report are Siemens AG (Germany), Schneider Electric S.A. (France), ABB Ltd. (Switzerland), Ingersoll-Rand PLC (Ireland), Emerson Electric Co. (U.S.), Legrand S.A. (France), Crestron Electronics, Inc. (U.S.), Lutron Electronics, Inc. (U.S.), Control4 Corporation (US), and more.

Markets and markets [60] had a technology push approach to smart housing while the number of companies looking smart homes as a provider or enabler of amities or assistance for daily easy living of residents has not dominated the business. Assistance in cleaning, cooking, dish washing, laundry, etc. rely on smart technology by home appliance providers who seem not to be part of the above-mentioned study, but are global players in the market of smart technology.

Markets and markets [60] foresees the future of smart home technology promising as they conclude:

> *'Traditional home automation devices were designed to control systems within a house and within a limited range of connectivity; however, with recent developments across different areas of connectivity of appliances and devices, … including mobile connectivity features, an integral component of smart homes provided by device manufacturers; and compatible communication protocol and technology based products offered by Internet Service Providers. Even though the concept of smart homes has been in existence for a long time, the market has witnessed a profound growth, mainly, during the last five years. The smart homes market is highly fragmented and expected to get consolidated. The smart homes service providers are beginning to gain traction in the marketplace, and we could see more smart homes technology players entering the market before an inevitable consolidation occurs.'*

This promise of Markets and markets [60] can be considered as an opportunity for startups competent in cloud technology. At time being, they are foreseen to be responsible for the ongoing future digitalization of electronic service sector in a wide sense comprising smart homes but by no means overlooking the skills and potential of startups to perform in any sector.

In many countries, third sector works for developing and implementing technology for independent living. In the Netherlands, both the elderly organizations as the smart house associations have worked for encouraging use of smart house technology within elderly housing, since 1990s (cf. Expert Center for Smart Technology and Smart Living (www.smart-homes.nl); previously Smart Homes Foundation. Still, van Berlo [11] concludes that we know how to build smart houses and about AAL technology (Ambient Assisted Living, cf. AAL [1]), but we may not understand why older people are not necessarily buying it.

Growing numbers of elderly population has pushed forward the development of smart housing. The European Commission has strongly encouraged the use of technology for independent living both by various policy initiatives: the EIP-AHA (European Innovation Partnership on Active and Healthy Ageing) initiative or the Digital Agenda action plan 2013, and for example such funding instruments as the EU Horizon 2020[13] Health, Demographic Change and Wellbeing Programme or the AAL Joint Programme [1] implemented by both the EU and EUREKA Network (www.eurekanetwork.org).

In addition, the Eureka Eurostars programme[14] (www.eurostars-eureka.eu) provides funding for SMEs (small and medium-sized enterprise, European Commission [38]) without limiting the subject and thus, it funds projects on smart housing and gerontechnology among others. Similarly, within the EU Horizon 2020 the MSC programme[15] does not limit the study subject and comprises scientific projects which has focused recently relatively often on smart housing. Smart housing is also part of the calls related to energy efficiency and smart city (cf. Sections on Energy Efficient Building and Smart Grid).

13.1.6 Energy Efficient Building

As mentioned Foster + Partners, as well as leading engineering and architects' offices have developed their design towards energy efficiency or sustainability which rely on advanced integrated technologies, in building automation in

[13]The European Commission Framework Programme for Research and Innovation 2013–2020.

[14]The Eurostars Programme (2008–2013) and the Eurostars-2 Programme (2014–2020) Co-funded by EUREKA member countries and the EU Horizon 2020 Research and Innovation Programme.

[15]The Marie Skłodowska-Curie actions (MSCA) provide grants for all stages of researchers' careers—be they doctoral candidates or highly experienced researchers—and encourage transnational, intersectoral and interdisciplinary mobility (cf. http://ec.europa.eu/programmes/horizon2020/en/h2020-section/marie-sklodowska-curie-actions).

particular. This trend has been supported by the development of energy certification by the commercial energy labels such as BREEM, LEEDS, Green Office or Gold Power by WWF, etc. (e.g. the Energy Labelling Directive (2010/30/EU), Himanen et al. [47], as well as, governmental initiatives, global bodies as for example the United Nations Industrial Development Organization (UNIDO) and the funding schemes of the European Commisision (EU) Framework Programmes for research and innovation (FPs), National Science foundation in USA (NSF [66]) and many others corresponding programmes.

Projects resulting in integrated commercial-scale solutions with a high market potential, in the field of energy, transport and ICT (Information and communication Technology) have been conducted since the EU Seventh Framework Programme (the EU FP7), starting in 1998. They focus on large experimental projects with advanced metering and experimental installations. They focus on study occupant experience when technology is in operation in true built environment, or in the context of Smart Cities Light House projects. Similarly, the EIT ICT Labs[16] promote smart energy efficiency, especially the action line Smart Energy Systems (SES). For example, one of the latest EIT Digital project calls was titled 'Intelligent Integrated Critical Infrastructures for smarter future Cities' (I3C).

The subject of energy efficiency has been in focus of the EU Framework Programmes from very beginning in 1984 and especially focused on the smart building and cities for example in

- the Programme of Energy-efficient Buildings (EeB) since 1990s till the current EU Framework Programme, Horizon 2020 where the implementation of the EeB programme has been executed since 2014 by the Executive Agency for Small and Medium-sized Enterprises (EASME)[17]
- the Earlier Intelligent Energy—Europe (IEE) programme in 2003–2007 (European Commission [37]).

13.1.7 Total Building Performance

Total building performance is an approach to integrated building service systems and keeping smart systems as enablers. The concept is in line with thinking behind the original intelligent building concept. The digital economy highlights the importance of the total building performance concept while demanding systems design of data flows and operational chaining across traditional limits in operational units.

---

[16]Founded in 2008, the EIT ICT Labs one of the Knowledge and Innovation Communities (KIC) at the European Institute of Innovation and Technology (EIT), www.eitdigital.eu. The EIT ICT Labs aims to create synergies between education, research, and innovation targeting results to be ready for commercialization.

[17]Set-up by the European Commission to manage on its behalf several EU programmes.

Assoc. Prof Lee Siew Eang is the pioneer in the total building performance studies. The Centre for Total Building Performance (CTBP) is a joint research centre of the Building and Construction Authority of Singapore (BCA) and the National University of Singapore (NUS). It is hosted by the Department of Building, School of Design and Environment.

The vision of the CTBP is to champion Total Building Performance R&D (Research and Development) and support the quest towards a quality and productivity driven construction industry. The research roadmaps are Research Programmes on

- Green and Energy Efficient Building
- Indoor Environmental Quality
- Building Performance Integration and Innovation
- Building Maintainability
- IT Design Decision Support Systems.

Interestingly in line with the idea of total building performance, the comparison of the intelligent office buildings with the other type high quality buildings showed stochastic remarkable difference in favour of intelligent buildings when the users evaluated the building performance although the differences in implementing the ICT or building automation between these two types of buildings cannot be found to be very clear (Himanen [45]). The survey covered several systems in addition to those based on ICT such as indoor air quality, indoor design, building mass, travelling, etc. The author suggests that the importance of the holistic approach in the intelligent office buildings design was a reason to the result favouring them, but further studies to prove the hypothesis is needed. As well, it turned out to be obvious that

- The research on intelligent office buildings have proved that neither one intelligent feature makes the building intelligent nor can any good quality substitute and even mitigate bad quality of building installations and properties.
- There is a reality of co-effecting factors telling to take care of good quality of all features for gaining the satisfactory results.
- Multidisciplinary[18] and inter-sectorial (or transdisciplinary[19]) innovation comprises both 'hard' technological disciplines and design criteria based on 'soft' sciences.

High-rises or tall buildings demand latest in any solution: structures, elevators, etc. Building process of them needs special attention in various phases of design as well as organization of property management for entire building life time. They can

[18]A problem is approached from several scientific or professional fields.

[19]A collaboration spanning multiple partners, both academic and non-academic as, to solve a common problem. Non-academic partners may include city officials, (non-) governmental agencies and offices, charitable organizations, companies, civil society, grassroots movements, etc. A synonym is inter-sectorial. Non-academic partners may include city officials, NGO's, companies, civil society, grassroots movements, etc.

be considered intelligent as such. Further information on tall building cases is found from:

- The Skyscraper Center, the Global Tall Building Database of the CTBUH (the Council on Tall Buildings and Urban Habitat, cf.
- List of tallest buildings and structures in the world can be found for example on Wikipedia.

Such new elements as commissioning or life-cycle procurements cover the whole building process as well.

If commissioning is quality control from the clients' point of view, the total building performance concept can be understood as the same as the inner quality control in manufacturing industry—that has been practiced there from the 1960s. Not the client as in the case of commissioning but the operators themselves within the building industry gain tools for improving their performance.

With the concept of integrated facilities management is understood that building automation is combined to the facilities management. Life-cycle procurements process covers the process from feasibility studies to the facilities management service provisions. The idea is to gain feedback that enables improvement in the performance of industry and makes it possible to serve their clients, the occupants of any built environment better than earlier.

13.2 Smart Integrated Urban Development

The current innovation of smart city has two focus areas: (1) the open city with public service accessibility to be discussed first and (2) the Smart Grid discussed that after. In addition, national information highways to enable the effective and smooth technological implementation of smart city innovations are discussed in Chapter of Future Potentiality. All these approaches have an end-user-oriented approach.

13.2.1 Smart City

Cities are the main driver of change in economic development and growth, knowledge and creative generation of production, innovation and overall liveability. The study of the European Environment Agency EEA [30] shows consistent evidence of

- A positive association between urban wealth and the presence of a vast number of creative professionals,
- A high score in a multimodal accessibility indicator, the quality of urban transportation networks (cf. Thulin [84]),

- The diffusion of ICTs (most noticeably in the e-government industry), and
- The quality of human capital.

Urban areas both face important sustainability challenges in social, economic and environmental issues and at the same time form an ideal setting to generate solutions. The Digital Agenda[20] proposes to better exploit the potential of ICTs in order to foster innovation, economic growth and progress. Among others, Caragliu et al. [17] have defined smart cities profiling the competitiveness of cities.

Smart cities are expected to become a sizable market with a large spending on smart cities technologies (cf. also Belissent [12], Bowerman et al. [14]). Real estate experts predict that smart cities will in the future be attractive to the educated, high salary knowledge work force and will therefore become profitable locations for real estate investors. The built environment is the core of the concept of smart city, however, the smart city is a wider concept

- According to the Smart City Council it comprises topics as Smart People (uses technology to make its citizens' lives better), Universal (e.g. open spaces for public use), Built Environment, Energy, Telecommunications, Transportation, Water and Wastewater, Waste Management, Health and Human Services, Public Safety, Smart Payments and Finance (Anon [5]).
- Mapping Smart Cities in the EU Study defined that a Smart City is one with at least one initiative addressing one or more of the following six characteristics: Smart Governance, Smart People, Smart Living, Smart Mobility, Smart Economy and Smart Environment (Manville [58]).

The ENSUF[21] [33] as well as, the European Innovation Partnership for Smart Cities and Communities (EIP) defines that smart city refers to cities in which ICT is increasingly pervasive and ubiquitous. They are cities whose knowledge economy and governance is being progressively driven by innovation, creativity and entrepreneurship. There digital technologies can be used to efficiently and run effectively by public services provision. Urban challenges cannot be separated from the regional and local setting in both the actual spaces and the institutions (the processes, practices and formal and informal rules). ICTs and data help to 'smartly' organise information flows in peoples' everyday life concerning e.g. energy, water, materials, food, goods, ideas, and in politics. Especially in building sector, the use of smart energy efficient applications are looked to reduce high energy consumption and green house gas emissions causing bad air quality. Mitigation of congestion by intelligent transport have similar effect.

[20]One of the seven pillars of the Europe 2020 Strategy which sets objectives for the growth of the European Union (EU) by 2020.

[21]The ERA-NET Cofund Smart Urban Futures (ENSUF) was established by the Joint Programming Initiative (JPI) Urban Europe in order to initiate a transnational joint call for RDI proposals developing our knowledge of the urban condition and sustainable development through creation and testing of new methods, tools, and technologies required to overcome current economic, social, and environmental challenges. ENSUF is supported by the European Commission and funded under the Horizon 2020 ERA-NET Cofund scheme.

Social and open innovation focuses on transdisciplinary co-creation[22] of smart urban development which is understood to be about the connectivity, accessibility and integration of various systems, sectors, services, infrastructures and public institutions. Increased urban liveability is gained by linking academic, practical and local knowledge.

These positive associations clearly define a policy agenda for smart cities, although clarity does not necessarily imply ease of implementation. Lists of smart or smarter cities have appeared. The progress is rapid and smart cities can be found in any location. Next, a couple of them are shown as examples, especially while they also refer to the guidelines in the area (cf. PAS 181:2014).

Manville et al. [58] concluded that there are Smart Cities in all EU-28 countries (after the selection of factors for successful Smart Cities that have been detailed in the report), but these are not evenly distributed:

- Countries with the largest numbers of smart cities are the UK, Spain and Italy, although the highest percentages are found in Italy, Austria, Denmark, Norway, Sweden, Estonia and Slovenia.
- Smart City initiatives are spread across several characteristics, but most frequently focus is on Smart Environment and Smart Mobility. Geographically, there is also a fairly even spread, although Smart Governance projects are mainly seen in the Older Member States of France, Spain, Germany, the UK, Italy and Sweden. Also noteworthy is that some characteristics typically occur in combination, such as Smart People and Smart Living.
- How the cities perform in the context of their country's national priorities and political and socioeconomic circumstances, led to the selection of the six most successful cities for further in-depth analysis: Amsterdam (the Netherlands), Barcelona (Spain), Copenhagen (Denmark), Helsinki (Finland), Manchester (UK) and Vienna (Austria). In each of these, a number of initiatives were assessed, showing focus on transport, mobility and Smart Governance, including building technologies.

Cohen [19] created the Smart Cities Wheel with a global advisory committee in order to charge on the smartest cities in the world and listed up to 400 potential indicators of which 62 were selected for assessment. Cohen [20], [21] listed the 10 smartest and innovative cities in 2014

- in Europe: Copenhagen, Amsterdam, Vienna, Barcelona, Paris, Stockholm, London, Hamburg, Berlin, Helsinki and
- in the world: Vienna, Toronto, Paris, New York, London, Tokyo, Berlin, Copenhagen, Hong Kong, Barcelona with strong candidates which are runners-up in this first ranking, including Amsterdam, Melbourne, Seattle, São Paulo, Stockholm, and Vancouver.

---

[22]An approach where heterogenous actors collaborate to produce knowledge, instruments, technology, artefacts, policy, know-how, etc.

Mass deployment of smart city concept on national or international perspective will not be possible without widely accepted standards. The widely used technology standards are not enough in city- or region-wide context. Standardization at smart application level is just in infancy phase (cf. Section Standardization of User-driven Smart Environment). A much stronger effort should be dedicated in this area in the future.

13.2.2 Smart Cities Digital Ecosystem

Generally speaking, the Smart City infrastructure is a digital ecosystem consisting of networked sensors and actuators, mobile phones, wearable devices and other embedded devices of general and/or specific purpose, and, by extension, of all *smart* devices with capabilities of interconnection, computation, and a sort of interaction with their environment (Schaffers et al. [80]).

Majority of the elements for improved thinking of the performance of smart environment and technology are existing. Advanced information or energy efficiency, public eHealth, advanced teaching and learning, etc. are in pipeline accordingly, for example, under the innovation in the Smart Cities and Communities lighthouse projects (the Horizon 2020 Work Programme 2014–2015 and 2016–2017).

Social impact of the research and innovation and ethical aspects has grown in importance within the European Research Area. The need for business planning within the projects or the gendered innovation has popped up (Expert group "Innovation through Gender" et al [39]) as well as the open science initiative. As well, the innovation on platformization asks the end-user involvement for example by the Internet of Things Focus Area (European Commission [35]). The ambition is to foster the take up of IoT in Europe and to enable the emergence of IoT ecosystems supported by open technologies and platforms. It will be addressed through a complementary set of activities structured around Large-Scale Pilots (European Commission [36], cf. Airaksinen and Kokkala [3]).

A good number of various functions and automated processes associated with smart environment generate vast amounts of data. Availability of valuable data has created insights of making analytics part of a growing marketplace exist. Data acquisition, normalisation, prioritisation and action are issues associated with data analytics—comprising a critical step in creating a trustworthy and functional smart built infrastructure and society.

More recently, development of smart urban environment has started to take into account the emergence of platforms for Big data[23] and IoT, applications on artificial

[23]Big data is high-volume, high-velocity and/or high-variety information assets that demand cost-effective, innovative forms of information processing that enable enhanced insight, decision-making, and process automation.

intelligence and their impact on smart buit environment. For example, the projects (e.g. FIWARE[24]) and platforms (e.g. SOFIA2,[25] PlanIT OS^{TM}, City OS, SOUL, City Protocol, LifeEngine, etc.) for smart cities and buildings focus on interoperability of technologies:

- Advanced platforms for middleware, standards, protocols, interfaces, etc. are under development for building e-infrastructure.
- Virtualisation of user experiences by connected environment with increasing virtual elements, augmented realities[26] and (hyper) connected (linked data) artificial intelligences that are embedded into the built environment around us.
- Safe and easy adaptability and occupancy-based control of digital ecosystems are targeted.
- 3D data visualisation techniques as surveillance based on video or machine-vision, and monitoring and control.
- A safe software development environment for testing new applications or a large-scale simulation framework for supporting the strategic and tactical decision process on smart cities are needed in innovation.
- Ontologies coping with heterogeneity and large-scale infrastructures for efficient aggregation of devices and compose/utilize complex virtual devices (De et al. [23]).
- Scalability while the number of objects connected to IoT increases exponentially for example, via smartphones, PCs, tablets, connected cars and wearable devices. IoT will become the largest device market in the world including hardware, software, and installation and management services. For example, Google processes more than 24 petabytes of data per day, an equivalent volume to thousands of times all printed collection of the US Library of Congress. At such scale of data, the technological capabilities, our capability of exploiting the data and, more importantly, our capability of reasoning about these data will be overflowed (Escolar [34]).

---

[24]The FIWARE Acceleration Programme promotes the take up of FIWARE technologies among solution integrators and application developers, with special focus on SMEs and startups.

[25]SOFIA2 is a middleware that allows the interoperability of multiple systems and devices, offering a semantic platform to make real-world information available to smart applications (Internet of Things). It is multi-language and multi-protocol, enabling the interconnection of heterogeneous devices. It provides publishing and subscription mechanisms, facilitating the orchestration of sensors and actuators in order to monitor and act on the environment. Cross-platform and multi-device through its SDK, APIs and extension mechanisms that allow integration with any device. A software development kit (SDK or "devkit") is typically a set of software development tools that allows the creation of applications for a certain software package, software framework, hardware platform, computer system, video game console, operating system, or similar development platform.

[26]A form of virtual reality augmented reality is a direct or indirect view of a real-world environment in real time which elements are augmented by computer-generated sensory input as graphics, sound and video. The user's view is enriched by virtual objects usually to provide information about the real environment. Typical hardware components are: processors, input devices, display, sensors and smartphones.

According to the App Economy Forecasts 2013–2016 report the global app economy accounted for 18 % of the combined application (app) services and handset market. It is estimated that the value migrates from handsets to apps by 2016, when the app market will rise to 33 % of the combined market (Vision Mobile [90]). An effect of this growing trend is the demand of application developers that in 2013 was estimated to be 12.6 % of the global developer population and predictably, will continue growing. The developer tools provided by most of the mobile platforms as Android or iOS offer a simple yet powerful programming environment for writing code and facilitate that both developers and users may program applications themselves. Apps can be published and stored in some repository (e.g. GooglePlay) from which they can be downloaded. Software as a Service (SaaS) is an alternative to apps for delivering applications in the cloud.

13.2.3 Smart Grid

Smart Grid technology is expected to efficiently manage supply and demand of electricity and modernize the technologies for grids, distributed generation (including microgrids) and improve grid reliability. A smart electricity supply network uses digital communications technology

- Firstly, for detecting and responding digitally to quickly changing electric demand and increasingly intermittent electricity production, even in real time, and
- Secondly, for two-way communication between the utility and its customers.

The Smart Grid consists of controls, computers, automation, and new technologies and equipment working interoperable together. The transition towards increasingly renewable energy systems or hybrid energy supply systems in the building, for example, calls for novel techniques of operation and control in response to the changing power transmission and distribution networks. The Smart Grid is not just about utilities and technologies; it is about giving the consumer the information and tools he or she needs to make choices about energy use in a similar way as already we manage activities on cloud such as buying tickets.

Adviser in Electric Networks at Finnish Energy Industries, Ms Ina Lehto tells that the Smart Grid is a two dimensional matter.

- Firstly, by definition the electrical grid as such is built of the transmission and distribution lines and digital technology which allows monitoring the power load, and secures energy transmission under stable and secure conditions. Primarily, the grid has simply been built smart from the needs of power supply side. This can be understood as a Smart Grid—representing an unprecedented opportunity to evolve the energy industry into increased reliability, availability and efficiency that will contribute to our economic and environmental health.
- Secondly, these smart technologies can be employed as a platform for consumer services of various kinds; from provision of the informative electricity bill to

cooperation with the customer as an energy supply provider, for example, of extra energy yielded from solar panels or industrial processes. In this context, the electrical engineering for Smart Grid has best to offer to smart built environment technologies.

Further on, Lehto adds on that the grid can be made sensing along the load and use the smart metering technology for the two-way communication, as done in Finland and Sweden, for example. Smart metering is indeed a key to two-way communication that provides customers with possibilities to use and produce electricity intelligently while ensuring efficient operation of the electricity markets.

At time being the Smart Grid is evolving, piece by piece, before all the technologies will be perfected, equipment installed, and systems tested to work fully on line. Research and innovation projects within the Horizon 2020, for example, will generate full scale use cases while the calls ask proposals of solutions that will be demonstrated in large-scale pilots and validated in real life conditions, or on real data by simulations that will take a long enough period of time for ensuring credibility and consistency of conclusions. A set of technologies and solutions are expected to be demonstrated in an integrated environment; to enable demand response, Smart Grid, storage and energy system operating under stable and secure conditions, rather in the context of an increasing share of renewable energy sources in the electricity grid.

The EIT ICT Labs are active in Smart Grid innovation within action line Smart Energy Systems (SES). Similarly, the Eurostars joint programme projects comprise innovation in Smart Grid. The perspective of introduction the outcome of the Horizon 2020 innovation actions in the market is expected to happen in the coming years after the project. The time to market of the Eurostars project results is in maximum 2 years after the project completion.

During the transition period, it will be critical to carry out testing, technology improvements, consumer education, development of standards (e.g. CEN-CLC-ETSI M/490) and regulatory environment for privacy, data protection, cyber security, as well as, Smart Grid deployment, infrastructure and industrial policy (cf. the Smart Grid Task Force and its Experts Groups in the field of Standardization on http://ec.europa.eu/energy/en/topics/markets-and-consumers/smart-grids-and-meters/smart-grids-task-force).

The projects in the consumer response and cooperation may deal with

- Mechanisms and tools allowing consumers to participate actively in the energy market and in demand response schemes such as smart metering programs, light controllers, PLCs,[27] SCADA systems,[28] weather station or a platform

[27]PLC is a digital computer used as control of machines, in many industries and designed for multiple arrangements of digital and analog inputs and outputs, extended temperature ranges, immunity to electrical noise, and resistance to vibration and impact.

[28]SCADA (Supervisory Control and Data Acquisition) is a system for remote monitoring and control that operates with coded signals over communication channels (using typically one communication channel per remote station).

incorporating an optimisation-based Building Energy Management System for improving the energy efficiency, for example, in residential microgrids.
- The application for use cases will vary from, for example
  - in terms of energy generation, storage (e.g. batteries, fly wheel, etc.), and loads as well as the strategies of operation, addressing both thermal and electrical energy management
  - concerning thermal inertia of buildings and building services such as water boilers or heat pumps, as well as home appliances (while taking into account their comfort and preferences), or
  - technological developments for hydrogen production and storage addressed in the frame of the Fuel Cell and Hydrogen JU.
- Demonstration and validation of new business models for a combination of distributed energy resources, self-consumption and storage with optimised utilisation of distribution networks from all energy carriers (cf. Fig. 13.1).

13.3 Soft Engineering—Towards Connected City

The user involvement in many areas is and has been a self-clear phenomenon, especially in service sector or in architecture, interior design, clothing. For example, functionalism has targeted to help building occupants to work effectively and dwell comfortable with spaces and furniture for easy household and housekeeping. But experts decided or studied what is best for people in labs with relatively limited samples. Just in the dawn of intelligent buildings was realized for example that all in reach of hand within the office desk might not be the healthiest and thus the most effective way of working. Currently, studies on true usage environment with large number of users is acceptable.

13.3.1 Intelligent Building Concept—Organizational Approach

The European Intelligent Building Group has had a good start for applying the organisational approach to the concept of intelligent building.

Emeritus Professor Derek Clements-Croome started the provision of a Master course on intelligent buildings at the Reading University. Their resent definition for intelligent building is activation based and comprises the user involvement:

> *'An intelligent building is a dynamic and responsive architecture that provides every occupant with productive, cost effective and environmentally approved conditions through a continuous interaction among its four basic elements: places (fabrics; structure; facilities): process (automation; control; systems): people (services; users) and management (design; construction; performance) and the inter-relationship between them.'*

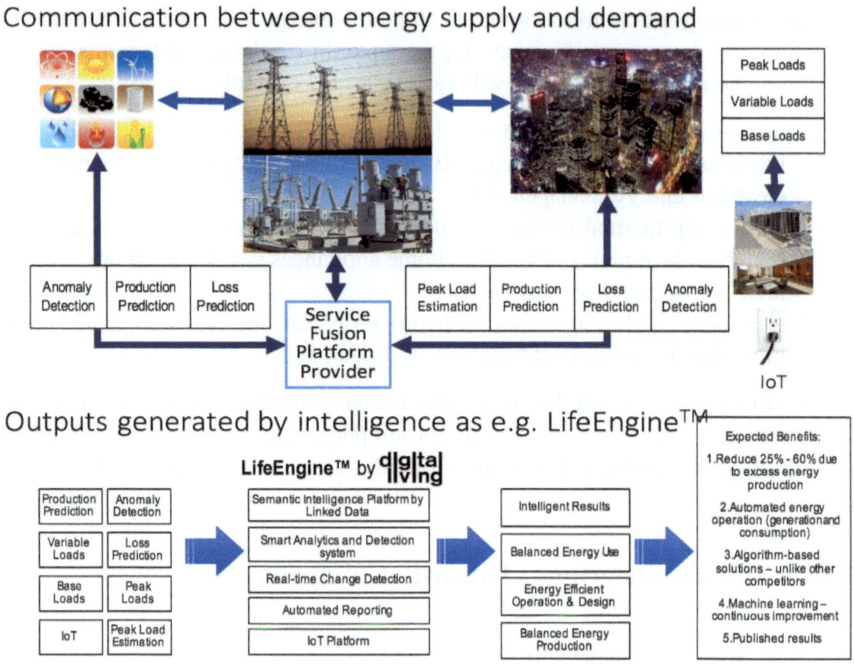

Fig. 13.1 Expected benefits of the Smart Grid by an example of semantic intelligence and linked data platform (LifeEngine) drafted by Ass. Prof. Oswald Chong [18] at Arizona State University

13.3.2 Buildings have Senses

In addition to the technological, organisational and activity-based definitions there are intelligent building definitions which approach the metaphor between man and machine. In science (source unknown) have popped up the idea that we build environment and related technology or artefacts that are alike human, similar to us or resembles ourselves. Also the metaphor between human and building or technology can be found; understanding how we copy ourselves—senses and cognition—into artefacts. The Swedish company T.A.C who later was merged with Schneider Electric introduced the concept of smart building as a metaphor between human senses and building automation. Automation components corresponding the human senses when placing on a level the human senses integrated into a person and the integration of building systems into the building (Himanen [45] p. 59).

T.A.C. together with NCC, a construction company from Sweden, was the key player of the EU FP6 project EBOB Energy Efficient Behaviour in Office Buildings (EBOB [29]). The combination of the human and social perspective with advanced modern control and ICT solutions called 'forgiving technology' was in focus (the concept initiated by the representatives from NCC Sweden and T.A.C., Himanen et al. [48]). The concept was targeted by the use case of making energy efficient behaviour natural, easy and intuitively understandable for the end-users, and at the same time to achieve the most energy efficient solutions while improving standards on indoor comfort in refurbished and new office buildings. Subjects on energy efficient end-user behaviour are still topical in the EU H2020 research for Energy Efficiency.

13.3.3 Buildings Borrow Cognition

The first intuitive expert guess of the author for the definition of the intelligent building concept was based on the idea of human intelligence in personal mental growth of the occupant as 'Intelligent Buildings are to be built after the needs of a person growing wise' (Lehto et al. [57]).

The building intelligence was then approached from two angles of cognition, First, search after the properties of building intelligence by a field study in real use environment with a sample of 12 buildings of which a half was intelligent and other advanced high class office buildings. Second, placing on a level the forms of human intelligence defined by Howard Gardner [42] and the hypothesis of the forms of building intelligence following the human approach (Fig. 13.2). The empirical sub-studies enabled to conclude that the designers had embedded human intelligence into technology in the forms of: ambient, dynamic, logic, connectivity and self-recognition. The classification was found from the seven forms of human intelligence defined by Gardner [42]: logical mathematical, linguistic, musical, visual-spatial, bodily kinesthetic, interpersonal and intra-personal.

The author's definition differs from others remarkably as other features (properties) than logic has been included in the concept of building intelligence: the spatial (ambient)—kinesthetic for adaptations (dynamic)—building knows its condition (self-recognition)—building informs of its status (connectivity). This type of holistic view could be worth of thinking for example in personalised computing or in role, event- and content-based applications.

The Intelligent Technology Framework describes the knowledge transformation process from design into usage of building and the role of the end-user involvement

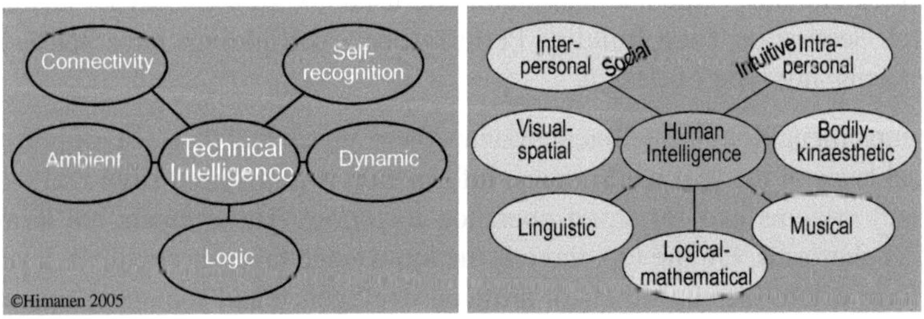

Fig. 13.2 The forms of building intelligence [45]

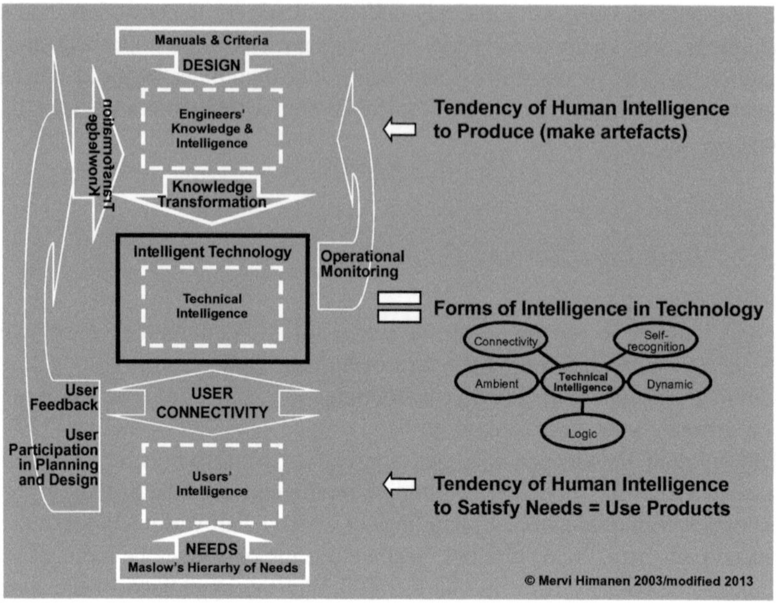

Fig. 13.3 The intelligent technology framework [45]

in it (Fig. 13.3) which at time being was not at necessarily clearly visible (cf. explicit) in design or soon that after. The learning cycle of Nonaka and Takeuchi [69], in Tuomi [87], p. 323) was used to understand the various forms of knowledge transformation and explain the process especially the tacit knowledge role in design (cf. implementation of the Nonaka and Takeuchi learning cycle in development of apps in Himanen et al. [47]).

The author has yielded from inductive reasoning after her empirical research on Intelligence of Intelligent Buildings (2003) that

- tacit knowledge of the designers which in general is difficult to comprehend consciously in the form of explicit knowledge is manifested in user feedback of the design and
- the influence of tacit knowledge on the buildings can be tracked by studying the explicit outcome—the intelligent building itself—with the end-user feedback of it (cf. Section on Total Building Performance and Buildings have Senses). This concerns any type of design.

In implementation of intelligent systems there are several layers starting from the physical layer of the real world items. Information is transferred from real world to the smart systems and for smart operation to follow. These layers are turning to cloud as platforms. On the top, there is the knowledge layer or cognitive layer, it is a platform where the algorithms of artificial intelligence and semantics operates by analytics or linked data technology for linking data, information or knowledge in order to enable operations is need. Actually, smart technology is a set of knowledge management tasks, mainly for decision-making, design and digitalization. The more sophisticated technology the better user involvement possibilities the algorithms in intelligence of technology allow.

The concept of intelligent building can be understood as a process of the knowledge management. When the engineers design and plan, they transfer their intelligence to the building. As well anyone can transfer ones intelligence to any artefact e.g. in the fields of manufacturing and service provision. It is clear that the previously learned knowledge in design tradition which is in explicit form (in calculation rules, guidebooks, regulations, etc.) will be copied to the new plans. The multidisciplinary user studies showed that intelligent technology is an outcome of the combination of explicit knowledge and thinking of engineers representing tacit knowledge which has not necessarily clearly manifested in physical form but the users can sense it. In other words, tacit knowledge can be defined by studying the result of engineering.

This type of thinking has value for example when describing how the process of design takes place and what is included into the design from the end-user feedback and from the tacit knowledge of the designers before it becomes in written form in the design guidelines (cf. PAS Standards Section Standardization of User-driven Smart Environment).

13.3.4 User Involvement in Smart Innovation

Always building and technology has aimed to help people to lied good life. The basic requirements to build a shelter for humans, from dangers such as climate and weather or animals and hostile intercourse, are understood as self-clear in a modern society.

Already industrialism and currently, information or knowledge society has increased the requirements of highly qualitative built environment while availability of resources or affordability does not dominate the demand in market. People like to increase their quality of life and accordingly the quality of built environment, the venue for living, pleasure or working. Workplaces have turned to a factors of effective production from being a costly burden of the production. A productive and cost effective built environment was gained through optimisation of architecture, structures, building services and management, and the interrelationship between them.

The freedom of choice in accordance of one's own preferences is large in the case of owner-occupied private houses which the owner has had built her or himself. Motivation on developing user involvement further comes from

- Unsatisfied needs of the residents and other occupants and tied to the local real estate broker's offering or what is listed in the manufacturer's catalogue.
- Tenants' possibilities to influence one's housing conditions are limited.

- Home buyers have few opportunities for influencing their future homes. People have different ways of living but constructors are still largely stick to the concept of a family unit with two parents and two children. Individual and local needs are pushed toward while all families do not fall within this category.
- Occupant in public or commercial buildings has limited possibilities to influence the design of their workplaces which the company representatives or executives often select. They are seldom involved in the initial stages of building such as in user requirement definition, commenting design drafting phase.
- It is claimed that the building developers and contractors as well as the institutional building owners—are not particularly interested in offering many alternatives.
- Designers are afraid of failures when experimenting new solutions which is both costly and spoils one's reputation.

A key issue still left was how to organise the interplay between user and machine or the machine to customer relationship (M2C) so that one can

- avoid machine controlling over human as we like to be aware of the actions going on in house but interfere as little as possible to the automated tasks when the manual mode is not preferred,
- deal with complex problems while human brain is able to operate with 7 issues at time in average (LeDoux [54] pp. 271).

The user involvement in design or operation of smart built environment has a new mental thinking pattern with advanced driving forces. They are (1) competiveness of various concepts and brands, (2) ecological sustainability, and (3) social impact of new building innovations or applications in facilities management and housing related e-services. Especially, serving certain user groups with solutions for their needs has popped up, as for example in housing among elderly and singles who are interested in housing concepts where they can choose between their own privacy and the common spaces, because loneliness is one of their major problems. In offices the way of working dominates the needs of space. In industrial practice, the building managers' possibilities to influence the decision-making of spatial arrangement have been experienced low. The management of core processes of the business dominates over the most effective way of building management. It is not a common practice that the building manager is the member of the board directors.

Taking end-users into account is not always easy (as reported for example by Ms Sonja Frosti at Digital Living Finland Ltd in Himanen et al. [47]) or considered not to be necessary (van Berlo [11, Toivanen 85]).

The user interest dominates the commercial demand. Recently, we have learned to admire firms that experiment and pilot with customers for new service and business models or user-driven innovation (von Hippel [49]), and create open source and social network-based innovation. Von Hippel argues for 'consumer innovations'; traditional division of labour between firms as innovators and customers is breaking down; a large portion of innovations comes from markets and customers.

Companies such as AGC, IBM, Nokia and many others have used new tools on social media or, end-user-oriented market study methods to increase their sales even with millions of customers. Such methods have been popular as, e.g. crowd-sourcing,[29] living labs born in Massachusetts Institute of Technology (MIT) (cf. Tang et al. [82] and [83], ENoLL, the European Network of Living Labs (www.openlivinglabs.eu)) or caving[30]. They provide dialogue for identifying the strategic challenges and grounds for their innovations. Still, the traditional methods of interviewing work well too, in this new context.

Short user studies by questionnaires have been a norm in property and facilities management companies to gather end-user feedback. Those questionnaires have moved into Internet. Internet and social media make a powerful platform for co-design. On the other hand, some building managers have even argued that listening without following corrective actions is enough.

User involvement may reach towards the role of a developer. Informing of one's needs might be relatively passive while just answering the questionnaire or to interviewer's questions. Neither the questions nor the analysis does necessarily head to figure out the problem of the user that the product or service should solve. The focus easily remains limited to the properties of the current product or service, and slight changes to be made.

The Finnish Government's innovation strategy [10] highlights skilled people and close-knit innovation communities as crucial for the competitive edge in the world economy (Fig. 13.4). Innovations are most often the fruit of new combinations of competencies crossing industry and disciplinary boundaries.

Nonaka and Takeuchi [68] argue for distributed leadership where wisdom is embedded in every individual and collective practice and action. They discuss cases of major structural, social, institutional and economic transformation where citizens, public agencies and firms have together saved their cities or lakes in crisis and brought them back to prosperous economic and social progress and development path for future. The distributed leadership includes competence of grasping the essence of a problem and knowing how to draw conclusions and acting on them immediately. This is 'hands-on' leadership in touch with the reality. This also implies that we consciously act based on values such as goodness, beauty and truth; they are applied, tested and recreated together with other people in every action.

There has always been self made men and volunteering among village men for building homes. The future studies identify the trend of individualism which has been one of the drivers for new forms of user-driven building. Individualism

[29]The practice of obtaining needed services, ideas or content by soliciting contributions from a large group of people and especially from the online community rather than from traditional employees or suppliers.

[30]A form of virtual reality is Caving. Cave is a room where its walls are typically made up of projection screens, where a computer-generated world is projected on the walls. In this a virtual reality where the user can experience the design before realization.

Fig. 13.4 Basic choices and key development areas for the innovation strategy 'A national innovation strategy' [10]

involving men and women which is important to remember because gendered innovation has been recognized both as a lacking in current production and thus one of the key success factors (Expert group "Innovation through Gender" et al [39]).

In Finland, the new forms of primary end-user involvement in smart innovation are such as a modern town houses with low ecological footprint, cohousing, group housing. The townhouses have the properties of raw houses without a garden. The Finnish group housing stands rather for self-acting homebuilding than putting emphasis on shared space and facilitation for communal dwelling. The members of the group housing group share the expert knowledge and facilitation for building during the building process as well as the costs due these activities. That after they can be rather independent unless they have not wanted to share any buildings with their neighbours. Both concepts have also legal status.

Internationally cohousing is understood as a type of intentional community composed of private homes supplemented by common spaces and shared facilities (cf. Vestbro [89], CoHousing Cultures Handbook 2012). CoHousing Platform (on http://co-housing-cultures.net/) summarises such keywords of cohousing as self-organised, community-oriented and sustainable, integrating, non-speculative and open to the neighbourhood, affordable and socially designed homes. The community is planned, owned and managed by the residents—who also share activities which may include cooking, dining, child care, gardening, and governance of the community. Common facilities may include a kitchen, dining room, laundry, child care facilities, offices, Internet access, guest rooms and recreational features.

According to the Houser Study (Himanen and Korhonen, [46]), the modern scheme of resident-driven housing business is in an early dawning phase in Finland based on the very low numbers of operators (about 150 in 2012) and their minimal share of the total housing construction volume. Into the sample was included companies who were involved in planning or complementation of such new user-driven housing concepts as cohousing, group housing, house builders as self-made men, plumbing renovation, independent living. Using the rough estimation of 6,000,000 M€ as the turnover of housing construction in 2011, it can be concluded that the resident-driven housing business corresponds less than 0.4 ‰ of the total of housing construction. This innovative niche in housing business is dominated by recently established micro and small companies. However, already also some of the big companies (10 of them in the sample) take part in the business by developing further their well established business concepts that might have been established a few decades ago.

13.3.5 User-Oriented Intelligent Building

Among the first ones to introduce the end-user approach to office design were the Skidmore, Owens and Merrill Architects in the 1980s in USA (www.som.com). Large post-occupancy studies followed the creation of intelligent building concepts and the realisation of smart buildings: e.g. (1) the ORBIT studies carried out by the Harbinger Group of Connecticut (Davis et al. [22], Duffy [27]), (2) the Intelligent Building in Europe [9] and the Office Tenant Survey of the BOMA and the (3) ULI [8]. They examined current practice by the user feedback, user requirements and changing work patterns in relation to the productivity of the work environment.

These studies focused on interior design and its meaning to the working performance. Factors of impact on environmental quality comprise colours, adaptation of spaces or furniture, active structures as automated walls and doors, etc. As well, leading engineering and architects' offices, which are many, have developed their design office layout further from these principles to have knowledge work places. Further on, the idea to stimulate the human brains which is the only 'raw material' of knowledge work is originated from the very first intelligent offices. Odours, music, water elements, etc. were used.

The Danish Professor Fanger started the indoor quality studies; conducting field studies on various parameters of the air quality as well as studies in laboratory conditions on the interplay between the air quality and user's work performance. Currently, the studies following Fanger's footsteps comprise an integrated concept of indoor environment quality, which is a part of the Total Building Performance concept.

The statistical significance in results was found in favour of intelligent buildings within the field study with a sample of 12 buildings which compared intelligent and other high class office buildings in true use environment of 534 office workers and a large number of parameters both in questionnaire survey and measurements of the knowledge work environment (Himanen [45]). The survey had drawn much from the above-mentioned large post-occupancy studies as well as from the indoor air quality studies. In addition, such other factors as influence of clothing or travelling were in. The new buildings were at same age and located in Helsinki Metropolitan area. The main criterion was the end-user feedback of the work spaces and the performance of office and building automation and their meaning to the end-users' work efficiency and efficiency.

Today, for example, studies carried out by Social Networks Analysis cover wide range of parameters with wide stocks of data, i.e. big data (cf. IEN Innovation Ecosystem Network, www.innovation-ecosystems.org). The scope of studies on intelligent built environment has expanded into studies on the smart city phenomenon. The partners involved in research and innovation come from all walks of life not only from scientific disciplines (cf. S3 Smart Specialisation Platform, a research scheme conducted by quadruple helix comprising academia, industry, public sector and end-users).

In her study on electronic services in municipalities Toivanen [85] found the large influence of attitudes and leadership style of developers and providers when applying ICT into service provision. Toivanen [86] emphasized the meaning of providers' positive and warm hearted attitudes towards end-users. She identified the meaning of predictability in defining even unconscious needs and desires of the end-users. When co-creating new services the interaction need be cordial and warm between end-users and service provider and the delivery. A new product or service must the answer to the true needs and expectations of the end-user and not the ones the provider has extrapolated (cf. Himanen et al. [47] on more about Toivanen's and others' research on user-driven innovation).

A strong emphasis on the development of the end-user involved smart environment has been and is within the AAL Joint Programme for funding research and innovation on smart housing, independent living and well-being of elderly and their communities (family, caregivers, neighbourhood, service providers, care system, etc.). The innovations should include both a user-centred approach and pilots with a considerable number of end-users involved in order to demonstrate the benefits and added value necessary to make impact on the market.

The definition of end-users in the AAL Programme

1. Primary end-user is the person who actually is using an AAL product or service, a single individual, 'the well-being person'. This group directly benefits from AAL by increased quality of life;
2. Secondary end-users are persons or organizations directly being in contact with a primary end-user, such as formal and informal care persons, family members, friends, neighbours, care organizations and their representatives. This group benefits from AAL directly when using AAL products and services (at a primary end-user's home or remote) and indirectly when the care needs of primary end-users are reduced;
3. Tertiary end-users are such institutions and private or public organizations that are not directly in contact with AAL products and services, but who somehow contribute in organizing, paying or enabling them. This group includes the public sector service organizers, social security systems, insurance companies.

Common to these is that their benefit from AAL comes from increased efficiency and effectiveness which result in saving expenses or by not having to increase expenses in the mid- and long term.

This user classification could be modified to be used in other context as well.

Digitalization enables building of digital ecosystems where the entire information process of service provision or production will be renewed. It concerns raw material, data sources, organization, process schemes, etc. and their place in the system or consideration of their existence or disappearance.

13.3.6 Standardization of User-Driven Smart Environment

Instead of smart building standards next will be referred the end-user related building standards and smart city standards.

The ISO (International Organization for Standardization) and the CEN (European Committee for Standardization), as well as, the Cenelec (European Committee for Electrotechnical Standardization) have launched standards related to user involvement and some of them are specific for building sector:

- ISO 21542:2011: 'Building construction—Accessibility and usability of the built environment'
- CEN/TS 16118: 'Sheltered housing (Requirements for services for older people provided in a sheltered housing scheme').

Human-centred design overall has established practice defined by

- ISO 13407 new version of ISO 9241-210: 'Human-centred design for interactive systems',
- ISO 13407:1999 Human centred design processes for interactive systems,
- ISO/TR 16982:2002: 'Ergonomics of humansystem interaction—Usability methods supporting human-centred design'

The interplay between human and robots can benefit from:

- ISO 8373: 2012. Robots and robotic devices—Vocabulary. International Organization for Standardization, 2012. 38 p.

For time being, technology standards are just in infancy at smart application level such as OneM2 M (ETSI partnership) in the Smart Mobility area and possibly soon some W3C initiatives the European M/490 Smart Grid mandate (ETSI/CEN/Cenelec) in the smart energy domain and possibly soon some W3C initiatives. A much stronger effort should be dedicated in this area in the future.

The Publicly Available Specification (PAS) is a sponsored fast track standard driven by the needs of the client organizations and developed according to guidelines set out by British Standards Institution (BSI). The PASs are intended for city authorities and planners, buyers of smart city services and solutions, as well as product and service providers such as national and local government departments, utilities, healthcare providers, transport, construction companies, ICT solution providers, city planners and developers. Some of them are particularly relevant to national and local government departments, utility companies, healthcare providers, transport service providers, construction companies, network companies, city planners and developers, designers, and vendors of ICT solutions be they big players, SMEs, or their clients.

The PAS defines a smart city as one where there is 'effective integration of physical, digital and human systems in the built environment to deliver a sustainable, prosperous and inclusive future for its citizens' [71]. The Executive summary of the PAS Standard on 'Smart city framework—Guide to establishing strategies for smart cities and communities' defines the concept of smart city in detail (Fig. 13.5, [72]).

The relevant titles for this review in the PAS Smart Cities suite include

- PAS 180:2014, Smart cities—Vocabulary, which defines terms for smart cities, including smart cities concepts across different infrastructure categories. To help build a strong foundation for future standardization and good practices PAS 180 provides industry-agreed understanding of smart city terms and definitions to be used in the UK for providing a common language of smart cities for developers, designers, manufacturers and clients.
- PAS 181:2014, Smart city framework—Guide to establishing strategies for smart cities and communities, which gives guidance on a good practice framework for decision-makers in smart cities and communities (from the public, private) to develop, agree and deliver smart city strategies that can transform their cities' ability to meet future challenges and deliver future aspirations.

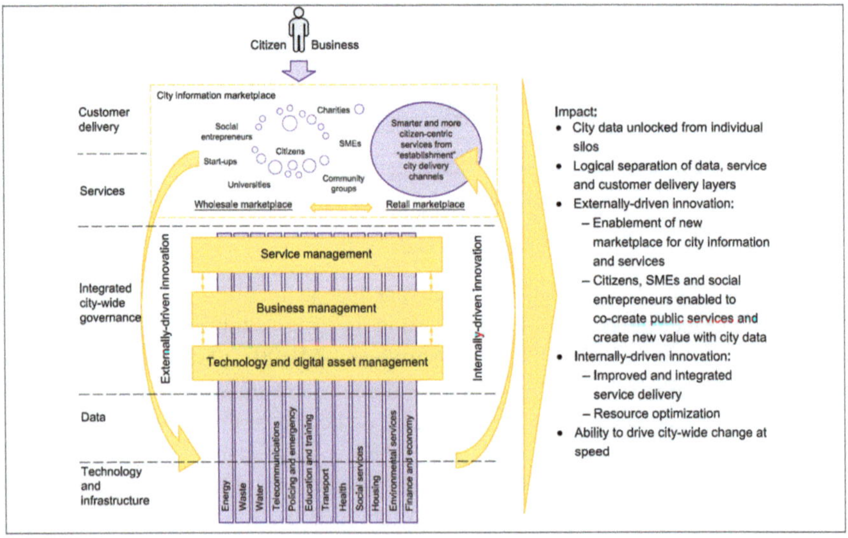

Fig. 13.5 New integrated operating model: in transforming cities to smart cities (PAS 181 [72])

- PD 8101:2014, Smart cities—Guide to the role of the planning and development process, which gives guidance on how the planning and implementation of development and infrastructure projects can equip cities to benefit from the potential of smart technologies and approaches (© The British Standards Institution 2014 and voluntary sectors). The guide is relevant to major developments, infrastructure projects, refurbishment programmes and improvements to public spaces. It considers how each stage of the planning and development process could support smart city opportunities and sets out what needs to be done at each stage.
- PD 8100:2014, an overview document that will provide guidance on how to effectively communicate the value of smart cities to key decision-makers. It gives guidance on how to adopt and implement smart city products and services in order to facilitate the rapid development of an effective smart city. It describes in detail the potential benefit of smart city strategies, provides recommendations on how to identify the first steps towards making the city smarter and covers the role of technology and data in providing the tools in this process.
- PAS 182:2014, Smart city concept model—Guide to establishing a model for data interoperability, which provides a framework that can normalize and classify information from many sources so that data sets can be discovered and combined to gain a better picture of the needs and behaviours of a city's citizens (residents and businesses). It gives guidance on how to promote data sharing across sectors in a city and help bridge the differences in data analysis between sectors like health, education and transport. It is intended to facilitate discussions between decision-makers and the specialists who build and design the systems and services that enable a city to function. The guidance addresses the fact that service providers do not always have the expertise to analyze the data they accumulate, that different sectors use a different language when describing data and offers a model that can be used by a variety of sectors.

In the context of smart city, the BSI refers also standards for

- Quality of life and services in cities: BS ISO 37120:2014 Sustainable development of communities. Indicators for city services and quality of life.
- Research on smart infrastructure projects: PD ISO/TR 37150:2014 Smart community infrastructures. Review of existing activities relevant to metrics.
- A specification for KPIs for smart infrastructure projects: PD ISO/TS 37151:2015 Smart community infrastructures. Principles and requirements for performance metrics.

13.4 Future Potentiality

Keywords of opportunities for construction in smart built environment can be listed: Apps for building on smart phones—Artificial Intelligence—Augmented Reality—Building Automation on Cloud and SaaS—Big Data BIM—Caving—

Clouding—Digitalization—Drones for scanning[31]—3D Printing—robotics—Telepresence[32] or Holoportation[33]—Integrated FM—Horizontal Platform—Internet of Things—Machine Guidance—Open data—Virtual Reality.

13.4.1 BIM—Building Information Management

The need to extend computer aided building design to construction and to real estate management phases of building has been on agenda from the very beginning of the born of smart building concept but still not reality today in mainstream building.

Nordic countries are often cited amongst the five strongest regions in the world regarding BIM implementation. Developments have taken place due to an awareness of the great potential for enhanced efficiency and productivity, not as a result of the demands of public construction client organisations, which have been the main driving forces in Norway and Finland. In Finland already in 2007 Senate Properties required information delivery of public building design in the form of BIM and prepared a BIM manual for documentation of new construction development (www.senaatti.fi/en). The shift from drawing-based design into the information-based one occurred. Instead of CAD drawings (Computer Aided Design)—which are familiar with all design software providers—the public real property developers started to require object-oriented databases to be carried out by BIM-based software on IFC standard (Industry Foundation Classes developed by the International Alliance for Interoperability (IAI)).

Just recently, BIM can be defined both as a technology and as a method applied to create, communicate and analyze building information models. The UK and Sweden are used as examples of this progress recently done in the UK and intended in Sweden.

In the UK, a national BIM strategy for the building sector has been initiated, in which the government, the private sector, the public sector, research institutes and academia are collaborating. The general aim of the strategy is to simplify and

[31]Drones are more formally known as unmanned aerial vehicles (UAV). Essentially, a drone is a flying robot. The aircraft may be remotely controlled or can fly autonomously through software-controlled flight plans in their embedded systems working in conjunction with GPS. UAVs have most often been associated with the military but they are also used for search and rescue, surveillance, traffic monitoring, weather monitoring and firefighting, among other things.

[32]As a form of virtual reality, telepresence is term used for a set of interactive technologies such as high definition video or audio that permit the users to feel or appear as if they are present (and able to influence and operate) in a location in which they presently are not physically not located.

[33]The HoloLens, Microsoft's much-hyped new augmented reality (AR) headset—a solution of telepresence. A series of cameras are set up around a room, tracking shapes and movement and stitching a 3D model together in real time. Speaking and interacting with remote friends, family, and work colleagues could become almost as natural as it is when you are face-to-face—except for the fact that everyone involved has to have a heavy AR device strapped to their head.

expedite BIM adoption throughout the sector and establish uniform requirements for BIM implementation in government-funded projects. Effective information management leads to business efficiency and profitability. Without the standardized approach to authoring of both geometrical and non-geometrical objects, any outputs from the model will be inconsistent and will not return valid results for schedules and other information-related queries. Following the Government Construction Strategy in 2011, the UK Government required fully collaborative 3D BIM (with all project and asset information, documentation and data being electronic) as a minimum by 2016 (NBS [64]). In 2015 NBS could conclude that they have clearly moved on from the time when 3D CAD could be mistaken for BIM:

> *'At NBS we have been working to deliver increasingly sophisticated, and standardized, levels of information into the federated information model through the timeline. This began with our innovative specification product, NBS Create, and then developed through the creation and growth of the NBS National BIM Library. We were able to fully integrate these products together through plug-ins, allowing information to be co-ordinated between the specification model and the geometry model. 2015 sees the next stage in this trajectory of development. Part funded by Innovate UK, and produced in partnership with the industry, we have released the NBS BIM Toolkit. This free to use toolkit offers a digital Plan of Work tool, and a new unified classification system. It provides support to define, manage and validate responsibility for information development, as well as its delivery, at each and every stage of the construction life-cycle.'*

Since 2015, in UK the standardization of BIM comprises such standards as (NBS [65])

- PAS 1192-5 Security,
- PAS 1192-3:2014 Information exchange—COBie—BS 1192-4:2014,
- PAS 1192-2:2013 Information management process,
- PAS 91:2013 Construction pre-qualification questionnaires,
- BS 8541 series Library objects for architecture, engineering and construction,
- BS 8536:2015 Facilities management briefing for design and construction,
- BS 7000-4:2013 Design Management Systems,
- BS 11000: Part 1 2010 and Part 2 2011 Collaborative Business Relationships, and
- Guide to managing design in construction.

Hamil (in NBS [64]) concluded that in the UK, the foundations for a digital construction future are now being put in place. These foundations include consistency in classification, standardized information requirements and guidance to help make the correct decisions quickly. In addition,the move to the cloud is changing the way teams collaborate through data access anddata sharing. With this in place, what will the construction industry look like in 2020?

In Sweden, Andersson et al. [4] summarize that BIM involves analysis of factors such as material strength, power consumption, noise, indoor air quality, constructability, occupational health and safety, accessibility, architectural design, cost, time and resource planning, supply chain management, operational optimisation and use of space. In other words, it concerns all the information that will be used at different stages in a facility's life cycle. Above all it involves simulating and optimising many of these factors from a long-term perspective. BIM entails the utilisation of information in a systematic manner being consistent with three-dimensional designs and explicit classification of information concerning the facilities. Thereby BIM facilitates all the possibilities that the technology provides in relation. BIM applications are usually divided into three main stages; (1) Three-dimensional models for visualisation and interaction; (2) Integrated analysis; (3) Automation linked to industrial processes.

By industrial processes Andersson et al. [4] mean that the processes are standardized and that platforms, products and information support are disconnected from construction projects and property management, and are developed independently from a life cycle, sustainability and customer perspective, then applied during construction and facility management. The recurring activities of construction and facility management are standardized, and the best overall solutions, such as inputs, subsystems and complete modules, are developed as separate platforms. Platforms and products are then offered in specific market niches.

The US and Singapore are other leading countries in usage of BIM.

13.4.2 3D Printing

In addition to new digital capabilities, 3D printing (cf. Additive Manufacturing (AM)[34]) allows the localisation of production, easy design modification and customisation as well as the introduction of technical, user centred (pls read: personalized consumer ready products) and aesthetic capabilities that are new to traditional industrial terrain.

As the organic architecture,[35] 3D printing can totally change the thinking of building. It is not any more fully clear for example where the wall transforms into roof while the printer structures a solid surfaces that serve multiple purposes as Prof. Arto Kiviniemi at Liverpool University has pointed out. The relations between objects in CAD and BIM tooling need radical redefinition.

By utilizing 3D printing in place of conventional construction techniques, the building's engineers project significant savings in time and costs, with estimated 50–70 % reduced production times, 50–80 % reduced labour costs and 30–60 %

[34]Additive Manufacturing refers to a process by which digital 3D design data is used to build up a component layer by layer using materials which are available in fine powder form. The term of professional production technique '3D printing' is increasingly used as a synonym for Additive Manufacturing. The technology has especially been applied in conjunction with Rapid Prototyping and now being used increasingly in Series Production.

[35]Organic architecture is a line of architecture which promotes harmony between human habitation and the natural world. This is achieved through design approaches well integrated with a site that buildings, furnishings, and surroundings become part of a unified, interrelated composition. Architect Frank Lloyd Wright used to describe his approach to architectural design by this concept.

reduced construction waste (Nield [67]). One of the biggest potential impacts of 3D printing technology is expected to be in creating low-cost and environmentally friendly housing. The world's biggest 3D printer can make houses out of mud, clay, water, dirt, and natural fibres, avoiding the expense and environmental consequences of cement.

Authorities in Singapore have announced plans to provide residents with 3D printed houses, and are now conducting a feasibility study to figure out how to get it Done, Nield reports [67]. If the proposal gets the go-ahead, house storeys will be printed independently and then assembled on site in the style of Lego bricks. When it comes to the Singapore project, not every part of the house would be 3D printed but just the main structural components. Any elements that cannot be printed cost-effectively could still be put together using traditional methods. The aim is to use machinery to build homes for Singapore's elderly population without relying on foreign labour, although the printing technology for concrete elements which is at the centre of the scheme is still in the development phase.

The UAE National Innovation Committee has proposed for construction a temporary headquarters for the facility's staff of Museum of the Future as the world's first 3D-printed office building in Dubai, Dockrill reports on ScienceAlert.com [25]. It will occupy close to 200 square metres of land. The completion date of the project has not yet been announced. It is set to be printed by a 3D printer measuring some 6.1 m tall, with the individual components subsequently assembled on site 'in a matter of weeks'.

13.4.3 Open Data—towards Collaborative Economy

With Open Data is meant the opening of information resources free of charge, in machine-readable format and with transparent conditions of use to businesses, citizens and society as a whole by the end of the decade. The goal is to create favourable conditions for new business activity and innovations, strengthen democracy and civil society, enhance administration in general and digitalization of it, and diversify the information resources available to education and research. The objective of Open Data initiatives includes on one hand, new business ideas with the aid of open data and better utilization of information resources, and on the other, the rationalisation of existing practices. The Government Programme also seeks to strengthen knowledge-based decision-making and openness.

The concept of collaborative economy is also used. It is defined as initiatives based on horizontal networks and participation of a community. It is built on 'distributed power and trust within communities as opposed to centralized institutions', blurring the lines between producer and consumer (Botsman and Rogers [13]). Transactions are facilitated via community-based online services. Also the term of shareconomy is used referring to a hybrid market model (in between owning and gift giving) of peer-to-peer exchange.

The EU's PSI (Public Sector Information) Directive on the reuse of public information resources aims to common practices and structures standardizing and to support the systematic opening of information resources. It is recommended that the standard open and internationally interoperable licence Creative Commons Attribution 4.0 (CC BY 4.0) be adopted in the reuse of the public sector's open information resources. A number of public authorities have adopted the recommended licence.

The second step of this open government process is to offer Open API to developers. Here it is possible to retrieve data but also to send some information or request. Open API are documented web services that allow programmers to interface directly with cities IT infrastructure without going through a human (like when going through a hotline) or a fixed interface (like filling a form on a web site). The goal is to offer the opportunity for third parties to develop applications on the top of city platform and create new services at better price for the benefit of the citizens. Some cities have even already launched their own app store. These Open APIs unleash a wide range of new services that would never have been considered by the city itself by lack of budget, creativity, insight, etc. Most of the applications are targeting mobile platforms (thus benefit from these platforms sensors and features like geolocalisation) and need to be integrated in social networks.

The European Data Forum (EDF) is a meeting place to discuss the challenges of Open Big Data and the emerging Data Economy and to develop suitable action plans for addressing these challenges (www.europeandataportal.eu/en). The European Commission Vice President for the Digital Single Market Andrus Ansip has stressed the importance of data to 'be able to move freely, across national borders' and 'any unnecessary or unjustified barriers should be stopped'. The European Data Portal does that by harvesting the metadata of Public Sector Information available on public data portals across European countries. Information regarding the provision of data and the benefits of reusing data is also included. Going beyond the harvesting of metadata, the strategic objective of the European Data Portal is to improve accessibility and increase the value of Open Data.

Examples of international cooperation are participation in the EU's SharePSI project, which seeks to develop international best practices in open data, and the Nordic Open Data Week 29 May–7 June 2015.

For example, according to the Finnish Open Data Programme 2013–2015, set up by the Ministry of Finance, open data will become a regular part of administrative activity. It has been prepared as part of the planning of the central government spending limits and as part of the general government fiscal plan. Every year, the Ministry of Finance has requested from the ministries plans outlining which information resources will be opened in the administrative branches and what the economic and social impacts will be. A growing number of municipalities are also opening their data while in Finland the central government and local government data are under separate authority: Ministry of Finance and the Association of Finnish Local and Regional Authorities.

Opendata.fi, an open data and interoperability service, was launched in 2014. The service aims to provide information about opened data resources as well as interoperability descriptions and guidelines for centralised use. Currently (2016), the service offers information diversely, on more than 1,400 opened resources, comprising the material from terrain data to weather, climate, sea, transport, financial, statistical and cultural data, the legal databank Finlex, owned by the Finnish Ministry of Justice, etc.

13.4.4 National Architecture for Digital Services

We are heading the era of digital economy and economy of platforms. Artificial intelligence and semantic web technology are realised to solve the problem of siloed systems which have dominated, e.g. the social service provision or private service sector, industrial digitalization, etc. As mentioned, cities actively look after solutions for city platforms that comprise integration and interoperability of social services for all, public transport, energy efficient building, efficient city administration.

The solution to siloed systems, however, is not only technological while the organisational and economic obstacles dominate the decision-making on the usage of the technological potential. The interests collide. In this context, technology alone is discussed.

A firm basis for developing nationwide governmental or municipal e-services for all citizens is aimed by the National Architecture for Digital Services as targeted and pioneered in Estonia, Finland or Denmark, for example. The National Architecture for Digital Services will be a compatible infrastructure facilitating information transfer between organizations and services. It carries potential for increasing the use of technology for all citizens' welfare being compatible, and personalized as well as event or content-based.

Architecture of for national technology is aimed to be personalized and user friendly, as well as, simultaneously relatively cost-effective and sophisticated which is enabled by the currently mature enough technological readiness level due to the previous software development. Still, the greatest benefits also in economic terms will be yielded by the benefits after the implementation the technology. For example, in Finland by electric public services is targeted [61]

- To simplify and facilitate transactions by citizens, companies and organizations with the authorities and to improve security,
- To promote openness in public administration and to improve the quality of public services
- To enable cost-efficiency in online services,
- To improve shared use of information and the compatibility of information systems,
- To promote corporate opportunities for leveraging public administration databases and services,

- To support the national economy by making public administration more efficient and by creating new business opportunities in the private sector.

Today, one cannot deny that the National Architecture for Digital Services in Finland for example is not yet completely planned (cf. Frosti [41], Harald et al. [43])—not to mention implemented, while in addition to current plans, to employ the full potential of the national architecture, compatible inter-city service fusion platforms are needed or an Enterprise Service Bus (ESB) architecture, which take care of the systems integration and data interaction between the public and the city services.

The Data exchange layer of the Finnish National Architecture for Digital Services has been demonstrated in the context of the social and health care in the city of Espoo [6]. It turned out to be suitable for the social and health care service provision and to improve access to data and interoperability, but not yet complete. The data transfer requires similar quality criteria as the KanTa technology (National Archive of Health Information, www.kanta.fi/en). All the data have to be encrypted and digital signature need to be in use in order to guarantee a trusted operation.

There is currently no well-functioning government-wide data network in Denmark [52], but still Denmark is a good example of the governmental commitment on nationwide interoperability of all. 'Digitalisér.dk' is the central repository of information on data interchange standards for the public and private sectors and a collaboration tool for the development of information society in Denmark. Launched by the Danish Ministry of Science, Technology and Innovation in October 2008, as a successor to the Infostructurebase (ISB), it is a key strategic element in the country's eGovernment architecture. Its main purpose is to support the exchange and reuse of data related to public and private service delivery, including cooperation, business reengineering and alignment of related services. An important part of the content is the standards approved by the Danish e-Government IT architecture and XML committees.

'Digitalisér.dk' also provides an uncomplicated basis for debating common public digitisation by using intuitive web which will be based on interaction rather than formal processes.

13.4.5 Connected Society

Economic growth and globalisation increase in complexity, during the information age in particular. Knowledge society trusts on transparency making life less complex thanks to good data, information and knowledge management on cloud. Artificial intelligent engines help human mind to cope with living and work and decide on personal life or operation at work. Virtual reality helps to comprehend quick and easy. Qualitative knowledge becomes a key factor in economy, politics, in operational decision-making. Service chains become ever more smooth going and shorter in time due to dropping unnecessary phases, or material as paper, for example, becomes ever more virtual.

Prof. Mischa Dohler from Kings Collage trust on the promise of Big Data (2013; www.linkedin.com/in/mischadohler). CEO Pirkka Frosti from Digital Living International Ltd (www.digitalliving.fi) has since 2008 shared the same idea of taking advantage of Big Data in digital economy ending up virtual ecosystem building of real-world items on cloud, using a smart knowledge management engine for platform of linked data semantics. They both foresee that the hype on Big Data will be realised right after the data will be employed for the benefit of connected society. It needs applications easing daily activities, enabling decisions that are based on valid knowledge and working in role and content-based automated environments.

CTO Jan Byfors at NCC Construction Sverige AB concluded (2016) in his speech on 'Opportunities of the digitalization for construction' that '*Digitalization is the solution! Construction is lagging behind! ... but catching up. The sector has already started its journey towards digitalization ....and that is BIM. BIM means Collaboration—a way of working. There will be new business logics. Sharing is the new norm. Digitalization is a game changer. Digitalisation is moving fast*'.

Byfors [16] urged the building sector: '*We have to adapt to the technology!*'

He foresees the IoT as a tool for realisation of smart homes, the implementation of the Integrated FM (facilities management) on the Horizontal Platform and Big Data. As well, IoT enables Smart City models linking real-world, land usage, levels, properties, roads and clients.

13.5 Discussion

Smart city and smart society carry a potential for advancing the beneficial and profitable use of smart technology. Among other things, the author's experience in adoption of information technology proves of advantages that will be achieved beyond the digital ecosystem. New technology has usually been implemented in the first place in certain parts of the entire operational ecosystem. Consequently, many advantages might be foreseen but only limited number of them can be directly realised. A wider digital ecosystem or business landscape is needed for indirect advantages to avoid the frustration of unfilled customer expectations or market promises (e.g. in interest of investors or funding bodies).

For example, an integrated home automation can save energy, but why invest in the all properties that smart housing technology provides, which might not be fully beneficial or, e.g. for safety can be purchased a mobile applications. The house automation and living can produce extra energy but before the smart metering and Smart Grid technology it cannot benefit the home owner. The unrealised vision of the integrated digital ecosystem from CAD and BIM via digitalized construction site management to smart integrated building automation and real estate property and facilities management has lived long in commercial building sector.

The requirement of integrated technology in the 1980s has turned out to be too high at that time, but the correct direction for the development was set. It has bear fruit and the readiness for smart housing, intelligent workplaces, smart cities, smart everything is good today. The best suited technology for applications in smart environment such as the linked data technology or semantic intelligence have just recently become promising enough to fulfil the requirements of integrating building technologies. But meanwhile, also new demand for integration or interoperability with smart city and Smart Grid concepts has risen. They are highly relevant and in need for better energy efficiency.

The original need of integration of technologies has been taken over by Internet which has reached all types of buildings and especially homes. Systems run after pre-decided algorithms on cloud. User operations are limited to pushing on or off. In homes entertainment and edutainment run on web by personal computers, smart phones and pads as well as via smart television. The same goes for example with safety and security applications. The original idea of integration in smart homes included the control of the home appliances, entertainment and energy supply and housekeeping facilitation. At time being, these systems run separately in the majority of homes but are going to be accessed on apps on smart phones, for example.

In the 1980s Intelligent Building Ass. defined the integration of technologies as the aim of the intelligent building research and development. This aim formed both the starting point and the driving force for the progress of the technology in concern. The implementation has followed accordingly and spread towards the concept of smart city. Due to the original aim in smart housing, the user requirements have not been paid special attention world wide which has hold up the creation of the smart services, especially those that are related to daily living and could interest the consumers or other primary end-user groups.

Nevertheless how highly in need in the dawn of digitalization the technology push approach has caused the lack of service-oriented approach in application development. The purpose of the technology has been lost. Technology has been added on without proper idea what for. Accessibility can be poor either due to lack of user friendliness or trained users. The end-user motivation to use smart solutions depends on how beneficial the solution is for the user and if it has the added value of making things easy-going or comfortable (the EU Elderathome survey in Himanen [44], the EBOB project in Himanen et al. [48]). The usage is dependent on limitations and possibilities. The possibilities are offered by technology, but the limitations can be both technical and social and psychological.

As from the technological point of view the smart building would be rather made of active structures and technologies than passive collection of preset technologies.

Some point out that there is no technological limitation for digital ecosystem building such as that of the smart environmental development, for example. The holistic approach needs experts with skills in several disciplines and their cowork. However, human brain has been proven to be able to manage maximum from 5 to 9 issues at time (Miller [62]). How difficult it might be to virtualize our experiences, lot of advanced science is going on new methods to observe human cognition and to reach data or information of the transcendence or understand the essence of human wisdom.

The user involvement is problematic while engineering education does not touch it almost at all. It is important to understand that user involvement needs special expertise on it and that for example the serious gaming technology is not necessarily enough in scoping the user needs and desires. The issues are not made easier while the experts in psychology or social scientist do not necessarily speak same language with technocrats.

Nevertheless, exclusively trusting on end-users alone and either neglecting or forgetting the expert knowledge from traditional developers or suppliers can be rather problematic. While the user involvement in design and operational solutions has been intensified, the focus tends turn strongly on the end-user feedback and the interplay between the users and the experts, as well as, the role of the expert knowledge seems sometimes been forgotten.

The problem of the quality of sample has popped up. For example, or some living lab methods as such are considered good but trust on end-user feedback can be questioned when–they have been used by soliciting contributions on ideas of product or service innovations or content provision, from a large group of people without any special selection criteria, and especially from an online community. The interaction that takes advantage both form user and expert knowledge can yield to a win-win solution—superior over one-sided approach from any source.

The design of smart user-oriented build environment starts from the understanding the processes of the occupants' activities and related information flows with big data from sensors, in-house registers, weather information, etc. The ability to adjust to constant change forms the base of design and control.

As innovation on intelligent building all technological advancement can benefit from linkages of (big) data, applications, digital ecosystems of several disciplines but not necessarily integration algorithms by one operator. Failed attempts to produce a killer application to smart building market have appeared. The use of open source licensing in intelligent building has based on consortia of companies operating in a certain market segment of smart building technology. Even this strategy has not led to a worldwide de facto standard. Smart built environment accessible on web or compatible with web has postponed and the stabilized business has been kept and retarded the digitalization of smart built environment which comprises a major part of society.

A potential for future competitiveness in the smart technologies lies in interoperability and ability of real estate and building sector to follow up the progress in digital economy. Digitalisation is dramatically impacting way of working and the whole operation and business environment. Neglecting digitalisation will cause a risk of losing the game in the highly competitive markets. The business will change not lost. Digitalisation can bring new business opportunities, business models, change the roles of operators in a value chain, and end existing business.

There is a need to extended thinking beyond technological approach towards interconnected daily activities at home and our duties in working life as well as in service fusion overall.

It is not enough to limit digitization referring to the conversion of analogue data (as images, video, and text) or information in oral or paper form into digital form. This transformation, refers to the ability to turn existing products or services into digital variants, and thus offer advantages over tangible product and changes associated with the application of digital technology in all aspects of human society.

13.6 Conclusion

No wider systematic review study on how the implementation of intelligent building concept has progressed in building sector has been found when the concept is considered in its wider holistic content. However, there are yearly granted awards highlighting the best built environments after various criteria.

Although several organisations have attempted to establish a universal definition for smart building or smart city, there are a multitude of definitions with different levels of detail and varying degrees of emphasis on various aspects of building intelligence.

Still, since 1980s the progress in software development has been excellent, while at the starting point even any commercial Internet was not here yet. The technology push type progress of smart built environment has taken advantage of this development.

A study on the user feedback has proved intelligent office buildings superior over other high quality office buildings.

The smart building has been the promise of the future. Engineers have not given up of this approach. Today digitalization and platformization are the promise which can be expected to have results in the very near future.

The IT giants are interested in digitalization of societies together with startups and spin-offs that have popped up from the unsatisfied user needs in smart built environment.

Digitalizing societies include also the human dimension into the concept of effective use of ICT. This progress concerns both smart city and intelligent buildings. Connected society has it all and is ready for collaborative and digital economy.

Despite the non-existence of systematic review study on how the implementation of intelligent built environment sector has progressed a few reasons for the fact that the concept of smart built environment has not yet been fully employed in the practice can be summarized

- The starting point in the integration of technologies has dominated and promoted technology push approach which as such has been very good.
- Internet has taken the role of integrator and decreasing the meaning of sector dominant solutions. Consequently causing frustration and diminishing business opportunities of closed solutions. Utilisation of cloud-based application has been slow.

- Although buildings have always been built as a shelter for people, in general, the user-driven building sector in the modern sense is in infant stage and consequently, the smart technology alone does not appeal customers if provision of services on smart technology for built environment is lacking.
- Municipalities' role in digitalization of the society is remarkable, but some of them might have ignored the importance of end-users' in the development of e-service provision.
- A common ICT architecture for all seems to be in favour or in interest of noboby including a large client pool of real estate and building sector.
- The market of smart housing has not started fly general as expected, and for example in the elderly housing despite the long lasting work while they do not buy. On the other hand, studies show that elderly are by no means overlookers of new technology but they buy only when the technology is satisfying their customer expectations.
- Despite the existence of user friendly cloud-based solutions representatives of building sector keep on favouring siloed ICT solutions.

References

1. AAL Active and Assisted Living Programme (2014) Horizon 2020 Joint Programme; Article 185 www.aal-europe.eu. Brussels: European Commission (EU)
2. ABI (2012) SaaS in the Building Automation Market. ABI research. www.abiresearch.com/market-research/product/1013772-saas-in-the-building-automation-market/. Accessed 2 May 2016
3. Airaksinen M, Kokkala M (eds) (2015) Smart city—research highlights12. Espoo: VTT Technical Research Centre of Finland Ltd, pp 112–113. ISBN 978-951-38-8288-4, www.vtt.fi/publications/index.jsp. ISSN 2242-1181 (Online)
4. Andersson R, Engström D, Samuelson O, Stehn L (2014) Smart Built Environment Processes and information management in construction and facility management. Stockholm: IQ Samhällsbyggnad, the Swedish Centre for Innovation and Quality in the Built Environment, 30 p. www.iqs.se/~/media/IQS/Files/Projekt/SIO_2014/SIO_agenda_webb.ashx
5. Anon (2015) Smart Cities Readiness Guide. USA WA: Smart Cities Council LLC. http://readinessguide.smartcitiescouncil.com/scc-readiness-guide-online. Accessed 2 May 2016
6. Anon (2014) Espoon palveluväyläpilotti. Yhteenveto 2.5.2014. Helsinki: Kuntien Tiera Oy. www.kunnat.net/fi/palvelualueet/projektit/akusti/akustiprojektit/palveluvayla/Documents/AKUSTI_palveluv%C3%A4yl%C3%A4selvitys loppudokumentti_FINAL_HM_puolitoista.pdf. Accessed 14 Jan 2014 (in Finnish)
7. Anon (2016) Tron project. Wikipedia, the free encyclopedia. http://en.wikipedia.org/. Accessed 2 May 2016
8. Anon (1999) What office tenants want? BOMA/ULI office tenant survey report. Washington: the Building Owners and Managers Association (BOMA) International & the Urban Land Institute (ULI), 102 p
9. Anon (1992) The intelligent building in Europe. executive summary. A multi-client study. London: DEWG (Duffy Eley Giffone Worthington), Milan: Technibank in cooperation of The European Intelligent Building Group, 32 p
10. A national innovation strategy (2008) Government's communication on Finland's national innovation strategy to the parliament. Helsinki: The Ministry of Employment and The Economy, 42 p. www.tem.fi/en/innovations

11. van Berlo A (2014) The market for smart houses and services. A perspective on funding schemes and market breakthrough. The Netherlands: Smart Homes—Dutch expert centre on smart housing & smart living Eindhoven. http://ec.europa.eu/research/innovation-union/pdf/active-healthy-ageing/van_berlo_2014.pdf. Accessed 2 May 2016
12. Belisent J (2010) Getting clever about smart cities: new opportunities require new business models. Forrester Research. November 2
13. Botsman R, Rogers R (2010) What's Mine is Yours : The rise of collaborative consumption. Harper Business. ISBN-10: 0061963542 and ISBN-13: 978-0061963544. 304 p
14. Bowerman B, Braverman J, Taylor J, Todosow H, von Wimmersperg U (2000) The vision of a smart city. In: 2nd International Life Extension Technology Workshop. Paris
15. van Berlo A (2001) What's a smart home? Eindhoven, the Nederslands: Foundation Smart Homes. http://www.smart-homes.nl. Accessed 3 Apr 2001
16. Byfors J (2016) Opportunities of the digitalization for construction. Helsinki Finland: Tekes, Finnish Funding Agency for Innovation (Finland), Vinnova Sweden's innovation agency (Sweden). Eurostars Match. ICT solutions for smart cities. 5.4.2016. www.b2match.eu/eurostarsmatch/participants/145. Accessed 2 May 2016
17. Caragliu A, del Bo C, Nijkamp P (2009) Smart cities in Europe. Košice. In: Slovak Republic: 3rd Central European Conference in Regional Science—CERS (Centre for Research in Sociology), 2009, pp 45–59. http://www.inta-aivn.org/images/cc/Urbanism/background%20documents/01_03_Nijkamp.pdf. Accessed 14 Jan 2014
18. Chong O (2016) Slide show on smart grid connectors. In: Arizona State University: Confidential Research and Innovation Proposal (INTE-GRID)
19. Cohen B (2015) The smartest cities in the world 2015: Methodology. Fast Company, Co.Exit. http://www.fastcoexist.com/3038818/the-smartest-cities-in-the-world-2015-methodology. Accessed 2 May 2016
20. Cohen B (2014a) The smartest cities in the world. Fast Company, Co.Exit. http://www.fastcoexist.com/3038765/fast-cities/the-smartest-cities-in-the-world. Accessed 2 May 2016
21. Cohen B (2014b) The 10 smartest cities in europe. Fast Company, Co.Exit. www.fastcoexist.com/3024721/the-10-smartest-cities-in-europe. Accessed 2 May 2016
22. Davis G, Becker F, Duffy F, Sims W (1985) Orbit 2. Executive overview. In: USA: the Harbinger Group Inc. on the behalf of DEWG (Duffy Eley Giffone Worthington) and FRA (Facilities Research Associates), 51 p
23. De S, Barnaghi P, Bauer M, Meissner S (2011) Service modelling for the Internet of Things. In: 3rd Workshop on Software Services: Semantic-based Software Services (WoSS 2011)
24. Digital Agenda (2013) One of the seven pillars of the Europe 2020 Strategy for the growth of the European Union (EU) by 2020. The former the European Commission's the European eGovernment Action Plan 2011–2015. Brussels: European Commission (EU). http://ec.europa.eu/digital-single-market/en/europe-2020-strategy. Accessed 2 May 2016 and former http://ec.europa.eu/digital-single-market/en/european-egovernment-action-plan-2011-2015. Accessed 2 May 2016
25. Dockrill P (2015) Dubai announces world's first 3D-printed office building, and it looks amazing. The printer being used is 6 metres high! ScienceAlert.com.au. 1 Jul 2015. www.sciencealert.com/. Accessed 2 May 2016
26. Dohler M (2013) Smart Cities–The Untold Story: Mischa Dohler at TEDxLondon City 2.0. You Tube, 20th Dec 2013. www.youtube.com/watch?v=xUFUp-ylfC4. Accessed 2 May 2016
27. Duffy F (1983) The Orbit Study DEGW (Duffy Eley Giffone Worthington), EOSYS. Information Technology and Office Design, London, 103 p
28. Duhon B (1998), It's All in our Heads. Inform 12(8):8–13
29. EBOB (2004) Energy efficient behaviour in office buildings. The EU sixth framework programme project NNE5/2001/263; 2002–2004. www.ebob-pro.com
30. EEA (2004) The 2004 Urban audit data set (Eurostat). Brussels: European Environment Agency. http://www.eea.europa.eu/data-and-maps/data/external/urban-audit-database

31. EIP-SCC (2016) The European innovation partnership. Brussels: European Commission (EU). http://ec.europa.eu/eip/smartcities/. Accessed 2 May 2016
32. EIP-AHA (2015) The European innovation partnership on active and healthy ageing. Brussels: European Commission. http://ec.europa.eu/research/innovation-union/. Accessed 22 Jan 2015
33. ENSUF (2016) Joint call for proposals. Brussels: ERA-NET Cofund Smart Urban Futures (ENSUF). p 52. Smart cities ensuf_call_text_final2.pdf. Accessed 25 May 2015
34. Escolar S (2016) CitiSim, smart city 3D simulation and monitoring platform. A ITEA label winning research proposal for the ITEA 3 Call 2. Confidential project proposal
35. European Commission (2016) Internet of things call, H2020-IOT-2016-17. Brussels: European Commission (EU). http://ec.europa.eu/research/participants/portal/desktop/en/opportunities/h2020/calls/h2020-iot-2016-2017.html#c,topics=callIdentifier/t/H2020-IOT-2016-2017/1/1/1/default-group&callStatus/t/Forthcoming/1/1/0/default-group&callStatus/t/Open/1/1/0/default-group&callStatus/t/Closed/1/1/0/default-group&+identifier/desc. Accessed 2 May 2016
36. European Commission (2015) H2020-EE-07 Behavioural change toward energy efficiency through ICT. Call text. Brussels: European Commission. http://ec.europa.eu/research/participants/portal/desktop/en/opportunities/h2020/topics/ee-07-2016-2017.html. Accessed 2 May 2015
37. European Commission (2008) Intelligent Energy Europe. IEE programme. Brussels: European Commission (EU). http://ec.europa.eu/energy/intelligent/about/iee-programme/. Accessed 2 May 2016
38. European Commission (2003) Commission recommendation of 6 May 2003 concerning the definition of micro, small and medium-sized enterprises (Text with EEA relevance) (notified under document number C (2003) 1422). Official Journal L 124, 20/05/2003, pp 0036–0041
39. Expert Group "Innovation through Gender", Schiebinger L (Chair), Klinge I (Rapporteur) (2013) Gendered innovations. How gender analysis contributes to research. Brussels: 2013 Directorate General for Research & Innovation EUR 25848. Luxembourg: Publications Office of the European Union, 2013 ISBN 978-92-79-25982-1. doi:10.2777/11868. 2013. 144 p
40. Foster (2016) Foster + Partners. www.fosterandpartners.com/projects/hongkong-and-shanghai-bank-headquarters/. Accessed 2 May 2016
41. Frosti P (2016) Integration and data storages. Implementation of real time assets. Helsinki: Ministry of Finance. Taltio workshop, 7.3.2016. Accessed 2 May 2016
42. Gardner H (1983) Frames of mind. New York, USA. Basic Books. IBSN 0465025080; 9780465025084; 0465025099; 9780465025091, 440 p
43. Harald B, Frosti P, Ylipekkala H (2016) Interconnecting ecosystems. Helsinki: Aalto University. The annual Real-Time Economy conference: Innovations for Europe. 26–27.5.2016
44. Himanen M (2005) Towards the new criteria of elderly housing by the model of independent mobility. In: Kähkönen K, Sexton M (ed.). 2005 The 11th Joint CIB International Symposium Combining Forces. Advancing Facilities Management and Construction through Innovation. Proceedings of 11th Joint CIB (International Council for Research and Innovation in Building and Construction) International Symposium. June 13th—16th 2005. Helsinki: RIL (Association of Finnish Civil Engineers), VTT (Technical Research Centre of Finland), pp 2697. 978–952-5004-62-5952-5004-62-7. www.irbnet.de/daten/iconda/CIB6138.pdf. CIB DC6138; 2008(01):1000880; NLCIB, pp 662–676
45. Himanen M (2003) The intelligence of intelligent buildings. The Feasibility of the Intelligent Building Concept in Office Buildings. Espoo: VTT, p 497 ISBN 951 − 38 − 6038 − 8 VTT Publications 492. ISSN 1235 − 0621. http://www.vtt.fi/inf/pdf/publications/2003/P492.pdf http://aaltodoc.aalto.fi/handle/123456789/2505
46. Himanen M, Korhonen IE (forthcoming) Weak signals of the finnish resident-driven housing business—A stakeholder analysis. A manuscript to be published
47. Himanen M, Frosti P, Frosti S, Varjo T (2014) Knowledge management of open innovation in digital ecosystem building—role-based, situation-aware personalisation in smart real estate

business. The Interdisciplinary Studies Journal (ISJ). Special Issue on How to integrate open innovation and ICT technologies in service design and delivery within a Smart City, vol 3, No 4. 44–72. http://www.laurea.fi/dokumentit/Documents/ISJ_vol%203_no%204_web_Smart%20Cities.pdf
48. Himanen M, Brissman J, Lundberg S, Vastamäki R (2005) Forgiving technology in automated office buildings. In: Kähkönen K, Sexton M (2005) (ed). The 11th Joint CIB International Symposium Combining Forces. Advancing Facilities Management and Construction through Innovation. Proceedings of 11th Joint CIB (International Council for Research and Innovation in Building and Construction) International Symposium. June 13th–16th 2005. Helsinki: RIL (Association of Finnish Civil Engineers), VTT (Technical Research Centre of Finland). 2697 pp 978-952-5004-62-5952-5004-62-7. http://www.irbnet.de/daten/iconda/CIB6490.pdf
49. Hippel, E von (2005) Democratizing innovation. Cambridge, Massachusetts and London, England: The MIT Press, p 220, ISBN 0-262-00274-4
50. I.B.I (1988) The intelligent building definition. first edition. In: Chicago: Intelligent Buildings Conference. Intelligent Buildings Institute Foundation (I.B.I.), 18 p
51. ISO 37120 (2014) Standard. Sustainable development of communities. Indicators for city services and quality of life. Switzerland Geneva: International Organization for Standardization, p 116. ISBN 978 0 580 80716 9, http://shop.bsigroup.com/
52. Joinup.eu (2015) eGovernment in Denmark, January 2015, Edition 17.0. Brussels: the European Commission under the Interoperability Solutions for Public Administrations (ISA) in Europe Programme
53. Kruus M, Kiiras J, Hämäläinen A, Sainio J (2006) Managing the design and delivery processes of building services under construction management contracts. In: Eindhoven University of Technology (TU/e): International Conference On Adaptable Building Structures, Eindhoven the Netherlands 03–05 July 2006. pp 2–128. www.irbnet.de/daten/iconda/CIB10814.pdf. Accessed 2 May 2016
54. LeDoux J (1998) The emotional brain. the mysterious underspinnings of emotional life. Touchstone, New York, p 384
55. Lehto M (1990a) TRON House—datorstyrt småhus. VVS-FORUM. 12. Dec 1990. ss. 20–24. ISSN 0346 – 4644 (in Swedish)
56. Lehto M (1990b) TRON House—PC–styrt bostadshus. VVS värme- och sanitetsteknikern. 53. årgången. 3/1990. ss. 11–15. ISSN 0356 – 6439 (in Swedish)
57. Lehto M, Talonpoika R, Huovila P, Jantunen J (1993) Älykäs asunto—tietoyhteiskunnan koti. Espoo: Valtion teknillinen tutkimuskeskus. 177 s. (VTT Tiedotteita—Meddelanden—Research Notes 1457). ISBN 951 – 38 – 4351 – 3, ISSN 1235 – 0605 (In Finnish)
58. Manville C, Cochrane G, Cave J, Millard J, Pederson JK, Thaarup RK, Liebe A, Wissner M, Massink R, Kotterink B (2014) Mapping smart cities in EU study. Brussels: European Union, Directorate-general for Internal Policies. Policy Department A, Economic and Scientific Policy, p 196. ISBN: 978.92.823-4761-4. www.europarl.europa.eu/studies. Accessed 2 May 2016
59. Markets and Markets (2015a) Smart Building Market by Building Automation System (Physical Security, BEMS, Building Communication, Parking Management, Plumbing & Water Management, Elevators & Escalators Management), by Application & by Region—Global Forecast to 2020. India. www.marketsandmarkets.com/Market-Reports/smart-building-market-1169.html. Accessed 2 May 2016
60. Markets and markets (2015b) Smart Homes Market by Product (Energy Management System, Security & Access Control, Entertainment Control, and HVAC Control), Protocol and Technology (Protocol, Cellular Technology, and Communication Technology), Service (Installation, and Customization), and Geography (North America, Europe, APAC, and ROW)—Trend and Forecast to 2020. India: Markets and Markets, p 179. www.marketsandmarkets.com/Market-Reports/smart-homes-and-assisted-living-advanced-technologie-and-global-market-121.html. Accessed 2 May 2016

61. MF (2016) National architecture for digital services. Helsinki: Ministry of Finance. http://vm.fi/en/national-architecture-for-digital-services. Accessed 15 Jun 2016
62. Miller A. The magical number seven, plus or minus two some limits on our capacity for processing information. Psychological review. Am Psychol Assoc 101(2):343–352
63. MSAH (2012) Socially sustainable Finland 2020. The Strategy for social and health policy. Helsinki: The Ministry of Social Affairs and Health. Publications of the Ministry of Social Affairs and Health 2011; 6 (ISSN 1797-9854), p 24 ISBN 978-952-00-3136-7. [retrieved 14.9.2012]: http://urn.fi/URN:ISBN:978-952-00-3136-7. Accessed 15 Jun 2016
64. NBS (National Building Specification) (2013) National BIM Report 2013. Newcastle: Autodesk Revit, Bentley, Tekla and Vectorworks, 23 p. www.thenbs.com/knowledge/nbs-international-bim-report-2013. Accessed 15 Jun 2016
65. NBS (National Building Specification) (2015) National BIM Report 2015. London: RIBA Enterprises Ltd. NBS and RIBA are members of the BIM Technologies Alliance supporting the UK Government's Construction Strategy BIM Working Group, p 40. www.thenbs.com/knowledge/nbs-national-bim-report-2015. Accessed 15 Jun 2016
66. NSF (2016) National Science foundation. NSF Funding & Research Community. Virginia 22230, USA. www.nsf.gov/funding/. Accessed 2 May 2016
67. Nield D (2016) Singapore just launched a plan to fill the city with 3D-printed homes. The future is 3D-printed. ScienceAlert.com.au. 19 Feb 2016. www.sciencealert.com/. Accessed 2 May 2016
68. Nonaka I, Takeuchi H (2011) The wise leader: how CEOs can learn practical wisdom to help them do what's right for their companies—and society. Harvard Business Review (HBR), May 2011, Reprint R1105B
69. Nonaka I, Takeuchi H (1995) The knowledge creating company: how Japanese companies create the dynamics of innovation, Oxford University Press, New York, 284 p. ISBN 978-0-19-509269-1
70. Onaygil S, Güler Ö (2009–2010). Intelligent building systems. Istanbul University: Lectures on Intelligent Building Systems. Chapter 1. 2009–2011. http://web.itu.edu.tr/~onaygil/ebt614e/presentation_2.pdf. Accessed 20 Apr 2016
71. PAS 180 (2014) Standard. Smart cities. Vocabulary. The UK: A Publicly Available Specification (PAS) under British Standards Institution. A Publicly Available Specification (PAS), p 38. ISBN 978 0 580 81874 5. http://shop.bsigroup.com/. Accessed 20 Apr 2016
72. PAS 181 (2014) Standard. Smart City Framework. Guide to establishing strategies for smart cities and communities. The UK: A Publicly Available Specification (PAS) under British Standards Institution, p 60. ISBN 978 0 580 81856 1. http://shop.bsigroup.com/. Accessed 20 Apr 2016
73. PAS 182 (2014) Standard. Smart city concept model. Guide to establishing a model for data interoperability. The UK: A Publicly Available Specification (PAS) under British Standards Institution, p 64. ISBN 978 0 580 84320 4. http://shop.bsigroup.com/. Accessed 20 Apr 2016
74. PD 8100 (2015) Standard. Smart cities overview. Guide. The UK: A Publicly Available Specification (PAS) under British Standards Institution, p 40. ISBN 978 0 580 88061 2. http://shop.bsigroup.com/. Accessed 20 Apr 2016
75. PD 8101 (2014) Standard. Smart cities. Guide to the role of the planning and development process. The UK: A Publicly Available Specification (PAS) under British Standards Institution, p 56. ISBN 978 0 580 85247 3. http://shop.bsigroup.com/. Accessed 20 Apr 2016
76. PD ISO/TR 37150 (2014) Standard. Smart community infrastructures. Review of existing activities relevant to metrics. Switzerland Geneva: International Organization for Standardization, p 124. ISBN 978 0 580 83691 6. http://shop.bsigroup.com/. Accessed 20 Apr 2016
77. PD ISO/TS 37151 (2015) Standard. Smart community infrastructures. Principles and requirements for performance metrics. Switzerland Geneva: International Organization for Standardization, p 70. ISBN 978 0 580 87092 7. http://shop.bsigroup.com/. Accessed 20 Apr 2016

78. Realcomm Staff Writer (2002) Network Appliances—Connecting Buildings & Processes to the Net! USA CA: Realcomm Conference Group, LLC: Internet Appliance, vol 1 no 5. 04.16.2002. www.realcomm.com/advisory/advisory.asp?AdvisoryID=13
79. van Rooijen M (2013) A vision for the Dutch health care system in 2040. Towards a sustainable, high-quality health care system. Geneva Switzerland: the World Economic Forum in collaboration with McKinsey & Company Amsterdam the Netherlands, p 24. www.mckinsey.nl; http://www.weforum.org
80. Schaffers H, Komninos N, Pallot M, Aguas M, Almirall E et al (2012) Smart Cities as Innovation Ecosystems sustained by the Future Internet. Technical Report 2012, pp 65
81. Sinopoli JM (2010) Smart buildings systems for architects, Owners and Builders. USA: Elsevier, p 225. ISBN-13: 978-1856176538, ISBN-10: 1856176533
82. Tang T, Wu Z, Hämäläinen M, Ji Y (2012a) From web 2.0 to living lab: an exploration of the evolved innovation principles. J Emerg Technol Web Intell 4(4):379–385
83. Tang T, Wu Z, Hämäläinen M, Ji Y (2012b) Internationally distributed living labs and digital ecosystems for fostering local innovations in everyday life. J Emerg Technol Web Intell 4 (1):106–115
84. Thulin S (2016) Mobility-as-a-Service for the Networked Society. Helsinki Finland: Tekes, Finnish Funding Agency for Innovation (Finland), Vinnova, Sweden's innovation agency (Sweden). Eurostars Match. ICT solutions for smart cities. 5.4.2016. www.b2match.eu/eurostarsmatch/participants/145. Accessed 20 Apr 2016
85. Toivanen M (2006) Sähköisten asiointipalvelujen kehittäminen kunnissa. Tampere: Tampere University, Acta Universitatis Tamperensis. p 1156. Acta Electronica Universitatis Tamperensis; 533. (In Finnish with English summary)
86. Toivanen M (2013) Miten kansalaisten sähköisten palvelujen käyttöönoton kynnystä voidaan madaltaa? In seminar: Tulevaisuuden sähköiset sosiaali- ja terveyspalvelut kunnissa. Helsinki: The National Institute for Health and Welfare (THL) and the Association of Finnish Local and Regional Authorities. URL:http://tapahtumakalenteri.thl.fi/tapahtuma/13945 (In Finnish)
87. Tuomi I (1999) Corporate knowledge: theory and practise of intelligent organizations. University of Helsinki, Metaxis. Helsinki, Finland, 453 p. ISBN 951-98280-0-1
88. UNIDO, Year unknown. Energy efficiency in buildings. Austria Vienna: the United Nations Industrial Development Organization (UNIDO), Sustainable Energy Regulation and Policy-making for Africa. Module 18, 120 p. Available at www.unido.org/fileadmin/media/documents/pdf/EEU_Training_Package/Module18.pdf
89. Vestbro D U (2010) Living Together – Cohousing Ideas and Realities around the World. Proceedings from the International Collaborative Housing Conference. Stockholm: Royal Institute of Technology
90. Vision Mobile (2012) App market forecasts 2013–2016. www.visionmobile.com/blog/2013/07/developer-economics-app-market-forecasts-2013-2016/. Accessed 20 Apr 2016

Chapter 14
SPHERE: A Sensor Platform for Healthcare in a Residential Environment

Przemyslaw Woznowski, Alison Burrows, Tom Diethe, Xenofon Fafoutis, Jake Hall, Sion Hannuna, Massimo Camplani, Niall Twomey, Michal Kozlowski, Bo Tan, Ni Zhu, Atis Elsts, Antonis Vafeas, Adeline Paiement, Lili Tao, Majid Mirmehdi, Tilo Burghardt, Dima Damen, Peter Flach, Robert Piechocki, Ian Craddock and George Oikonomou

14.1 Introduction

It can be tempting to think about smart homes like one thinks about smart cities. On the surface, smart homes and smart cities comprise coherent systems enabled by similar sensing and interactive technologies. It can also be argued that both are broadly underpinned by shared goals of sustainable development, inclusive user engagement and improved service delivery. However, the home possesses unique characteristics that must be considered in order to develop effective smart home systems that are adopted in the real world [37].

The home is the quintessential personal space and, therefore, there is a greater expectation of privacy at home than in public spaces. People are likely to behave differently when they are at home, in the knowledge that information about their behaviour belongs only to themselves and perhaps others who share this environment with them. But these lived experiences in real-life settings are of great interest to many research fields, in particular those aiming to improve well-being and healthcare provision. Studies conducted in living labs and prototype smart homes can produce information concerning system functionality and usability, but they represent a compromise in terms of the complexity of everyday life. The shift of smart home technologies into real-life contexts has begun to expose a number of

P. Woznowski (✉) · A. Burrows · T. Diethe · X. Fafoutis · J. Hall · S. Hannuna · M. Camplani · N. Twomey · M. Kozlowski · B. Tan · N. Zhu · A. Elsts · A. Vafeas · A. Paiement · L. Tao · M. Mirmehdi · T. Burghardt · D. Damen · P. Flach · R. Piechocki · I. Craddock · G. Oikonomou
Faculty of Engineering, University of Bristol, Merchant Venturers Building, Woodland Road, Bristol BS8 1UB, UK
e-mail: p.r.woznowski@bristol.ac.uk

V. Angelakis et al. (eds.), *Designing, Developing, and Facilitating Smart Cities*,
DOI 10.1007/978-3-319-44924-1_14

barriers to a wider adoption of these systems, including high cost of ownership, inflexibility, poor manageability and difficulty achieving security [9].

In reality, homes are very dynamic environments. They vary in typology as well as layout, and can change sporadically as residents refurbish interior spaces. This can create a variety of challenges ranging from the initial system installation to signal propagation and data analytics. Homes contain material possessions, but they also have personal meaning for the people who live in them. They can be home to a single person or to multiple people, with different individual characteristics. If we think about several generations living under a single roof, it is likely that each individual has different abilities and motivations in terms of interacting with a smart home system. Previous research about how people create a smart home identified three roles of household members in relation to their smart homes: home technology drivers, who were primarily involved in planning and maintaining the technology; home technology responsibles, who did not deal with the technology directly but wanted it installed and would outsource necessary repairs and adjustments; and passive users who were removed from any phase of the home automation, but had learned the basics about the system by using it [36]. Overall, people have a greater expectation of control over technology at home than in public spaces. The home can also be a place of conflict and negotiation between the people who live there, which can affect how domestic technologies are adopted and used.

The need to restructure healthcare services is widely acknowledged and has led to the home being viewed as a key setting for health and care. Perspectives on the role of the patient in preventing and managing chronic illness have shifted from the self-management approach of conventional medicine towards an approach that sees patients, healthcare professionals and the wider community working together to develop holistic and personalised care plans. Within this context, smart home technologies have emerged as one way to empower patients to actively engage in the management of their well-being. Achieving a system that is technically feasible and clinically effective requires a multidisciplinary approach that combines the expertise of various stakeholders, including end users who will steer the design towards an acceptable outcome. The domestic end users of such a smart home system may comprise healthy individuals, individuals who are living with one or more chronic health conditions, and individuals who experience an acute disease or injury. Furthermore, people's health and care needs change suddenly or progressively throughout their lives.

The challenges of developing smart home technologies for health and care become evident as we begin to break down the various facets of the home and the diversity of its residents. This type of rich understanding of potential users has been used to develop inclusive design resources to inform the development of appropriate smart home technologies for health and care [10]. The remainder of this chapter details the development of a smart home system, which was designed to be retrofitted into real homes. We begin with a description of the SPHERE project, under which this research was conducted.

14.1.1 Overview of SPHERE

SPHERE (Sensor Platform for Healthcare in a Residential Environment) is an EPSRC-funded interdisciplinary research project, led by the University of Bristol in collaboration with the University of Reading and the University of Southampton. The overall aim is to develop a smart home platform of non-medical networked sensors, capable of gathering and integrating multiple types of data about the home environment and the behaviours of its residents to better understand a range of healthcare needs. Rather than targeting subsets of the population based on demographics or health conditions, the project takes an inclusive approach with a view to generating rich data sets. We anticipate that analysis of these data sets will generate evidence-based insights into factors that affect health and well-being, thus informing more appropriate and effective interventions.

The system comprises various sensors that can be broadly grouped into three categories: environmental sensors, which monitor temperature, humidity, luminosity, noise level, air quality, room occupancy, electricity metering and cold and hot water consumption; vision sensors, which are able to track people and provide information about quality of movement; and wearable sensors, primarily a low-power wrist-worn device that uses accelerometers to measure patterns of movement. A prototype of this system is installed in a two-bedroom Victorian residential property in Bristol, which serves as a test house for short- to long-term user studies. We felt it was important to test the system in a realistic setting, in a familiar and otherwise unremarkable domestic environment. Among other things, this has allowed researchers to experience some of the technical challenges of retrofitting this system into real homes. Once this system has been thoroughly tested and iterated, we plan to deploy it in up to 100 homes in Bristol for long-term and 'in the wild' studies.

14.2 Enabling Technologies

This section briefly describes the different sensing modalities used by the SPHERE system, namely, the wearable sensor, the environmental sensors and the video monitoring system.

14.2.1 The Wearable and Environmental Sensors

The SPHERE architecture monitors the residential environment using a custom environmental sensor board, named SPES-2. SPES-2 (shown in Fig. 14.1) is a battery-powered sensor board that is based on the Texas Instruments CC2650 System-on-Chip (SoC) for processing and wireless communication. The CC2650 is a multi-standard 2.4 GHz wireless system; it supports Bluetooth Low Energy (BLE) and IEEE 802.15.4. A meandered 2.4 GHz monopole antenna is printed on

Fig. 14.1 The SPHERE environmental sensor

the board. For processing, the CC2650 incorporates an ARM Cortex-M3 microcontroller unit (MCU).

The SPES-2 incorporates a series of sensors for monitoring the residential environment. These include: a temperature and humidity sensor (HDC1000); a light sensor (OPT3001); a barometer (BMP280); a passive infrared motion sensor (EKMB1101); and a microphone (SPH0641LU4H-1) used for noise level sensing. Moreover, SPES-2 exposes an interface for connecting external sensors. Any low-power analogue or digital 3.3 V sensor is compatible. The board is powered by a 3 V CR2477 coin cell battery (typical capacity of 1000 mAh). The physical dimensions of the board are 75 × 75 × 1.6 mm, enclosed in an off-the-shelf casing (dimensions 85 × 85 × 25 mm). In addition to the in-house developed environmental sensors, the SPHERE system also incorporates a commercial electricity monitoring system by CurrentCost.

The SPHERE infrastructure also incorporates a custom activity tracker. The SPW-2 (Fig. 14.2) is a wearable sensor board, mounted on the user's wrist. Similar

Fig. 14.2 The SPHERE wearable. *Left* While charging. *Right* Physical dimensions

to the environmental sensor board (SPES-2), SPW-2 is based on the CC2650 multi-standard 2.4 GHz SoC. Similarly to its predecessor [15], it employs two ADXL362 accelerometers. The accelerometers are separated by 30 mm, and are aligned with the user's limp in such a way, so that differential measurements provide rotational information on the limp's movements [57]. Additionally, the board incorporates a low-profile 2.4 GHz inverted-F antenna, and a flash memory of 512 MB for temporary data storage, whilst the user is out of the house.

The board is powered by a 100 mAh Lithium–Polymer (Li–Po) rechargeable battery (3.7 V). A Qi compatible inductive contact charging circuit [27] is incorporated to wirelessly charge the battery. Commercial off-the-shelf charging pads are compatible. The physical dimensions of the board are 39 × 20 × 1.6 mm, enclosed in a custom wristband (dimensions 41 × 22 × 8 mm).

The final component of the SPHERE infrastructure for environmental and body sensing is a mains-powered gateway, named SPG-2. SPG-2 employs two CC2650 sub-systems, which can be used either for simultaneous BLE and IEEE802.15.4 support or for implementing antenna diversity on the same standard to improve wireless coverage, as proposed in [17].

The gateway board incorporates a subset of the sensors of the environmental sensor. These include: a temperature and humidity sensor (HDC1000); a light sensor (OPT3001); a barometer (BMP280); and a microphone (SPH0641LU4H-1) used for noise-level sensing. The board is mains powered via USB and the same interface can also be used for programming and software debugging. Its physical dimensions are similar to those of the SPES-2.

14.2.2 Video Monitoring

The video monitoring component of the SPHERE system is a real-time multi-camera system, which is tasked not only with tracking individuals navigating their home environment, but also with providing continuous quality of movement information. Tracking information is provided in the form of 3D bounding boxes in world coordinates and quality of movement information is currently in the form of the log-likelihoods of a particular movement being 'normal' [43].

In the SPHERE platform, integration with other sensing modalities, user acceptance and deployment budget are primary considerations. To make deployment into the local community financially viable, it has been necessary to limit hardware selection to the low cost consumer RGB-D camera, Asus Xtion.[1] This camera needs to be coupled with a machine with suitable processing capacity, minimal intrusion to the user and minimal cost. The Intel Next Unit of Computing (NUC) with an i5 processor and 8 GB of RAM fulfils these requirements (Fig. 14.3).

[1]https://www.asus.com/3D-Sensor/Xtion/.

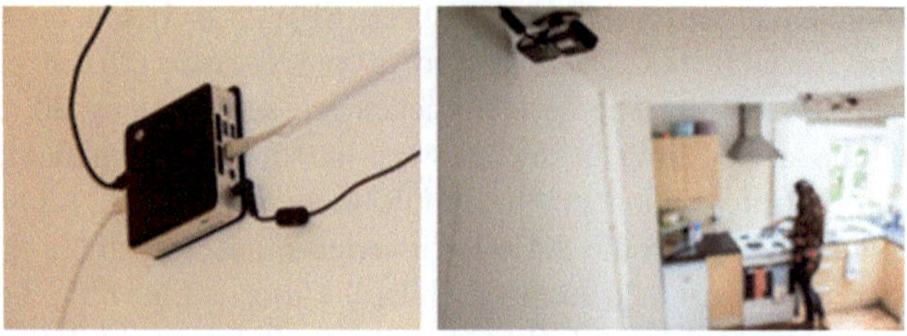

Fig. 14.3 NUC and Asus Xtion sensor in the SPHERE house

The small size and relatively low cost of the NUC, when compared to most workstations, allows it to be strategically placed in close proximity to other sensors. Another key feature of the NUC is its four USB 3.0 ports, which provide enough bandwidth to capture from four depth cameras simultaneously. This gives the system far more flexibility in deployment. Specifically, a range of configurations is possible: one NUC operating all of the cameras, one NUC per camera, or some configuration in between, depending on user preference as well as the specific circumstances of each individual deployment.

Another critical consideration for all of the SPHERE sensors and particularly the video subsystem is installation overhead and long-term reliability. Consider that SPHERE currently plans to deploy the system into up to 100 homes in the local community. With up to three NUCs and ASUS Xtions per home, the setting up and management of such a large enterprise rapidly becomes intractable without streamlined installation protocols and extremely reliable subsystems.

In light of these requirements, each of the NUCs runs GNU/Linux with the video system configured as a service, which is automatically launched when the machine boots. This makes the video system resistant to temporary power loss. A standardised system image is used to configure the system so that a single NUC can be unboxed and setup ready for deployment in around 10 min.

As mentioned previously, software reliability is critical to facilitate extended operational periods without being physically accessed. To aid this, we have used object-oriented design principles and common software design patterns where appropriate.

To support the collection of video data we have designed a pair of classes which capture all of the functionality needed in the video system: the Camera class and the CameraObserver class. The first of these encapsulates camera functionality providing depth, colour, bounding box and skeleton information. Instances of the CameraObserver class register themselves with a Camera to be notified when a new frame is available, with its associated data. The observer is then free to choose what to do with the data. This provides the system with a unified method to implement specific video experiments without understanding the underlying hardware configuration.

To integrate the video system into the SPHERE platform, we implemented a subclass of the CameraObserver which takes the video data, excluding the frames themselves, and serialises them into a JSON string. These strings are then transmitted over MQTT (Message Queue Telemetry Transport) protocol on the appropriate topic. This allows the video system to act as an IoT device to the rest of the SPHERE sensing platform.

Even considered in isolation, the output of the video system is a rich source of information. The velocity, aspect ratio and location of the bounding boxes provide important clues for measuring activities and behaviours. For example, when combined with environmental contextual information, such as the location of kitchen appliances, activities such as cooking, washing up and watching television can potentially be identified. Moreover, it may also be possible to determine general activity levels including the amount of time spent sedentary. Individuals are tracked using state-of-the-art algorithms such as [38]. The bounding boxes obtained may be further exploited to yield coarse estimates of human pose. For example, an approximately square bounding box is indicative of sitting, whilst a vertical elongated rectangle implies standing.

Quality of movement assessment is based on the skeletons provided by the PrimeSense[2] middleware. These are normalised for the global positioning and orientation of the camera and height variation. The relatively high dimensionality of the normalised skeletons is reduced using a modified version of Diffusion Maps [11], where Gerber's [22] method for addressing outliers in Laplacian Eigenmaps is exploited. The resulting high level feature vector, obtained from the normalised skeleton at one frame, represents individual poses and is used to build a statistical model of normal movement. Abnormal movement patterns are detected by their deviation from this model.

14.3 Overall System Architecture

The SPHERE system will be deployed in up to 100 homes in and around Bristol for long-term and 'in the wild' studies. Each deployment will consist a number of environmental, wearable and video sensors (Sect. 14.2), accompanied by a number of devices required for (i) data storage, (ii) network connectivity among sensors, and (iii) system management and monitoring.

The SPHERE system also consists a back end (called the "SPHERE Data Hub"), which is made up of a number of virtual machines, servers and storage devices physically situated at the University of Bristol. The Data Hub is used for data analytics and it also provides a system administration dashboard. It will lastly be used for long-term storage of all data collected from the 100 properties.

Each deployment property will be connected with the Data Hub through a secure Virtual Private Network (VPN) over a 3G, 4G, or fixed broadband link. On the deployment property, this VPN tunnel will terminate on a device called 'SPHERE Home Gateway', an Intel NUC PC identical to the one used by the video monitoring subsystem (Sect. 14.2.2). The overall system architecture is illustrated in Fig. 14.4.

[2]http://www.i3du.gr/pdf/primesense.pdf.

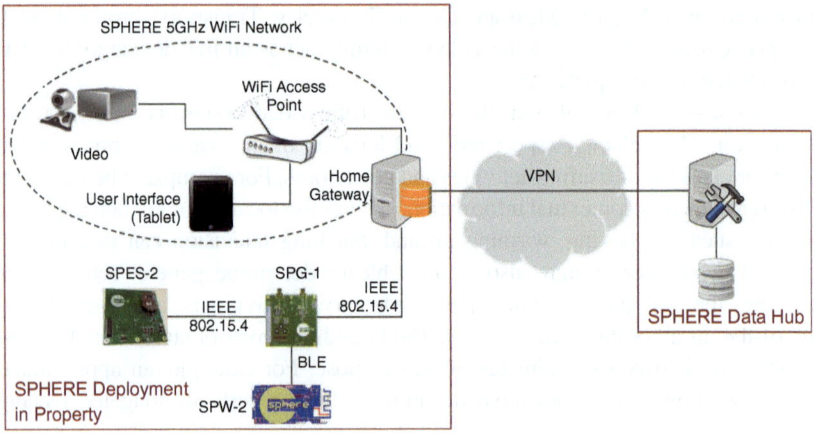

Fig. 14.4 Overall architecture of the SPHERE system

In addition to terminating the VPN tunnel between the University and the property, the Home Gateway serves a number of purposes:

- It provides a reliable, redundant and secure data storage medium for sensor data collected by the wearable, environmental and video sensors,
- It provides a time synchronisation source for all other SPHERE systems in the property,
- It hosts a dashboard that can be used by participants to visualise data, monitor and manage (e.g. start or stop) the system. The dashboard is presented to participants over a web-based interface through a pre-configured tablet computer.

The SPHERE system generates data that falls within one of two categories:

- **Sensor Data**: Sensor measurements collected by the wearable, environmental and video sensors. This category, for example includes wearable acceleration data, ambient light levels, environmental temperature, water or electricity usage, presence detection data and video bounding boxes.
- **Control and Monitoring Data**: Data used to monitor the system's overall state and to manage its individual components. This category includes, for example, network statistics, version of the software running on each device, device uptime, battery levels for battery-powered devices and more. In terms of managing the system, data within this category can be used to, for example, stop and restart the system or to install software updates.

Noteably, *Sensor Data* are stored on a Mongo database hosted at the Home Gateway, but—due to the sheer volume and also for security and data privacy purposes—they are *not* transmitted to the Data Hub over the network. Conversely,

Control and Monitoring Data are transmitted to the Data Hub over the VPN link, in order to allow remote system management, administration, monitoring and generation of alerts to notify about system malfunctions.

In terms of network connectivity, a SPHERE deployment is conceptually broken down into two logical segments within each property. The wearable and environmental sensors communicate among themselves and with the Home Gateway using a combination of BLE and IEEE 802.15.4 wireless links. Both of those technologies operate in the 2.4 GHz radio frequency band. A 5 GHz WiFi network is used to ensure communications among the Home Gateway, Video NUCs and the tablet that provides users with the SPHERE User Interface. The choice of using the 5 GHz band was made for two reasons: (i) increased network performance, which is required to accommodate the large volume of video data; and (ii) prevention of interference with the 2.4 GHz low-power networking used by the wearable and environmental sensors. The entire SPHERE system is thus fully isolated from the user's own home network.

At the application layer, the SPHERE system makes extensive use of the MQTT protocol for data collection as well as for system monitoring, with an MQTT broker installed on the Home Gateway at every property. An MQTT client at the Data Hub is used to collect monitoring data and to issue control commands to all systems. Lastly, an MQTT client is used on every Video NUC in order to: (i) publish sensor data to the Home Gateway, (ii) publish monitoring data to the Home Gateway and the Data Hub, and (iii) receive management commands from the Home Gateway or the Data Hub.

For data collection from and management of the environmental sensor (SPES-2), we use the Constrained Application Protocol (CoAP) instead. Translation between MQTT and CoAP is achieved with a SPHERE-developed application layer proxy running on the Home Gateway. Lastly, the SPHERE wearable uses BLE to communicate with the rest of the infrastructure via an SPG-2 device. This SPG-2 encapsulates wearable sensor and monitoring data into CoAP packets before submitting them to the Home Gateway for processing and storage. The inverse process is followed in order to send control requests to the wearable.

14.4 Data Analytics and Interpretation

Ambient Intelligence (AmI) spaces process large quantities of *sensor data* and require robust and accurate Activity Recognition (AR) strategies. This is needed to infer activities of interest to monitor health, well-being or other personal benefits such as fitness level. A typical approach to the problem is to define a hypothesis to test which informs experimental design. Quantification of the hypothesis can be done either by simulation or by processing real sensor data. Experimental design also involves the selection of appropriate data collection methods, ground truth acquisition methods and annotation strategies.

14.4.1 Ground Truth

The validation of AR strategies involves comparing their output against ground truth/benchmark data. But before this can take place, this data has to be acquired. Ground truth data acquisition involves three stages: collecting data upon which to base the ground truth, deciding what labels are appropriate to describe the data and applying these labels annotating the data to obtain the ground truth. These stages are often developed as a result of an iterative process to determine the best data collection methods, most appropriate labels to use, and a suitable way to annotate the data to produce consistent and informative ground truth.

Obtaining annotations from self-reported diaries is imperfect as they rely purely on participant's compliance and their subjective perception and memory, which in general, becomes inaccurate with time. This is particularly the case with annotation which requires accurate temporal precision in order to be maximally effective, so it is unrealistic to expect detailed activity diaries with the exact timings of the activities. Researchers approach this problem in many different ways. Allen et al. [1] collected unsupervised activity data in the home using a computer set up to take participants through a routine. Input from the user was in the form of a button press from which the data was annotated. Van Kasteren et al. [52] asked participants to wear a Bluetooth headset that used speech recognition to label ground truth data. This method is inexpensive, but is of limited utility because it does not capture enough detail and contextual information.

Another strategy is for the researcher to record the activity and context during data collection [35, 42]. Pärkkä et al. [42] adopted a semi-overseen approach to collecting data for AR classification based on realistic activities. A single researcher followed the participant during the experiment and used an annotation app to record the activities. Even with this approach it was noted that there were annotation inaccuracies that were most likely correlated to predictive inaccuracy.

Methods using video recordings provide an objective reflection of participant's activities enabling a far more accurate and detailed activity ground truth data, however, these require additional attention in the form of ontology. For example, Atallah et al. [2] used video to annotate activities during laboratory experiments to train AR classifiers. Data can also be collected in an unobserved environment, encouraging natural behaviour; however, it can also be perceived as intrusive and will only capture actions with no room for participant interpretation.

Video annotation can be costly and time-consuming. Active learning is a technique that can reduce the amount of annotated data needed for training a classifier. In this approach, classifiers are to be trained with a minimal set of annotated data. Active learning algorithms attempt to quantify the utility of obtaining labels for new instances tensioned against the financial cost of querying an oracle for the true label. Only the instances that are deemed to yield maximal utility are selected. In particular, when coupled with transfer learning techniques (i.e. transferring knowledge from different contexts to a new context), active learning can dramatically reduce the quantity of labelled data required [12]. Hoque and Stankovic [26] employed a clustering technique to group smart home environmental data into activities and the user labelled the clusters. Another application for active learning, is to update classifiers or personalise them [33]. While attractive, these methods rely on a collaborative effort on the part of the user.

There are a number of available software tools which are suitable for video annotation, such as the ANVIL video annotation tool [29] or ELAN [8] developed by the Max Planck Institute for Psycholinguistics, The Language Archive, Nijmegen, The Netherlands.

The labels used to annotate data is another annotation consideration. Labels are often application specific, e.g. [42] used a hierarchical list of labels, aimed at capturing the context of the activities, whereas [49] focused purely on a specific disease and the associated symptoms. Logan et al. [32] presented a detailed activity ontology for the home using a custom tool that enabled annotators to label foreground and background activities for when the participant's attention is focused on another activity, addressing the fact that humans naturally multitask. Roggen et al. [44] used a four 'track' annotation scheme for annotating human activities based on video data including tracks for locomotion, left- and right-hand activities (with an additional attribute that indicates the object they are using), and the high-level activity.

In the SPHERE project, machine learning algorithms are initially trained and validated against recordings from a head-mounted video camera worn by participants. The data originates from the three different sensing modalities (depth cameras, wearable accelerometer and environmental sensors) deployed in a real house, which constitutes the testbed [56, 58]. The same SPHERE ontology of ADLs, underpins system-generated activity data and the controlled vocabulary used in video annotation.

14.4.2 *Activity Recognition*

In order to make instantaneous or longitudinal inferences about the health status of individual residents, a necessary first step is to be able to recognise normal Activities of Daily Living (ADL). To this end, we need a framework that allows us to take the inputs from multiple heterogeneous sensor sources, such as those described in Sect. 14.3, and make informed decisions that are tailored both to the individual and the context. Naturally, the SPHERE setting presents many sources of uncertainty. First, we are dealing with multiple sensor modalities (environmental, body-worn, video), each of which will have different noise profiles and failure modes. Second, as described in Sect. 14.4.1, we are dealing with a situation where annotated or labelled data is expensive and intrusive to acquire, and the resulting labels are potentially noisy and inaccurate (indeed in some cases there may be no *ground truth* in the classical sense, and we need to resort to modelling annotator disagreement explicitly). Lastly, patterns of human behaviour are subject to many factors (internal and external) that may or may not be attributed to the particular health context of a given individual.

Faced with such a situation, the most sensible approach would be to use *white box* modelling methods where possible. Model-based machine learning [6, 54] attempts to follow this ideal by encoding assumptions about the problem domain explicitly in the form of a model. Indeed, the model can be viewed simply as this set of assumptions, expressed in a precise mathematical form. These assumptions include the number and types of variables in the problem domain, which variables affect each other, and what the effect of changing one variable is on another variable. The result is that any decisions made by the system can be inspected, so that if the model is performing poorly, the solution is to re-examine the assumptions being made.

In the Bayesian paradigm, degrees of belief in states of nature are specified through the use of probabilities, which through the construction of probabilistic graphical models [30] allow us to apply a principled mathematical framework of the quantification of uncertainty to perform model-based machine learning. On the basis of the models we build, Bayesian decision theory tries to quantify the trade-off between various decisions, making use of probabilities and costs [3, 4].

A typical problem that we face is that the differences between individuals are too large to be captured by a single model. Hierarchical Bayesian models [21] allow us to simultaneously generalise over communities of residents whilst also learning personalised models. In addition, they allow us to be more flexible with our priors, by specifying *hyper-priors*, and then performing inference over the priors instead.

Adapting to multiple operating contexts

However, transparently dealing with degrees of belief does not solve all modelling challenges posed by the SPHERE project. Our models and inferences have to be applied in multiple contexts, and indeed any given context is liable to both gradual and abrupt shifts. Let us consider the example of modelling daily patterns of behaviour. A common approach for coping with the temporal aspect of daily patters is to introduce an *hour of day* feature to the classification model [28, 51]. However, when summarising the temporal nature of an activity into a coarse feature such as this, not only is information lost after discretisation, but also the strength of the periodicity of the action is ignored. Bayesian approaches have been ascribed to such periodic data [13] and these models can not only capture the complex multi-modal aspects of periodic patterns, but the resulting posteriors are interpretable and may be studied to increase practitioners understanding of the nature of their data.

In such situations, it will be crucial that we are able to trust the probabilities coming from the system. A machine learning system is well *calibrated* if the predicted probabilities it gives correspond to observed frequencies. This is natural in forecasting where we would expect it to rain in 60 % of days where a weather forecaster predicts a 60 % chance of rain [39] but carries over to machine learning as well. If a system is poorly calibrated then it suggests a problem either in the model (such as an overly restrictive assumption) or in the inference.

Different operating contexts call for different performance metrics, which perhaps incorporate a different notion of expected loss [25]. If the goal is to minimise loss, for example for the case of classification, a systematic approach would be that given a model, threshold choice methods that correspond with the available information about the operating condition should be applied, followed by comparison of their expected losses.

The operating contexts of smart environments can be affected by many factors. For example, it is well known that when multiple occupants reside in a smart home, models that do not adapt to these contexts will generally yield poor predictions. This happens because 'confusing' sensor data can arise when co-occurring activities are performed in different locations of the home. However, if the topology of the residence can be learnt, not only will predictive performance be boosted, but predictive confidence and calibration will likewise be improved [16, 50].

Different classification performance metrics such as F-score also imply a different notion of calibration [18]. More generally, the choice of the performance metrics in use should be seen as another modelling assumption rather than being independent from the model. Given that we expect the end users of our systems to include medical professionals as well as the residents themselves, we can easily see how the types of decision we would want to surface should be adaptable.

Explicitly modelling context change also favours domain adaptation and model reuse. We are building on the results of the REFRAME project,[3] which developed a general methodology for model reuse in machine learning called *reframing* (Hernández-Orallo, Prudêncio et al. (in press) [24]). The setting is exemplified by the recent ECML-PKDD'15 Discovery Challenge *MoReBikeS: Model Reuse with Bike rental Station data,*[4] which encouraged participants to build predictive models for new bicycle rental stations making use of previously trained models on other stations (for which the training data was however no longer available).

14.4.3 Localisation

Research [31] has shown that human activities in residential areas are highly correlated with their corresponding locations. Activities pertaining to SPHERE's research interests mostly occur indoors, thus rendering the traditional global navigation satellite systems like GPS or Galileo is redundant. Instead, a localisation solution which can indicate the relative indoor position is essential for future research opportunities.

Academia and industry have both tried various approaches to tackle indoor localisation problem. GE[5] and Philips[6] offer the LED-based indoor localisation

---

[3]http://reframe-d2k.org/.

[4]http://reframe-d2k.org/Challenge.

[5]http://www.gelighting.com.

[6]http://www.lighting.philips.co.uk/systems/themes/led-based-indoor-positioning.html.

system for use in retail outlets or hospitals. Video systems, such as MS Kinect or Intel RealSense, can also provide tracking information when the target is recognised within the effective region, but only if proper light condition is preserved. In scenarios where accurate location information is not necessary, the outputs from passive infrared [40] or sound sensors [5] are used as an indication of the subject's presence in certain areas. In addition, accelerometer and gyroscope [23, 55] data are also used for keeping track of the subjects, if the initial location is known—this method is known as one embodiment dead reckoning. It cannot, however, be considered an ideal solution to provide indoor location information yet, much like GPS system is now, for outdoor applications. The above approaches are used for different purposes in different contexts, and as such have drawbacks and limitations. These include arduous installation and deployment, low accuracy, high cost, limited coverage, and depending on the mentioned differing contexts, intruding upon subject's privacy. Thus, in relation to SPHERE, other indoor localisation methods are considered.

Nowadays, with an ever-increasing use of wireless systems, radio frequency (RF) signals are present everywhere penetrating all living spaces, including residential homes. Modern RF receivers allow for the parameters such as time delay, power strength, profile distortion, and even experienced reflections, to be accurately measured. These parameters can be used to estimate the distance between the transmitter and receiver. The distance estimations from spatially differentiated locations then are used for localising the target by triangulation or multilateration. In Bose and Foh [7] and Wang et al. [53] received signal strength (RSS) based ranging methods are described in detail. Generally an RSS-based method requires a relatively simple propagation environment, in order to avoid signal superimposition caused by multipath propagation. Thus, a high-density residential area creates a very challenging scenario for the application of this approach. Time is another parameter that can be used to indicate the distance. In both, academia and industry, researchers have tried to extract timing by specially designed wide band signals, such as those shown in Sahinoglu et al. [1, 19, 45]. Günther and Hoene [20] and Ciurana et al. [34] introduced a round trip measuring method based on the ACK mechanism of the 802.11 protocol with standard commercial off-the-shelf (COTS) devices. This method avoids multipath propagation problems, but is still limited by low clock resolution and stability of COTS devices. Exel [14] presents an improvement to the clock resolution problem by extracting the wireless communications signals from a dedicated receiver—this, however, is much more expensive than using COTS devices. By now it is apparent, that the above mentioned methods are either too expensive or inaccurate for widespread deployment.

Regarding the applications and requirements of SPHERE, the technology which can leverage existing wireless signals and cheap COTS equipment are very much preferred. Also, differing from the industrial applications such as the storehouse or assembly line robotics, which have hard requirements on localisation accuracy and resolution, there are no exact firm requirements, when referring to human activity research. Therefore, the development of the indoor localisation system for SPHERE includes two stages: the *premier* stage and the *advanced* stage. The Premier stage takes advantage of resources already deployed in the house—for example Bluetooth (BLE) access points (APs) and PIR sensors in each room. This provides us with room-level location information given, for example the optimum amount of BLE AP's spread around the house. According to the SPHERE's signal propagation research, three access points in the house at any one time are sufficient for this purpose [48]. Room-level information can provide only limited data to differentiate between activities. For instance, the detection of presence in the living room is associated with activities such as watching TV, reading, or chatting, and excludes actions not normally associated with this location, such as making tea, washing or taking shower. However, room-level information is still not fine-grained enough to aid recognition of some ADLs. In the context of multiple sensor platforms used in the SPHERE system, the localisation can be further strengthened by fusion of the other sensors' data—for example the on-body accelerometer, video or even electrical and water metres. Another approach is to deploy extra RF sensors (BLE in the SPHERE's context) in the house in order to better calculate the location information (*advanced* stage). The aim for this stage is to provide between 1×1 m^2 and 1.5×1.5 m^2 resolution location information based on the high density of RF receivers.

The implementation of the premier and advanced stages of the indoor localisation in the SPHERE prototype testbed are as follows:

Premier Stage: The SPHERE localisation system in this stage includes one custom wearable [15] and three distributed receivers [17]. The wearable broadcasts BLE advertisements in channel 37, 38 and 39 with 4 dBm emission power. Each receiver is equipped with one horizontal polarised folded dipole antenna and one vertical polarised folded dipole antenna for error correction purpose. On each receiver, wearable BLE advertisements are sniffed, timestamped, and subsequently saved into the database. In the database, the RSSs of the same BLE advertisements captured by different receivers are extrapolated together using the timestamps. Not all BLE advertisements can be received successfully by all receivers due to the signal attenuation caused by extended distance and obstacles in the propagation path. Thus, the received RSS samples are resampled using Pandas library. The data used for localisation experiments was acquired through scripted data collection. This script consists of a representative sample of ADLs occuring in every location in the testbed house. Classification methods, such as k-nearest neighbours (KNN) and support vector machine (SVM), were applied to this dataset. These classifiers provide around $80 \sim 85$ % correct room-level localisation recognition with only three BLE receivers in the house [16]. Subsequently, the same data set was analysed using a Hidden Markov Model (HMM) approach which builds the time sequential relation between locations. HMM provides similar levels of correction rates. By comparing the characteristics of the errors in classification methods and the HMM, we surmise that the erroneous classifications occurs in a random spike manner while the errors in HMM method manifested themselves mostly as bursts. If we define the change of the location as one event, HMM shows much less spurious events than classification methods. Hence, mutual calibration between the classification and time sequential method is necessary to improve the localisation performance.

Advance Stage: In this stage, addtional BLE APs are deployed. Approximately two additional APs are installed in each room. Each AP is constituted by one Raspberry Pi and one COTS Bluetooth dongle. The APs are installed around the ceiling to mitigate the body shadow effect [41] and loitering personnel interference. The house is divided and labelled into 82 grids which are roughly $1 \times 1\ m^2$. As the grid size is very small, it is very difficult for classification methods to cope with slight RSS difference between neighbouring grids. Thus, the localisation in this situation is more reliant on the time sequential relation.

Passive Sensing

There are multiple ways of approaching the localisation challenge in a sensor-rich, multi-modal setup. Methods based on simple binary sensors (e.g. PIR) are limited to single occupancy scenarios and can only provide room-level accuracy. On the other hand, RF-based approaches relying on change in RSSI, not only depend on participant wearing an RF transmitter (e.g. SPHERE wearable) but also require high-density infrastructure of RF receivers. Thus, passive sensing [47] is considered as one of the most likely candidates to provide fine-grained location information in a residential context. It originates from radar technology, and can then be extended to civilian applications by taking advantage of the RF signal already present in residential areas. As demonstrated in [46], passive sensing technology can quantify the precise distortion of human reflected RF signal which is related to the object moving along its trajectory. By synthesising the quantified signal distortion into the predefined classifiers or machine learning modules, pose and location information can be extracted. The successful implementation of passive sensing will lead to a device free solution for collecting location and even activity information in spaces where the wireless signals are presented. Passive sensing is thus an important, albeit budding, ambition of academia and industry, and requires large amounts of testing and research to be fully utilised. Incomplete as it may be though, it is nonetheless a rising research topic due its potential performance and applications in healthcare, security and entertainment.

Acknowledgement This work was performed under the SPHERE IRC, funded by the UK Engineering and Physical Sciences Research Council (EPSRC), Grant EP/K031910/1.

References

1. Allen FR, Ambikairajah E, Lovell NH, Celler BG (2006) Classification of a known sequence of motions and postures from accelerometry data using adapted Gaussian mixture models. Physiol Meas 2006:935
2. Atallah,L, Lo B, Ali R, King R, Yang G-Z (2009) Real-time activity classification using ambient and wearable sensors. IEEE Trans Inf Technol Biomed Publ IEEE Eng Med Biol Soc 13(6), 1031–1039

3. Berger JO (1993) Statistical decision theory and Bayesian analysis, 2nd edn. Springer-Verlag, New York, p 1993
4. Bernardo JM, Smith AFM (2008) Bayesian Theory. John Wiley & Sons, Hoboken, NJ, p 2008
5. Bian X, Abowd GD, Rehg JM (2005) Using sound source localization in a home environment
6. Bishop CM (2013) Model-based machine learning. Phil Trans R Soc A
7. Bose A, Foh CH (2007) A practical path loss model for indoor WiFi positioning enhancement. Inf Commun Signal Process
8. Brugman H, Russel A (2004) Annotating multi-media/multi-modal resources with ELAN. In: Proceedings of the 4th International Conference on Language Resources and Language Evaluation (LREC 2004). Lisbon, 2004, pp 2065–2068
9. Brush AJ, Lee B, Mahajan R, Agarwal S, Saroiu S, Dixon C (2011) Home automation in the wild: challenges and opportunities. In: Proceedings of the SIGCHI Conference on Human Factors in Computing Systems, pp. 2115–2124. ACM
10. Burrows A, Gooberman-Hill R, Coyle D (2015) Empirically derived user attributes for the design of home healthcare technologies. Pers Ubiquit Comput 19(8):1233–1245
11. Coifman RR, Lafon S (2006) Diffusion maps. Appl Comput Harmon Anal 5–30, (Elsevier)
12. Diethe T, Twomey N, Flach P (2016) Active transfer learning for activity recognition. In: 24th European Symposium on Artificial Neural Networks. Bruges: ESANN
13. Diethe T, Twomey N, Flach P (2015) Bayesian modelling of the temporal aspects of smart home activity with circular statistics. Mach Learn Knowl Discov Databases, 279–294. Springer International Publishing, Porto
14. Exel R (2012) Receiver design for time-based ranging with IEEE 802.11b signals. Int J Navig Obs
15. Fafoutis X, Janko B, Mellios E, Hilton G, Sherratt S, Piechocki R, Craddock I (2016) SPW-1: a low-maintenance wearable activity tracker for residential monitoring and healthcare applications. Int Conf Wearables Healthc (HealthWear). EAI
16. Fafoutis X, Mellios E, Twomey N, Diethe T, Hilton G, Piechocki R (2015) An RSSI-based wall prediction model for residential floor map construction. In: Proceedings of the 2nd IEEE World Forum on Internet of Things (WF-IoT). IEEE
17. Fafoutis X, Tsimbalo E, Mellios E, Hilton G, Piechocki R, Craddock I (2016) A residential maintenance-free long-term activity monitoring system for healthcare applications. EURASIP J Wirel Commun Netw 2016, 23
18. Flach PA, Kull M (2015) Precision-recall-gain curves: PR analysis done right. In: Proceedings of the Twenty-Ninth Annual Conference on Neural Information Processing Systems. NIPS
19. Fontana RJ, Gunderson SJ (2002) Ultra-wideband precision asset location system. Ultra Wideband Systems and Technologies, Baltimore
20. Günther A, Hoene C (2005) Measuring round trip times to determine the distance between WLAN nodes
21. Gelman A, Carlin JB, Stern HS, Rubin DB (2013) Bayesian data analysis, 3rd edn. Chapman and Hall, London
22. Gerber S, Tasdizen T, Whitaker R (2007). Robust non-linear dimensionality reduction using successive 1-dimensional Laplacian eigenmaps. In: Proceedings 24th International Conference on Machine learning, pp 281–288. ACM
23. Harle R (2013) A survey of indoor inertial positioning systems for pedestrians. IEEE Commun Surv Tutor 15(3):1281–1293
24. José H-O et al, Reframing in context: a methodology for model reuse in machine learning. AICOM, (in press)
25. Hernández Orallo José, Flach Peter, Ferri Cèsar (2012) A unified view of performance metrics: translating threshold choice into expected classification loss. J Mach Learn Res 13 (1):2813–2869
26. Hoque E, Stankovic J (2012) AALO: Activity recognition in smart homes using Active Learning in the presence of Overlapped activities. In: Proceedings of the 6th International Conference on Pervasive Computing Technologies for Healthcare. IEEE, pp 139–146

27. Hui SY (2013) Planar wireless charging technology for portable electronic products and Qi. Proc IEEE 101(6):1290–1301
28. Kim E, Helal S, Cook D (2010) Human activity recognition and pattern discovery. Pervasive Comput. 48–53
29. Kipp M (2012) Annotation facilities for the reliable analysis of human motion. In: Proceedings of the Eighth International Conference on Language Resources and Evaluation (LREC). Istanbul, pp 4103–4107
30. Koller D, Friedman N (2009) Probabilistic graphical models: principles and techniques. MIT Press, Cambridge, Massachusetts
31. Lao L (2006) Location-based activity recognition. University of Washington
32. Logan B, Healey J, Philipose M, Tapia EM, Intille S (2007) A long-term evaluation of sensing modalities for activity recognition. In: Proceedings of the 9th International Conference on Ubiquitous Computing (UbiComp'07). Berlin: Springer-Verlag, pp 483–500
33. Longstaff B, Reddy S, Estrin D (2010) Improving activity classification for health applications on mobile devices using active and semi-supervised learning. In: 2010 4th International Conference on Pervasive Computing Technologies for Healthcare (PervasiveHealth), NO PERMISSIONS. IEEE, pp 1–7
34. Ciurana M, Barcelo-Arroyo F, Izquierdo F (2007) A ranging system with IEEE 802.11 data frames. In: IEEE Radio and Wireless Symposium. Long Beach
35. Maurer U, Smailagic A, Siewiorek DP, Deisher M (2006) Activity recognition and monitoring using multiple sensors on different body positions. Wearable Implant Body Sensor Netw. IEEE, Massachuset, pp 113–116
36. Mennicken S, Huang EM (2012) Hacking the natural habitat: an in-the-wild study of smart homes, their development, and the people who live in them. Pervasive Computing. Springer Berlin Heidelberg, pp 143–160
37. Mennicken S, Vermeulen J, Huang EM (2014) From today's augmented houses to tomorrow's smart homes: new directions for home automation research. In: Proceedings of the 2014 ACM International Joint Conference on Pervasive and Ubiquitous Computing. ACM, pp 105–115
38. Munaro M, Menegatti E (2014) Fast RGB-D people tracking for service robots. Auton Robots, pp 1–16
39. Murphy Allan H, Winkler Robert L (1984) Probability forecasting in meteorology. J Am Stat Assoc 79:489–500
40. Narayana S, Prasad RV, Rao VS, Prabhakar TV, Kowshik SS, Iyer MS (2015) PIR Sensors: Characterization and Novel Localization Technique
41. Obayashi S, Zander J (1998) A body-shadowing model for indoor radio communication environments. IEEE Trans Antennas Propag 46(6):920–927
42. Pärkkä Juha, Ermes Miikka, Korpipää Panu, Mäntyjärvi Jani, Peltola Johannes, Korhonen Ilkka (2006) Activity classification using realistic data from wearable sensors. IEEE Trans Inf Technol Biomed 10(1):119–128
43. Paiement A, Tao L, Camplani M, Hannuna S, Damen D, Mirmehdi M (2014) Online quality assessment of human motion from skeleton data. In: Proceedings British Machine Vision Conference 2014
44. Roggen D et al (2010) Collecting complex activity datasets in highly rich networked sensor environments. In: 2010 Seventh International Conference on Networked Sensing Systems (INSS). IEEE, pp 233–240
45. Sahinoglu Z, Gezici S, Guvenc I (2008) Ultra-wideband positioning systems
46. Tan B (2015) Wi-Fi based passive human motion sensing for in-home healthcare applications. In: IEEE 2nd World Forum on Internet of Things. Milan
47. Tan B, Woodbridge K, Chetty K (2014) A real-time high resolution passive WiFi Doppler-radar and its applications. In: International Radar Conference. Lille
48. Tsimbalo E, Fafoutis X, Mellios E, Haghighi M, Tan B, Hilton G, Piechocki G, Craddock I (2015) Mitigating Packet Loss in Connectionless Bluetooth Low Energy. In: 2nd IEEE World Forum on Internet of Things (WF-IoT). Milan: IEEE. pp 291–296

49. Tsipouras MG, Tzallas AT, Rigas G, Tsouli S, Fotiadis DI, Konitsiotis S () An automated methodology for levodopa-induced dyskinesia: assessment based on gyroscope and accelerometer signals. Artif Intell Med 127–135. Elsevier
50. Twomey N, Diethe T, Flach P (2016) Unsupervised learning of sensor topologies for improving activity recognition in smart environments. Neurocomputing
51. Twomey N, Flach P (2014) Context modulation of sensor data applied to activity recognition in smart homes. In: Workshop on Learning over Multiple Contexts, European Conference on Machine Learning (ECML'14). Nancy, France
52. van Kasteren T, Noulas A, Englebienne G, Kröse B (2008) Accurate activity recognition in a home setting. In: 10th International Conference on Ubiquitous Computing —UbiComp'08. New York: ACM Press, pp 1–9
53. Wang, Y, Yang X, Zhao Y, Liu Y, Cuthbert L (2013) Bluetooth positioning using RSSI and triangulation methods. Las Vegas
54. Winn John, Bishop Christopher M, Diethe Tom R (2015) Model-based machine learning. Microsoft Research, Cambridge, p 2015
55. Woodman, O, Harle R (2008) Pedestrian localisation for indoor environments. In: Proceedings of the 10th International Conference on Ubiquitous Computing (UbiComp'08)
56. Woznowski P, et al (2015) A multi-modal sensor infrastructure for healthcare in a residential environment. In: IEEE ICC Workshop on ICT-enabled services and technologies for eHealth and AAL. London: IEEE, pp 271–277
57. Tsai Y-L, Tu T-T, Bae H, Chou PH (2010) EcoIMU: a dual triaxial-accelerometer inertial measurement unit for wearable applications. 2010 International Conference on Body Sensor Networks (BSN), Singapore
58. Zhu, Ni, et al. "Bridging e-Health and the Internet of Things: The SPHERE Project." Intelligent Systems, IEEE (IEEE), 2015: 39–46

Erratum to: Making Onlife Principles into Actionable Guidelines for Smart City Frameworks and #IoT Policies

Nenad Gligoric, Christine Hennebert, Srdjan Krco, Carmen Lopez, Ignacio Maestro, Colin Ő Reilly, Michele Nati, Antonio Skarmeta, Rob van Kranenburg, Nathalie Stembert and Alberto Serra

Erratum to:
Chapter 3 in: V. Angelakis et al. (eds.), *Designing, Developing, and Facilitating Smart Cities*, http://doi.org/10.1007/978-3-319-44924-1_3

In the original version of the book, the author name "Nenad Gilgoric" has to be changed to "Nenad Gligoric" in Chapter 3 and Contributors in FM part, which is a belated correction. The erratum book has been updated with the change.

The updated online version of this chapter can be found at https://doi.org/10.1007/978-3-319-44924-1_3

V. Angelakis et al. (eds.), *Designing, Developing, and Facilitating Smart Cities*,
http://doi.org/10.1007/978-3-319-44924-1_15

Index

A
Accessibility, 27, 30
ADL recognition, 325
Ambient Intelligence (AmI), 323
Anonymous communication, 135
Architecture design, 86
Assumed user, 17, 19, 21–25, 29, 30

B
Big data, 49–53, 57, 58, 258, 261
Bottom up, 18, 24, 29
Business Innovation, 58

C
Citizen-centric, 39, 43
Citizen-driven, 28
Cloud computing, 195, 196
Cloud Internet of Things (Cloud-IoT), 169
Cloud-IoT Service Migration, 187, 188
Co-creation, 38, 43, 46
Co-existence, 9
Collective adaptive systems, 93, 94, 100, 108, 109
Communication Technologies, 141–143, 145, 152, 154, 158, 161
Confidentiality, 112, 115, 118–121, 123, 129, 130, 134, 135
Cyber Physical Systems (CPS), 55

D
Data protection, 94, 99–101, 103, 106
Demand response, 256, 259, 260, 262
Device of Virtualization, 67, 167, 183
Digital Signatures, 124, 125
Diversity, 18, 21, 22
Dynamic energy pricing, 243

E
eHeath, 270
Electrical optimization, 244, 256, 262
Encryption, 114, 120–123, 129–135
End-to-end security, 113, 114, 117, 129, 130
Energy markets, 256
Ethics, 38

F
Federated Cloud Computing Infrastructures, 169, 175
FIWARE, 195–199, 202–207

H
Historical perspective, 13

I
Inclusion, 30
Integrity, 112, 119, 123–126, 129, 130, 134, 135
Intelligence of technology, 267, 270, 289, 307
Internet of things (IoT), 35, 36, 38, 39, 42, 44, 46, 52, 53, 58, 63, 67, 80, 87, 102, 103, 105, 106, 108, 111, 113, 114, 116–118, 128, 130, 131, 195–197, 203, 208, 321
IoT-A, 195–197, 199, 203
IoT architectures, 63–68, 75, 84
IoT Urban Applications, 142

K
Knowledge management, 268, 290, 306

L
Legal, 94, 99–101, 103, 104
Local, 34, 35, 39, 43, 45–47

M
Machine learning, 259, 260
Mix networks, 132, 133
Multi-Cloud platforms, 175, 183, 196

V. Angelakis et al. (eds.), *Designing, Developing, and Facilitating Smart Cities*,
DOI 10.1007/978-3-319-44924-1

N
Network architectures, 142, 157, 159

O
Online OD estimation, 224
OPEX/CAPEX, 51

P
Personal data, 93, 95–105, 107, 109
Privacy, 94–98, 100–106, 108, 111, 112, 114, 116, 119, 120, 128, 130, 135, 136
Privacy by design, 64, 84, 86, 102
Privacy impact assessment, 96, 100, 102
Progress of smart technology, 309
Prosumers, 246, 249

R
Renewable energy, 51, 258, 259, 285, 286

S
Security by design, 64
Semantic Data Models, 179
Semantic Storage Systems, 178
Sensor networking, 149, 159
Sensors, 317–323, 325, 328, 329
Serious games, 257
Smart cities, 63, 78, 83, 84, 94–97, 99, 100, 103, 104
Smart ecosystem building, 282
Smart grid deployment, 246
Smart grids, 244, 252, 253, 259, 260, 262
Smart home, 315–317, 325, 327
Social, 94, 95, 97, 98, 100, 104, 107, 108
Social diversity, 12
Software architecture, 199

T
Total Cost of Ownership (TCO), 52, 54
Traffic control, 214, 231, 233, 237
Traffic data, 215–219, 221, 223, 230, 231
Traffic information, 214, 215, 231
Traffic management, 213, 214, 218, 219, 221, 230
Traffic prediction, 215
Trust management, 72

U
Unobservable communication, 135
Urbanization, 11
Urban sociology, 8
User involvement, 267, 269, 270, 283, 287–293, 296, 308

W
Wireless, 145, 152, 154, 157, 160

MIX
Papier aus verantwortungsvollen Quellen
Paper from responsible sources
FSC® C105338

If you have any concerns about our products,
you can contact us on
ProductSafety@springernature.com

In case Publisher is established outside the EU,
the EU authorized representative is:
Springer Nature Customer Service Center GmbH
Europaplatz 3, 69115 Heidelberg, Germany

Printed by Libri Plureos GmbH
in Hamburg, Germany